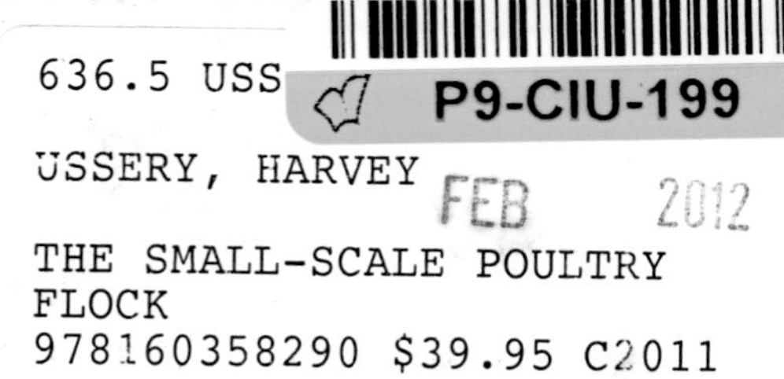

Praise for *The Small-Scale Poul*

"Harvey Ussery has spent a lifetime developing and showca...
that is ultimately carbon-sequestering, hygienic, neighbor-friendly, and food-secure. . . . This book is about a call to heritage, to the wisest of wise traditions in food security and relationships. Harvey brings the latest tools and practices within the grasp of any aspiring flockster. It is this functional spirit that will make this book a classic in the small-scale poultry rearing genre."
—From the foreword by Joel Salatin, Polyface, Inc.

"*The Small-Scale Poultry Flock* is the only complete guide available to using your poultry as an integrated part of a self-reliant farmstead—a topic not addressed at this depth and breadth in any other poultry book. Author Harvey Ussery combines his clear, down-to-earth writing style with creative strategies throughout. He comprehensively explores a wide range of topics including chicken behavior, anatomy, holistic health care, making your own poultry feeds and finding alternative home feeds, breeding your own poultry stock, butchering poultry, and much more. This book covers it all." —Elaine Belanger, editor of *Backyard Poultry* magazine

"Harvey Ussery delivers all the practical information you need to grow your own eggs-and-meat birds, in a style and format that will keep you interested and amused. Plus, he raises the larger question: What kind of world do we want to live in? One that treats animals as units of production, or one that honors all life, especially that farmstead marvel, the domesticated chicken?"
—Sally Fallon Morell, president, The Weston A. Price Foundation

"Ussery's outstanding book is certain to withstand the test of time both for its encyclopedic and practical information and for its acknowledgment that the future of our culture and our food security is in the hands of the small farmer and backyard producer. If you are starting out with your first flock, this is your book. And when you've been keeping poultry for 30+ years, this will still be your best book."
—Shannon Hayes, author of *Radical Homemakers*

"This book is packed with practical advice on raising poultry by someone who has not only done it all, but has learned from his broad experience and knows how to communicate that wisdom clearly and in a lively, readable style. Harvey Ussery has written one of the most comprehensive guides out there, but what places it above the rest of the crowd is that he shows you how to work with nature rather than against it in ways that will minimize work while ensuring the health and happiness of the flock. Whether you're a beginner or an old-time poultry farmer, you shouldn't go any further without this excellent manual."
—Toby Hemenway, author of *Gaia's Garden: A Guide to Home-Scale Permaculture*

"There is a revolution going on, and it is the popular return of keeping poultry to provide food for our home tables. This book helps lead the way by integrating the small flock with its natural environment: the homestead, or small farm. Nowhere else will you find such valuable information on putting poultry to work in the garden, producing much of their feed, and producing healthful food for ourselves."

—Don Schrider, author of the American Livestock Breeds Conservancy's *Chicken Assessment for Improving Productivity* and *Storey's Guide to Raising Turkeys (Second Edition)*

"Anyone interested in practical, experienced, insightful information about how to select, breed, care for, manage, feed, protect, process, eat, and/or market small-scale or personal poultry flocks for their own eating pleasure or selling to others—and have *fun*—should read this book."

—Frederick Kirschenmann, author of *Cultivating an Ecological Conscience*

"Here's the ultimate book for those who want to know everything there is to know about raising poultry. . . . I could not find—in this encyclopedic array of chicken know-how—one detail that I would quibble with."

—Gene Logsdon, author of *Holy Shit* and *The Contrary Farmer*

"*The Small-Scale Poultry Flock* is about establishing a free-range poultry flock fully integrated into a healthy homestead ecosystem. Based upon the author's decades of hands-on experience with many breeds and species, it covers all the basics about raising poultry, and fills some important gaps not usually covered well enough elsewhere, including chicken behavior, poultry breeding, raising chicks with broody hens, managing free-ranging, dealing with predators, using electric net fencing, feeding poultry with home-grown feeds, and integrating the poultry with soil mineral balance, gardens, lawns and pastures, orchards, worm bins, and soldier fly (larvae) production. If you want to raise chickens and can afford just one book, I recommend this one."

—Carol Deppe, author of *The Resilient Gardener*

"No other book on small-scale poultry provides so integrated an approach to issues of soil health, home economy, resource conservation, food quality, and animal welfare. Harvey Ussery's tireless passion for experimentation and empirical observation offers a wealth of information based on decades of first-hand experience. This is the big picture of poultry; no homesteader or backyard chicken enthusiast should be without a copy."

—Kate Hunter, of livingthefrugallife.blogspot.com

The Small-Scale Poultry Flock

The Small-Scale Poultry Flock

An All-Natural Approach to Raising Chickens and Other Fowl for Home and Market Growers

HARVEY USSERY

Foreword by Joel Salatin

CHELSEA GREEN PUBLISHING
WHITE RIVER JUNCTION, VERMONT

Project Manager: Patricia Stone
Developmental Editor: Makenna Goodman
Copy Editor: Laura Jorstad
Proofreader: Eileen M. Clawson
Indexer: Shana Milkie
Designer: Peter Holm, Sterling Hill Productions

All photographs by Harvey Ussery, unless otherwise credited.
Photo on page 198 by Auke-Bonne van der Weide, vanderweide01@yahoo.com.
Illustrations by Elayne Sears, unless otherwise credited.

DISCLAIMER: Information offered in this book is based on decades of research and practical experience. However, the author is not a trained professional in any health, environmental, or other field; neither he nor the publisher is responsible for the consequences of the application of any information or ideas presented herein.

Printed in the United States of America
First printing September, 2011
10 9 8 7 6 5 4 3 2 1 11 12 13 14 15

Chelsea Green Publishing is committed to preserving ancient forests and natural resources. We elected to print this title on FSC®-certified paper containing at least 10% postconsumer recycled paper, processed chlorine-free. As a result, for this printing, we have saved:

24 Trees (40' tall and 6-8" diameter)
11,374 Gallons of Wastewater
11 million BTUs Total Energy
721 Pounds of Solid Waste
2,522 Pounds of Greenhouse Gases

Chelsea Green Publishing made this paper choice because we are a member of the Green Press Initiative, a nonprofit program dedicated to supporting authors, publishers, and suppliers in their efforts to reduce their use of fiber obtained from endangered forests. For more information, visit www.greenpressinitiative.org.

Environmental impact estimates were made using the Environmental Defense Paper Calculator. For more information visit: www.papercalculator.org.

Our Commitment to Green Publishing

Chelsea Green sees publishing as a tool for cultural change and ecological stewardship. We strive to align our book manufacturing practices with our editorial mission and to reduce the impact of our business enterprise in the environment. We print our books and catalogs on chlorine-free recycled paper, using vegetable-based inks whenever possible. This book may cost slightly more because we use recycled paper, and we hope you'll agree that it's worth it. Chelsea Green is a member of the Green Press Initiative (www.greenpressinitiative.org), a nonprofit coalition of publishers, manufacturers, and authors working to protect the world's endangered forests and conserve natural resources. *The Small-Scale Poultry Flock* was printed on FSC®-certified paper supplied by RR Donnelley that contains at least 10-percent postconsumer recycled fiber.

Library of Congress Cataloging-in-Publication Data
Ussery, Harvey.
The small-scale poultry flock : an all-natural approach to raising chickens and other fowl for home and market growers / Harvey Ussery ; foreword by Joel Salatin.
p. cm.
Includes index.
ISBN 978-1-60358-290-2
1. Poultry. I. Title. II. Title: Small scale poultry flock.
SF487.U87 2011
636.5--dc23

2011021217

Chelsea Green Publishing Company
Post Office Box 428
White River Junction, VT 05001
(802) 295-6300
www.chelseagreen.com

DEDICATED TO
Heather
who started our first flock

and to
Ellen
love of my life
sine qua non

. . . how we eat determines,
to a considerable extent,
how the world is used.

Wendell Berry

CONTENTS

PART THREE: Working Partners

PART FOUR: Feeding the Small-Scale Flock

PART FIVE: Other Management Issues

PART SIX: Breeding the Small-Scale Flock

PART SEVEN: Poultry for the Table

APPENDICES

PART SEVEN: Poultry for the Table

APPENDICES

FOREWORD

"But can you feed the world?" This is by far the most common question I am asked everywhere I speak around the world. As much as the groundswell of locavores and biological farming advocates desire a localized and non-industrial food system, most still wonder if such systems can actually feed the world. After all, if this fundamental question cannot be answered positively, then we should all jump on the Monsanto bandwagon for the health and prosperity of our fellow man.

Most people in our culture still believe the industrial Green Revolution and the advent of factory farming saved us from certain starvation. This is why most people think chemical-petroleum farming, transgenic modification, and concentrated monospecies production are wonderful and righteous food production methods—together with increased globalization and food miles, and ever-growing, ever-more-centralized, industrial-sized production and processing facilities to achieve assumed economies of scale. The endorsement stems directly from the assumption that such farming techniques are necessary to feed us all.

Wading into this malaise of misperception is equivalent to venturing into enemy territory. Many people, and especially today's experts, think that an alternative to these assumed food trajectories does not exist. Any suggestion that localized, compost-driven, carbon-sequestering, biological food production systems are a credible alternative is met with derision, laughed to scorn by credentialed experts who assume they know better. Those of us who espouse such thoughts are treated condescendingly with a puppy-dog pat to the head and told to trot back to our play-farms that don't really count. Or we are accused of spreading a cult, a false hope, and jeopardizing real progress, like modern Luddites. Or even criminalized with accusations that our pastured livestock and non-vaccinated, non-medicated animals threaten science-based feedlots and factory farms, thereby threatening the planet's food supply.

Well now, wait just a minute. Let's zero in on just one monument to industrial food production: the factory chicken house. While your mind's eye zooms in on that football-field-sized monolith containing twenty thousand chickens or more, realize what is *not* in the picture. First, the hundreds of acres of grain that are plowed, chemicalized, harvested, dried, and trucked to the house. It does not stand alone; it has a support land base spreading out from it that is enormous. And then consider the manure disposal. Not being near the grain-producing land that could benefit from the manure, this historically normal blessing becomes a modern-day curse. Indeed, the manure is fed to cattle, spread excessively on hundreds of acres of land nearby, dehydrated and pelletized for lawn fertilizers, and even converted to biodiesel, all processes that use an excess of energy and fossil fuels, and create more waste and pollution down the line. The point is that these amazing factory houses stand on the shoulders of countless acres, tax-subsidized disposal networks, corporate welfarism, and cheap fuel. If all these chickens were grown on pasture, however; fed locally produced grain, supplemented with bugs, worms, and kitchen scraps; dropping their manure out on the same fields that are their salad bar, they would nest into the local ecology on a lot less energy and be a blessing on all points without requiring one acre of additional land.

And lest you think we are running out of land, consider that America has 35 million acres of lawn. And 36 million acres housing and growing feed for recreational horses. Last time I checked, 71 million acres is actually enough to almost feed every American without any farms—and we haven't even talked about golf courses. We send our young men and women to war around the world to ensure cheap energy to make fertilizer to grow more grass to mow with more petroleum to send carbonaceous grass clippings into landfills. This is insane. Why not fill our yards with food production, from vegetables to pasture to poultry? Looked at in this light, the local and biological food tsunami moving across America represents a credible alternative to industrial fare.

I was ten years old, in 1967, when my first fifty as-hatched, heavy-breed-special chicks arrived in the mail. A cardboard-box brooder in the basement provided ample protection and housing for the chicks until they moved into our backyard. Although this first foray into poultry yielded thirty-two roosters and only eighteen hens, it was enough to start a fledgling egg enterprise. The cockerels provided an early immersion into the art of poultry processing, and the concomitant ecstasy of delectable dining on transparently grown fowl.

This early love affair with poultry morphed into a serious pastured egg business during my teen years, providing me with spending money and entrepreneurial savvy. The poultry became a natural centerpiece of our pasture-based livestock farm in the early 1980s and remain so to this day. Both broilers and egg layers grace our pastures in portable structures, providing exceptional taste and nutritional integrity to thousands.

But in my perfect world, I don't think we should be growing all this poultry for sale. Except for perhaps the rabbit, I cannot think of another more compatible, more nutrient-dense food opportunity this close to a home. And rabbits don't lay eggs—despite what you may have heard about the Easter bunny. In reality, if every single kitchen, both commercial and domestic, had enough laying hens attached to it to eat and recycle the scraps and inedibles generated therein, the entire commercial egg industry would be obsolete. Who says we can't feed the world?

In a time when people fear for their food security, what is more secure—centralized egg factories inventorying and trucking millions of eggs a year to retailers 1,000 miles away, or millions of backyard and condominium-housed flocks generating eggs and meat a few feet away from the kitchen? Lest anyone recoil in horror that this would usher in a new age of poultry-induced disease and pathogens, I urge them to realize that we have many pieces of knowledge and infrastructure that did not exist from 1900 to 1950, when the early crowding and the industrialization of farming were instituted, without the necessary protocols or infrastructure.

In those times we didn't even have electricity, much less lightweight and portable predator-proof electrified poultry netting to permit easy movement and control. We didn't have refrigeration, stainless steel, or on-demand hot water. And we didn't understand a lot of principles surrounding hygiene and sanitation. We didn't have shredders for generating cheap carbonaceous "diaper bedding," or simple extruded steel arches and fifteen-year-stabilized laminated plastic for building warm winter hoophouses.

Utilizing all of this technological infrastructure and knowledge, Harvey Ussery has spent a lifetime developing and showcasing a truly viable home-scale poultry model that is ultimately carbon-sequestering, hygienic, neighbor-friendly, and food-secure. The answer to the abuses, both animal and environmental, in the factory poultry industry will not come from railing against governmental regulations. Instead, it is for thousands and even millions of people around the globe to catch Harvey's vision and bring to their own lives, in their own living space, the joys of poultry.

Unlike larger livestock like cows or sheep or llamas, chickens require only lightweight infrastructure. They can get your children off the TV and video games because their real-life antics will entertain far better than Hollywood. Bringing food responsibility and relationship to your doorstep can have profound character-building and spiritual consequences. Viscerally participating in the dance of death and regeneration, the sacrifice required to give life sustenance, can bring awe and reverence to the family that will yield a new generation of sensible leaders.

Perhaps the thing I appreciate most about this book is Harvey's gift, both to the home-scale poultry enthusiast and the small-scale producer like myself, of an incredibly practical, can-do attitude. These are not pets, although every flock has its individuals who endear themselves to the flockster in one way or another. But rather, this book is about a call to heritage, to the wisest of wise traditions in food security and relationships. Today's modern tools and knowledge revolutionize negative associations with poultry—the dirty, dusty drudgery of yesteryear—and Harvey brings the latest tools and practices within the grasp of any aspiring flockster. It is this functional spirit that will make this book a classic in the small-scale poultry-rearing genre. I know you'll find it as insightful and helpful as I did.

Joel Salatin
Polyface Farm, Swoope, Virginia
December 2010

ACKNOWLEDGMENTS

Thanks to Lottie and Marshall Ussery for teaching me that producing some of what I eat in the backyard is a good thing; and to Harry Feldman and Fritzi Grabosky, who would have found that a peculiar idea.

Thanks to poultry friends everywhere, both those locally with whom I swap ideas and stock and those I meet and correspond with online. Thanks especially to members of American Pastured Poultry Producers Association, American Livestock Breeds Conservancy, and Society for the Preservation of Poultry Antiquities, from whom I have learned so much.

Thanks to Joel Salatin for being mentor and inspiration for more than two decades, and for all he is doing to lead the way toward a saner agriculture.

Thanks to my editor at Chelsea Green, Makenna Goodman, for her patience and good judgment.

Don Schrider and Kate Hunter read the manuscript and gave helpful suggestions, in addition to contributing text and pictures. Leigh Glenn and Eva Johansson also read the manuscript and offered valuable input. Susan Burek of Moonlight Mile Herb Farm offered insights for the chapter on natural poultry health; Linda Anderson helped with information about guardian dogs; and Tim Koegel shared information about his eggmobile, his cost/profit spreadsheets, and insights into pastured poultry for markets.

Thanks to others who contributed material: Carol Deppe, Jean Nick, Sam Poles, Ellen Ussery, Bridget Chisholm, Steven Blake, Stephen Day.

Thanks to Mike Focazio, chicken buddy extraordinaire, especially for his photos that illustrate the chapter on poultry butchering; for help with soil issues in "A Question of Balance"; and for serving as philosophical collaborator in defining the perspective that underlies this book.

Jeff Mattocks of Fertrell generously responded to technical questions about poultry nutrition and feeding, out of his vast store of expertise on the subjects.

Federico Lucas helped with work on the homestead to give me more time at the keyboard. *Gracias.*

Thanks in memoriam to my good friend Reid Putney, who asked every time he saw me, "When you gonna write that book?" And thanks to Nani Power, whose writing classes helped get me off my duff.

Thanks to all the editors I've worked with at *Mother Earth News*, *Countryside & Small Stock Journal*, and especially to Elaine Belanger, my editor at *Backyard Poultry*; and to André Angelantoni of Post Peak Living (http://postpeakliving.com) for the opportunity to present my "Chickens 101" in this online venue.

Thanks to Bonnie Long for sharing her photos in this book—it has been a pleasure working together the past two years.

Thanks in loving memory to Marvin Grabosky for seeds he planted. Thanks to Melba Hendrix, and to Jo T. and his friends, for lending helping hands when I really needed them.

And finally, thanks to my daughter Heather, who was my "accountability partner" for the project, and to my wife Ellen, without whose support I would have never gotten past the third chapter.

INTRODUCTION

I have been raising poultry for almost three decades. Though I started as far too many beginners do—with a flock of chickens on a bare dirt run, fed commercial feeds exclusively—I came to see that the liberation of the flock from that dreary model not only made for happier chickens but also opened up possibilities for partnership with them in all parts of the homestead or farm.

Despite the fact that I live in a small rural village in northern Virginia, keeping poultry was at the time a rarity among my neighbors. My passion for poultry stood out, and before long I was known locally as "the chicken man." That epithet drew increasing numbers of people to my back door, people who had begun to aspire to more self-reliant ways of living. Almost ten years ago I began offering classes on poultry husbandry and tours of our working homestead. I found a hunger for information about the productive backyard flock that only seemed to grow as I tried to satisfy it. Soon I was making presentations in local farming and gardening venues. To reach out to more would-be flocksters, I created a website, www.themodernhomestead.us, which in turn opened up opportunities to write on poultry and other homesteading subjects for *Mother Earth News*, *Countryside & Small Stock Journal*, and especially *Backyard Poultry*, and to present my ideas to larger audiences farther afield, at conferences of farming, sustainability, and traditional foods organizations.

More and more of my readers and listeners asked me, "Why don't you write a book?" It seemed indeed that there was no book available that outlined the all-natural, all-inclusive care of poultry in the breadth and depth for which I was constantly reaching. And there was no poultry book in the field that took quite as seriously as I do the coming enormous changes in the national and global economies, with their threats to basic food security, and with the consequent need to find new ways to produce the food we eat.

This book is for keepers of small-scale poultry flocks who see their birds as respected partners in the homesteading, food-independence enterprise, rather than as pets—and who are not satisfied with the isolation of the flock in its own static little corner of the homestead or small farm. This does not mean that poultry husbandry isn't fun, nor that we cannot develop a profound relationship with our birds. Indeed, it is my assumption that we develop a more interesting and satisfying relationship with the flock as partners rather than as pets.

This book is definitely not for those whose passion is competitive showing of fowl. I enjoy the displays of the breeder's art in a large poultry exhibit as much as anyone; but I have neither the experience nor the interest to address this specialty subject.

Since the book is informed most of all by my own experience raising poultry in a rural setting, there may be a few special concerns of urban—and to some extent suburban—keepers of poultry I cannot address. But note that most of the ideas presented herein easily scale either up or down, and thus can be applied even by keepers with tiny flocks on small properties, or those contemplating stepping up to production for small local markets.

None of the above implies that this is not a book for beginners. While in some ways it targets an intermediate to advanced level of experience, it will prove a meaty, challenging resource as well for the beginner excited about creative, natural, and integrated ways to use the poultry flock. My ideal reader is the backyard homesteader, or small-scale farmer, whose goal is the production of *all* the family's eggs and dressed poultry (and potentially enough to provide to other nearby families), using methods that utilize the flock as an *integrated* part of the total food-independence enterprise.

My emphasis will be on chickens: I have most experience with them; they are the most common domesticated fowl; and they may be the easiest of all starter livestock. But this book is about homestead *poultry.* Other fowl species will be considered from time to time, usually when their management, needs, or habits diverge significantly from those of their chicken cousins. A chapter near the end of the book, "Other Domestic Fowl," will discuss in greater detail the other common gallinaceous species (guineas and turkeys) and the waterfowl (ducks, Muscovies, and geese). The more exotic, ornamental, and specialty fowl (pheasants, swans, pigeons) will not be covered in this book, except for a brief sidebar on growing game birds for market.

This book is not a set of recipes to be applied formulaically to your flock to meet your goals in your particular circumstances. It is about exploring possibilities, between here and the horizon. The truth is that all of us backyard flocksters are making it up as we go along. At one time most families, at least in the country and in small towns, either owned chickens or were closely familiar with families who did—they had useful, traditional models to guide them. Today, to work with a small-scale flock in a holistic way is to find our way back to paradigms that make sense, to create workable models ourselves, to work at the cutting edge of what is possible with *Gallus gallus domesticus* as partner. Isn't that exciting!

I cannot claim any special expertise, other than what has been hammered into my skull by hard-won experience. The truth is, I'm just an old hick with chickenshit on his boots, who's made a lot of mistakes. But one thing I can assure my reader: I quickly came to distrust the static, limited model for poultry husbandry with which I—like most beginning flocksters—began. I have constantly searched for more natural, inclusive ways to manage and use the flock—always experimenting, always taking my cue from my birds, whose happiness after all defines good husbandry.

If you join me in this alliance with our feathered friends, you will learn more from a similar willingness to heed your flock, and a spirit of adventure, than from anything said in this book.

PART ONE
Getting Started

1 | WHY BOTHER?

Eating with the fullest pleasure—pleasure, that is, that does not depend on ignorance—is perhaps the profoundest enactment of our connection with the world. In this pleasure we experience and celebrate our dependence and our gratitude, for we are living from mystery, from creatures we did not make and powers we cannot comprehend.

—Wendell Berry

Keeping a small-scale flock is almost certain to bring more joy than frustration, is great fun, and requires only minutes in a routine day. Most important from the perspective of this book, the home flock can make a major contribution to the effort to eat better and become more food-independent. But we do have to invest significant effort to set up the project; and we have to see to the flock's needs *every day*, even better several times a day, without fail. This serious commitment to living creatures dependent on our care could require significant changes in how we organize our lives.

So why bother? Especially when the alternative is so easy and so cheap?

The "Cheap Chicken" Alternative

Eggs and dressed broilers from the supermarket are among the cheapest protein foods you can buy. Indeed, I estimate that my backyard chicken and eggs cost me (considerably) *more* than I would pay for their equivalents in the supermarket. In order to understand such a peculiar choice, you need to know more about how eggs and dressed broilers are produced in the poultry industry.

As in other areas of modern industrial agriculture, the poultry industry operates on an almost unimaginable scale. Modern broiler houses contain between ten thousand and one hundred thousand birds. A typical one is 20,000 square feet and contains twenty-two to twenty-six thousand growing broilers—less space per bird than the size of a sheet of typing paper. The hapless broilers stand in the deepening accretion of their own manure, which accumulates at the rate of several tons per week in a typical broiler house.

Industrial broilers are highly selected for fastest possible growth to slaughter size. To counteract the stress of living in unsanitary, crowded conditions, broilers eat antibiotics as part of their feed[1] from hatch until slaughter—as early as forty-four days, no later than seven or at most eight weeks. Butchering—of millions of birds per week in a factory-style slaughterhouse—occurs on highly automated "kill lines," in which key processes, including evisceration, are handled robotically.

Laying hens are even more confined: The typical commercial layer house, which might contain up to a hundred thousand layers, is arranged in a system of stacked tiers, like those in figure 1.1, of battery cages, each about the size of a filing cabinet drawer (67 to 86 square inches per bird—a sheet of typing paper is 93.5 square inches) and holding eight to ten

Fig. 1.1 Layer hens in tiers of battery cages. PHOTO FROM FARM SANCTUARY

hens, dimly lit by artificial lighting. The hens cannot stretch or even fully stand up, to say nothing of engaging in natural behaviors such as foraging and dust bathing.

Since laying hens under such stress would naturally engage in crazed pecking of one another, chicks who will become layers have half their upper beaks chopped off immediately after hatch—a mutilation known as debeaking that may cause chronic pain. And the male chicks, produced by the hundreds of thousands per day industry-wide, every one of whom is by definition "useless" to the enterprise? They are conveyed alive into something like a sausage grinder, minced by high-speed blades, gassed, vacuumed through a series of pipes onto an electrified metal plate, or simply thrown into a bin to suffocate.[2]

Feeds for industrial poultry are mediocre at best, based on whatever by-products are cheapest on a given day—feather meal from slaughterhouses, soybean meal from oil extraction, cottonseed meal, wheat and other grain bran from flour refinement. Much of the fat content is oil from discarded fast-food deep fryers—which is in the industrial recycling bin to begin with because it is *stale*. But then *all* ingredients in the feeds are stale as a result of the heat and pressure of processing, and the resulting damage to the more perishable nutrients.

Food Safety

Every year [in the United States] 76 million cases of foodborne illnesses occur, leading to about 300,000 hospitalizations and 5000 deaths.

—*New York Times*, SEPTEMBER 4, 2010

In August 2010 the FDA recalled half a billion eggs from supermarket coolers for salmonella contamination. Half a *billion*? That's enough to furnish an egg to one out of thirteen people on the planet—an egg that could cause serious food poisoning (and, in the worst case, death). In a single recall. How can anyone trust a production system that has failed so catastrophically its first priority, food safety?

I too was amazed at the numbers—though my amazement was more that the recall was *only* half a billion. Now that you know the conditions under which industrial eggs are produced, you might join me in wondering why they aren't *all* contaminated.

The news is even more grim for eaters of supermarket chicken, who might reflect on salmonella and campylobacter, among the most common sources of foodborne disease (along with *E. coli* O157:H7). *Consumer Reports* published the results from a random sampling of fresh supermarket broilers in twenty-two states in the nation.[3] Sixty-six percent of the samples were contaminated, either with salmonella or campylobacter, or both. (Samples from large-scale certified "organic" production were free of salmonella but had a 57 percent contamination rate for campylobacter.) Most of the disease-causing bacteria detected were resistant to at least one antibiotic—the result of the feeding of antibiotics referred to above—so potential infections will likely be more difficult to treat.

It is a shock to most consumers to learn that the pristine, shrink-wrapped, government-inspected package of poultry in the refrigerated supermarket display is *routinely* (well over half the time) contaminated with bacteria that can make them seriously ill (fever, diarrhea, abdominal cramps, possibly even death)—but it is no surprise to anyone experienced in butchering poultry. Even the rankest beginner will take great care to remove the entire gastrointestinal tract, from esophagus to vent, in a single unbroken package, thus avoiding spillage of intestinal contents into the body cavity. The high-speed robotic kill lines in poultry processing plants, however, allow plenty of fecal spillage, as evidenced by the fact that evisceration is the most common source of contamination of supermarket chicken. That is to say: "There is shit in the meat."[4]

Corporate interests intone the mantra endlessly: The American Way of Poultry is necessary in order to "feed the world." In the wake of news of egg recalls and contamination of broiler meat, your response might well be: "Yeah, feeding the world is a great thing. I'll start with my children."

Food Quality

We're all too acquainted with the item in the supermarket produce section that *looks* like a tomato but ends up as a slice of tasteless pulp on the plate; or the beautiful imported strawberry—huge, deep red, and plump—with no discernible flavor. It shouldn't be surprising that eggs and chicken produced in equally mechanized, chemicalized systems of gargantuan scale are equally insipid. Anyone who assumes—as I emphatically do—that taste, color, and aroma reflect nutrient quality in a whole, unaltered food will give a failing grade to supermarket eggs, with their pale yolks and flabby whites, and supermarket chicken that has flavor only if accompanied by a sauce.

I don't know if you've seen the T-shirt that reads, MY TASTES ARE SIMPLE—I LIKE THE BEST, but that describes my wife Ellen and me precisely, at least as regards food quality. Call us food snobs if you like, but our "simple" tastes are a major motivation for producing our eggs and chicken in our own backyard—we *will not eat* the debased equivalents in the supermarket.

Apparently this intuitive sensory recognition of the good stuff nutritionally is obvious to most people, if not all: When we serve guests whose only experience of food comes from the supermarket and fast-food franchises, their faces light up with surprised pleasure as they taste our omelets, to say nothing of our roast chicken.

Lest you dismiss the above as mere "better than thou" snobbishness, know that a growing body of studies demonstrate the nutritional superiority of eggs and dressed poultry produced using some of the methods suggested in this book. If you too "simply" want the best for your family, may I suggest a flock of easy-to-keep poultry in your backyard?

Spillover

While the US industry touts its "cheap" chicken—ten billion broilers annually, 40 percent of global production—to "feed the world," the astute reader will have noted the massive feeding of antibiotics necessary to keep industrial flocks alive and growing, or laying. Perhaps you're troubled by excessive feeding of antibiotics[5]—and their release into the environment, with what long-term effects we haven't a clue—which causes resistance to antibiotics among pathogenic bacteria. That is: The routine feeding of antibiotics to livestock results in loss of the therapeutic use of whole classes of antibiotics for treating people. Troubling as well is the routine feeding of arsenic to the nation's broiler flocks. (Did he say *arsenic*?)[6]

Wendell Berry observed that industrial agriculture has a habit of taking a beautiful natural *solution* and neatly bifurcating it into two hideous, insoluble *problems*. An example is chicken poop, which I will argue in this work is "holy shit."[7] But its holiness as part of the great cycles of nourishment and decay and fertility is a reality only when small flocks are dispersed over grateful landscapes that can utilize the droppings as nature intended. We *cannot* manage or effectively utilize mountains of chickenshit from tens of thousands of birds per

Organic versus Conventional: What Should You Buy?

These days there are more and more "organic" foods in the supermarket. But can you be certain that eggs, broilers, and other poultry products that carry the organic label are safer, are of higher quality, and contribute to more sustainable farming and protection of the environment?

For years Big Ag and government agencies scoffed at the notion that "organic" farming produced better food while avoiding ecological harm. But then the market share of organic foods began to climb as consumers became more concerned about food quality; and Congress, in October 2002, established the National Organic Program, to be administered by the US Department of Agriculture. Complex and detailed protocols were put into place to accredit and certify production of foods allowed to be labeled "organic" in the marketplace. Do note that qualifier: The NOP is about establishing the right to *label* food as organically produced. No farmer, however conscientious about avoiding toxic chemicals in farming, using practices however sustainable or ecologically sound, may label foods for sale as "organic" that have not been so certified by an approved third-party certifying agency. Obviously, whether the resulting "organic" foods really are superior in nutrition, safer, and more ecologically benign has everything to do with how rigorously the certifying agencies ensure that producers are abiding by the organic standards.

Unfortunately, the USDA has been notoriously lax in its enforcement of the standards, leading to the accusation that the organic label since adoption of the NOP is more about allowing Big Ag and Big Food to reap profits from premium organic prices than about supporting a saner agriculture or more reliable foods.

I observed the result firsthand at a conference of the Virginia Association for Biological Farming, where I attended a presentation by the head of the organic subdivision of a national egg company that produces millions of dozens of eggs per year. The presenter proudly showed slide after slide of his showcase "organic" layer facility, containing seventeen thousand hens. While the hens were not confined to battery cages as in a conventional layer house, they were "free-ranging" inside it at a spacing of 1.75 square feet each—about two and a half times the size of a sheet of typing paper. Organic standards require that the layers be given "access to the outdoors," and that was provided by a 10-by-20-foot wire enclosure off one end of the facility, with a door opened onto it from ten o'clock in the morning until four o'clock in the afternoon, weather permitting. If the seventeen thousand layers rotated through their "outdoor access" in groups of five hundred at a time, each hen would be "outdoors" for about ten minutes, at a concentration of less than half a square foot each. Needless to say, there was not a blade of grass in the pen, which was ankle-deep in chicken manure instead. The resulting eggs were certified to be "organic"—though surely no consumer who ever bought a single dozen of them would have agreed, knowing the realities of their production.

Even Walmart, in the spring of 2006, began offering a number of certified organic foods, many of them produced in China, where the USDA has exerted almost no meaningful controls over the certification process. More and more, "corporate organic" has come to be something entirely different from the organic *movement* engendered by Sir Albert Howard, J. I. Rodale, and other critics of industrial farming. It has come to mean for the most part, in the case of industrial laying and broiler flocks, that the *feeds* are certified organic. Never mind that the corn and soybeans were grown in China or Brazil and unsustainably shipped here. Never mind that those feeds are fed to confined layer flocks of eighty thousand or even one hundred thousand who do not range on pasture eating green plants and insects.

As for broilers, as noted in the text, *Consumer Reports* found that certified organic supermarket broilers were superior to conventional broilers regarding salmonella contamination—though the contamination of more than half of them with campylobacter hardly inspires confidence in their "organic" label.

I do not suggest that the organic label is meaningless. With regard to our own food purchases, my wife Ellen and I seek out certified organic products we cannot buy locally—coffee, olive oil, and almonds, for example. That is, the greater the degree of *anonymity* in the purchase, the more an organic certification means to us—despite our knowledge of the shortcomings in the certification process. But we buy all our beef, pork, and lamb locally, face to face with farmers whose methods we know and trust. When the anonymity is removed from the transaction, we do not care whether the farmer chooses to be certified or not.

I recommend that you not blindly assume that the label "organic" automatically means that the food in the package really is more nutritious, safer, or more sustainably produced. If at all possible, seek out food with a face—food from a farmer you can query about production methods used, and thus can come to trust with your family's health and well-being. If you cannot buy directly from the source, pay some attention to the certifying agency that attests to the "organic" label—some are considerably more rigorous than others. And use common sense. "Organic" broilers selling for $2.50 per pound, or "organic" eggs selling for $1.75 per dozen, simply cannot be truly organic in any sense you would recognize as meaningful. One of my correspondents in the American Pastured Poultry Producers Association (APPPA), for example, told me that he spends $1.50 per dozen on feed costs *alone* to produce eggs that truly meet the organic standards.

Check out an important watchdog organization, the Cornucopia Institute, to inform yourself about the complex issues surrounding the National Organic Program—and to help find your way to organic foods that truly deserve the label.[8]

broiler house without leaching of their nutrients—which should be a priceless asset—into groundwater and natural water systems, where they function as pollutants rather than nutrients. In the Shenandoah Valley not far from where I live—the site of dozens of huge commercial broiler houses—poultry house litter stored in big open piles (because there was nowhere else to haul it) has been a large source of pollutants into the Shenandoah River. It has become necessary to haul the manure produced ever farther from its source to spread on fields (ones that have not already reached the limit of phosphates they can absorb without becoming effectively toxic).

The problems created by MegaPoultry involve profound quality-of-life issues—issues that are not just about being "greenie," caring for the environment, or embracing feel-goodism but are profoundly *practical* issues of sustainability. And to anyone who reads

Natural Eggs Are Better

In 2007 *Mother Earth News* published a comparison of the nutrient profiles of naturally produced eggs with those produced in industrial laying flocks.[9] The article concluded by citing mounting evidence that there are important differences between the two.

- In 1974 the *British Journal of Nutrition* found that pastured eggs had 50 percent more folic acid and 70 percent more vitamin B_{12} than eggs from factory farm hens.
- In 1988 Artemis Simopoulos, co-author of *The Omega Diet*, found that pastured eggs in Greece contained thirteen times more omega-3 polyunsaturated fatty acids than US commercial eggs.
- A 1998 study in *Animal Feed Science and Technology* found that pastured eggs had higher levels of omega-3s and vitamin E than eggs from caged hens.
- A 1999 study by Barb Gorski at Pennsylvania State University found that eggs from pastured birds had 10 percent less fat, 34 percent less cholesterol, 40 percent more vitamin A, and four times the omega-3s compared with the standard USDA data. Her study also tested pastured chicken meat and found it to have 21 percent less fat, 30 percent less saturated fat, and 50 percent more vitamin A than the USDA standard.
- In 2003 Heather Karsten at Pennsylvania State University compared eggs from two groups of Hy-Line variety hens, with one kept in standard crowded factory farm conditions and the other on mixed grass and legume pasture. The eggs had similar levels of fat and cholesterol, but the pastured eggs had three times more omega-3s, 220 percent more vitamin E, and 62 percent more vitamin A than eggs from caged hens.
- The 2005 study *Mother Earth News* conducted of four heritage-breed pastured flocks in Kansas found that pastured eggs had roughly half the cholesterol, 50 percent more vitamin E, and three times more beta-carotene.
- In 2007 *Mother Earth News* sent samples from fourteen pastured flocks in all parts of the United States to an accredited independent laboratory. The results showed they were nutritionally superior to eggs from the poultry industry. (See a summary of their report in appendix G.)

that statement and says, "Well yeah, that's all true, but there's nothing I can *do*," I would say: *One dollar, one vote.* The industry cares not a whit about our tender feelings for the environment: The dollar we plunk down at the supermarket checkout is first and foremost a *vote*—for more of the same. Fortunately, many of us have an alternative: taking the trouble to cast our votes for local suppliers of sustainably raised eggs and poultry.[10] Really lucky consumers may have an even better option: a flock of chickens in the backyard eating the way they evolved to eat and enjoying the lifestyle they evolved to live—and gracing the table with the best eggs and dressed poultry on the planet.

Maybe Not So "Cheap"

Every decade for the past 150 years, the percentage of our income that we spend for food has decreased. We now spend less than 10% of our income feeding ourselves compared to over 15% thirty years ago and over 50% before 1900. To put this in more concrete terms, in 1875 it took 1,700 hours of work to purchase the annual food supply for a family. Today it takes about 260 hours of work. Oh, and in 1875 we spent 1% of our annual income on health care. Today, we spend over 16%. Not surprisingly, health care costs have tracked the food cost trend, but in the opposite direction.

—Matthew O'Hayer,
Vital Farms, Austin, Texas[11]

"Cheap chicken" is an illusion hiding a number of costs not factored into supermarket price tags. The estimated annual cost of foodborne illness in the United States is about $152 billion—about $1,850 every time someone gets sick from food they ate. Many cases of illness go way beyond an upset stomach—those dollar figures merely hint at the enormous amount of suffering for victims and their families from salmonella and campylobacter in supermarket eggs and contaminated chicken.

Chicken and eggs are "cheap" to a great extent because their production is *subsidized*—by *you.* Taxpayer dollars help pay Big Ag's production of the mountains of commodity corn and soybeans without which the profitability of the egg and broiler industries would be impossible.[12]

A major reason chicken is "cheap" is that the poultry industry is almost never required to pay to prevent or clean up its spillover effects (waved aside by economists as "externalities"). Dollar costs of spillover are impossible to quantify, though a reasonable estimate runs into the trillions of dollars.[13] We experience these costs every day, though, even if we never make the connection. Public water treatment costs increase, for example, as a result of phosphate and nitrate contamination of water systems by runoff from industrial poultry production. Think of the billions of gallons of bottled water consumers buy, because of mistrust about the safety of what comes out of their taps.

The full scale of spillover costs, however—contamination of natural water systems including vast areas of ocean, loss of species to pollution and soils to destructive commodity agriculture, emergence of antibiotic-resistant diseases—cannot even be thought about in dollar terms, though they diminish us all—and the future of our children.

Karma

Discussing the way industrial broilers and layers are raised, I focused on the filth, mediocre feeds, and feeding of antibiotics and arsenic to suggest that industrial eggs and broilers *are not fit to eat.* But the reader must have felt as well a sense of horror: *Wait a minute, this is the way chickens are treated?* Perhaps you imagined yourself, trapped with others in a cage too small to stretch in, or shoulder to shoulder with others, ankle deep in your own wastes; eating nothing but junk food; never a breath of fresh air or glimpse of the sun; lacking all sense of social order—*for your entire life.* You don't have to be a hard-core vegan—I emphatically am not—to feel utterly outraged.[14] There is clearly a moral dimension to such abuse of God's creatures.

We might reflect on karma, an Eastern concept that to some sounds a little spooky, but simply means: Nothing takes place in a vacuum. What goes around comes around. Whatever you do, you live with the consequences, whether intended or not, whether you acted from greed or out of ignorance—or even for the noble purpose of "feeding the world." The suffering we cause these sad creatures in their hellish concentration camps comes around to all of us, a karmic retribution, in our own suffering: foodborne illness, contaminated water, threats to biological systems on which our lives depend. It will doubtless sound "woo-woo" to some, but I believe as well that when we eat eggs and meat from birds whose lives are psychic hell, we are *eating that anguish.*

Food Security

We live in an age of multiple crises: climate change; massive degradation of agricultural soils; peak oil; increasingly critical problems of fresh water supply and quality; a witches' brew of toxic chemicals in our environment; loss of species on a scale that possibly exceeds the rate of extinction in the thousand-year period following the impact of a 7-mile-diameter asteroid sixty-five million years ago (the one that did in the dinosaurs). Any one of these threats raises serious concerns about the stability of the global supply systems we've put into place—and they are converging, in some cases reinforcing one another's effects. Mostly that's for another book; but it is worthwhile focusing on a couple of emergent crises that should have us thinking about *food security.*

Peak Oil and Its Discontents

Any analysis of our economy and its prospects is worthless if it does not start with the fact that it is fundamentally and at every point dependent on fossil fuels—and most especially on petroleum. The elephant in the room that far too many economists (and corporate executives and politicians) ignore is that world production of the black gold will soon peak—indeed, some analysts believe it has already reached its peak—meaning that forevermore there will be less oil produced each following year. An economy based on endless growth cannot survive unchanged in an environment in which its lifeblood—cheap, abundant, readily available oil—is becoming more expensive and more scarce. (And this in a context where major players such as China and India are rapidly expanding their own demands on the market for crude.)[15]

There is no part of the economy in which we are more dependent on petroleum (and natural gas, also approaching peak) than in the way we produce our food: farming that is massively mechanized and powered by petroleum (source also of most agricultural pesticides) and natural gas (source of most artificial fertilizers); high-energy processing of most foods in the food supply; and a system of long-distance supply and just-in-time deliveries to the point of sale. (It is estimated that the average bite of food on the American plate moves 1,500 miles from field to fork.) There is no greater threat to the industrial food system than a shrinking petroleum supply.

Meltdown

The financial crisis in the fall of 2008 shook the global economy to its core—and left ordinary citizens like us feeling deeply threatened by the remote fall of dominoes we didn't understand and over which we had absolutely no control.

In spring of 2010 the global economy again had the jitters, this time because Greece was teetering on the brink of national default. Since Greece is such a minor player in the global economic system, I wondered about all the fuss. It would be different if one of the big guys failed to pay off its national debt—maybe the United States? Fortunately, all the pundits assured me, "of course" that wouldn't happen. Being reassured sometimes makes me nervous, so I checked some figures—a particularly impressive one being the level to which our national debt has ballooned. As of this writing, that figure is more than $14 trillion. Since I don't have any more idea what a trillion is than a chicken, this image helps: A stack of $1,000 bills 4 inches thick equals $1 million. A billion dollars is a similar stack, 358 feet high. A trillion-dollar stack would be 68 miles high. So $14 trillion is a stack of $1,000 bills—almost as high as the driving distance is long from Chicago to Boston.[16]

The current national debt amounts to more than $46,000 *per citizen* in this country. Less than a majority of citizens are taxpayers, however: Repayment of the debt *per taxpayer* comes to $129,000—*in addition to* normal levels of national, state, and local taxation to fund ongoing operations of government. Oh, and then there's the accumulation of interest on the $129,000 over the years our average taxpayer is heroically trying to pay it off.

I'm just a hick with no expertise in these dizzying realms—but a hundred and twenty-nine thousand? *It ain't gonna happen.* Which is to say that reneging on our national debt, either outright or by way of hyperinflation, is as inevitable for us as for Greece (or Spain, Portugal, Iceland, Great Britain, or any other cripple du jour in the global intensive care unit); and that the notion that we can keep borrowing our way into the future is a hallucination.

Access to Food

I don't want to waste too much time with doom 'n' gloom, but unless we recognize the coming constriction of declining oil supplies, together with the increasing fragility of national and global economic systems, and the impossibility that we are going to make our stupendous national debt go away—we will be planning for a future that is not going to happen.

Who knows how the near future will play out? But when complex, centralized systems fall apart, the first changes will likely be in where and how we get our food. Food production must become more local, more small-scale, more driven by human (and animal) energy, with a majority of families producing a portion of their own food out of necessity. Those with experience at growing their own food will be better able to provide for themselves—and to be of service to neighbors who don't have a clue.

A small-scale flock of poultry, creatively managed in imitation of nature, can be a key part of the homestead or small farm's food production efforts, helping us reach our goal of being less dependent on food from somewhere else. The time to enlist a poultry flock as partners and mentors is now. *Climb ye learning curves while ye may.*

2 | THE *INTEGRATED* SMALL-SCALE FLOCK

The poultry husbandry suggested in this book is based on allowing the flock to live as closely as possible to the way Ms. Natural Chicken would live on her own in the wild. And on the observation that, if we do so, the birds in return can help us improve the homestead or farm and make it more productive. One reason those ideas sound so good to me now is because that's not exactly how we started out.

How We Got Started

Ellen and I met in a Zen monastery (bet you weren't expecting that one!). True love blossomed when we discovered we shared a passion for—compost. After we married and lived for a couple of years far too close to Washington, DC, we followed the call of the compost westward and ended up in a two-hundred-year-old house on 3 acres in a crossroads rural village in northern Virginia—we called our new home "Boxwood" after the prominent border of mature boxwood out front. Though, as said, compost was foremost in our minds, my daughter Heather, who was living with us at the time, was thinking *Chickens*. We approved her project and, not long after our move (in the spring of 1984), we ordered twenty-five New Hampshire chicks through the mail. (Later that summer we ordered a mixed batch of bantams as well. Even at that early date, we were enamored with the idea of breeding our own chickens naturally, and we had read that bantams make great mothers. See chapter 4 for more about bantams.)

We brooded the chicks in a cardboard appliance carton and at four weeks moved them to a small shed that was on our place when we moved in. Though we lost most of that first group to a weasel and had to start over, it was not long before we had a busy flock in a wire run off the coop. After supper we would take dessert out on the grass by the run, enjoying the show. Our neighbors probably thought us peculiar—but we found the chooks a lot better 'n television. And not so long after that we had the thrill of every beginning flockster: our first egg.

But I was troubled by our project. Within two weeks of being allowed onto the grass in their wire-enclosed run, our scampering young chooks had voraciously eaten every last blade of grass. They didn't look quite so pretty against a patch of bare dirt festering with chicken poops—nor did they seem happy with the change. Each time I threw in a cricket, or grubs from the garden, or lettuce and cabbage trimmings, their obvious hunger for these foods made the dry stuff from the feed bag look ever more lifeless in comparison. Memories of my grandmother's flock, which she allowed to range freely over a 100-acre farm in the North Carolina piedmont, made my chicken run seem more and more a prison—or more accurately, a concentration camp.

Then Ellen told me she'd heard about a farmer in the Shenandoah Valley named Joel Salatin who was raising his chickens on pasture—growing meat birds in movable pens, following beef cattle with laying chickens eating fly maggots out of the cowpies.

Knowing that somebody was managing to "do chickens" somewhat as Granny had once done excited me—and crazy stuff like chickens eating out of cow poops appealed to the kid in me. I had to give this guy a call! I did so, clueless how much I was imposing on the time of a busy farmer and prolific writer, but Joel was both gracious and encouraging. He urged me to explore options for the Great Liberation—the exodus of the flock out of the dead-zone chicken run. He was pleased with my obvious enthusiasm for giving my birds a more natural life on grass. "There aren't many of us out there," he said, "but we know there's a better way."

I built a 10-by-12-foot Polyface-style mobile pen (Polyface being the name of Joel's farm) and began raising batches of broilers on our acre or so of pasture. Dinner guests remarked in astonishment, "Man, chicken was never like this!" Before long I discovered electric net fencing; its use allowed me to do what Granny did with her free-ranging flock, but within the limits imposed by close neighbors on either side.

Poultry husbandry ever since has been, for me, about giving the flock the most natural, happiest lives possible; and discovering all they can give as partners in return. Some key ideas have guided our progress.

Imitation of Nature

The great mistake of modern agriculture is assuming we can *control* the processes of growth and yield, and that problems or needs that arise can be dealt with using one-for-one solutions—most likely needing to be purchased. Preventing disease in flocks? *Antibiotics.* Soil fertility? *Easily applied supercharger chemical salts.* Crop-damaging insects? *Blast those guys with pesticides.* Runoff pollution? *Not my department.*

We have lost the intuitive understanding that natural systems are highly diverse and interconnected, one species supporting others and benefiting from them in return; and therefore the practice of agriculture must see "the whole problem of health in soil, plant, animal, and man as one great subject."[1] The key to success with the homestead flock is a return to that vision. The setting for the flock is not our backyard, but its total ecology—another way of saying that, if *Gallus gallus domesticus* evolved to a state of near-perfect health ranging the landscape and eating live, self-foraged foods, the key to success is a life as open and unconfined as our situation allows, with access to as wide an array of natural feeds as possible.

Imitation of nature means most of all fitting flock husbandry into the great cycles of reproduction and growth, death, decay, and renewal—in which nothing is squandered; no spillover from one sphere of activity poisons the whole; the vigorous pushing of a species' own agenda places necessary limits on the excesses of others; and the potential of the underlying foundation, the soil, to support ever more robust, abundant life *increases* with each round of growth and death.

"Efficiency"

Modern agriculture is said to be more efficient, as proved by the immense quantities of dressed broilers and eggs produced in its factories. But it's a peculiar notion of efficiency that touts production of 1 calorie of food energy for 10 calories of energy input—in contrast with a model in which much of the energy to produce eggs and dressed poultry comes directly from the nearest star. And a system that produces $1.29-per-pound chicken while polluting waterways is hardly a model for *efficiently* providing for human needs.

Biological efficiency offers itself anytime we're inclined to make use of it. Getting in one operation free fertilization of the garden, control of slugs, tillage of weeds, waste management control, with a dozen eggs thrown in as a bonus—what could be more efficient than that! The key is to avoid sticking the flock in some isolated corner all to themselves, and to make them happy by giving them good work to do.

Closing the Circle

This book is about patterns of poultry husbandry that are more independent of outside purchases from far away, and more dependent on home efforts and local resources. It sees the homestead or small farm as an energy system; and wise usage as seeking out all available energy resources on the home place, while avoiding energy leaks out of the system, thus creating a more closed and resilient whole. It assumes that local resources will become apparent as we stop looking to purchased inputs to fill every need. (See the caveat in the sidebar "Dependencies.")

Poultry are arguably the starter livestock par excellence—easier and cheaper to set up on a small scale, with perhaps more to offer as working partners than any other livestock option. And they easily reproduce, adding yet another dimension to the independence of outside inputs.

Putting It All Together

It is the organizing principle of this book and my approach to poultry husbandry that the ultimate efficiency is to find every opportunity to fit the flock into larger patterns of diversity, individual and ecological health, resource utilization, and recycling of "wastes." Let me tell you about my winter greenhouse flock, shown in figures 2.1 through 2.4, to illustrate the operative concepts, and give you an idea of what to expect in this book.

I *hate* confining my birds, but for years I kept the flock inside my main poultry house in winter because I saw no alternative—if released to the pasture as they are in the green season, the constant scratching of the chickens would destroy the dormant sod. The nightmarish result would be one I had seen all too vividly, early on in the life of our flock here: a bare stretch of frozen dirt, dotted with chicken poops just waiting for the next rain to run for every natural water system between here and the sea. I had already found the natural solution to best manure management, deep litter over an earth floor, but assumed that solution only worked within the walls of the poultry house—I saw no way to bring it outside.

One of our two garden areas is behind the greenhouse. A constant about my gardening practice has been the planting of more cover crops for soil improvement every year. Four years ago I had managed to establish a fine lush mix of cover crops over the entire

Dependencies

Becoming more independent of purchased inputs is always a work in progress. Simply to make a start with a home flock is to decrease our dependence on purchased eggs and poultry, even if we are buying all our feed for the flock. Practicing more natural husbandry will likely eliminate purchased medications entirely. However, most of us, especially at the beginning, will be heavily dependent on resources brought in from outside. Honest recognition of the nature and extent of our dependencies, rather than an illusion of independence, is more likely to provide the incentive to climb the necessary learning curves toward closing the circle on the homestead or farm, while we have the chance.

I am keenly aware of my dependence on purchased resources in three areas, and constantly try to answer the question of how my husbandry would have to change in the event of an economic calamity, personal or systemic, that prevented bringing in resources from outside.

Purchased stock: Every year I buy in several batches of hatchlings—chicks, ducklings, goslings—and occasionally adult birds, such as guineas. This is the dependency that concerns me the least, since use of natural mothers to produce new stock, discussed in detail in chapter 27, is a constant and fundamental part of my husbandry. With every species I raise, I hatch out at least some of my new stock using broody females who retain the instinct to incubate and to nurture young—I am thoroughly seasoned in the rhythms and requirements involved. Being able to buy stock from elsewhere is convenient, to be sure. However, if necessary, from one season to the next I could be totally independent of purchased hatchlings, using the services of proven broody hens (and ducks and geese).

Of course, in such an eventuality I would have to be concerned about maintaining productive capacity and genetic diversity in my birds. That is why the principles of conservation and improvement breeding (discussed in chapter 25) are so important.

Purchased feeds: Even though I make my own feeds, I am heavily dependent on the purchased primary ingredients—small grains, corn, peas, a few supplements—to feed my flock. It is my keen awareness of this dependency, as much as a concern about feed quality, that underlies my compulsion to increase the portion of my birds' diets supplied by home resources, and even to cultivate "recomposer" species to help feed the flock using conversion of organic "wastes." Indeed, you could say that almost every strategy in this book is about finding ways to minimize purchased feeds. Whatever crisis emerges to limit the ease of bringing in feeds from outside, dealing with it would simply be an extension of what I'm already doing to feed my flock on my own.

Electric net fencing: I rely heavily on electric net fencing and energizers, high-tech inputs I cannot duplicate on my own. This is the area in which I am most uncertain what management changes would be required should our agricultural supply system ever falter. I certainly do not envision closely confining my flock in any circumstances, but can imagine in a post-electronet situation making more use of "day ranging" and "compost runs." The big challenge would be protection from predators. Ranging strategies would key on the fact that most wild predators are nocturnal. A major exception is dogs, whose diurnal predatory tendencies are another matter entirely.

greenhouse garden—crucifers for their "cleansing" effect on the soil, peas to set nitrogen, small grains for their addition of root biomass. I knew the cover would be a great boost to the garden's soil when I sent in my tiller chickens come spring.

But why wait for spring? Suddenly I saw how to get an early start on the flock's tillage work—and to bring deep-litter practice outside.

I built two 8-by-8 pens for my mixed flock—chickens, ducks, and geese—in the far end of our 20-by-48-foot greenhouse. The project was easily done. The greenhouse itself provided shelter from the brutalities of winter; all that was required in addition was some chicken wire on light wooden framing to keep the flock out of the growing beds. I furnished the pens with the usual henhouse accessories: dustboxes for dust bathing, nestboxes mounted above floor level on the end wall of the greenhouse, two

Fig. 2.1 My 20-by-48-foot greenhouse offers perfect winter housing for my mixed flock.

simple 2-by-4 roosts, and hanging feeders. I left an earth floor in the pens and covered it 8 to 12 inches deep with oak leaves. I found it was okay to leave the waterers inside the pens if I had chickens only in them. When ducks and geese were part of the greenhouse flock, it was better to keep the waterers outside to avoid wet litter—they're pretty messy with their water. On especially cold nights, I moved the waterers to the basement to prevent freezing.

On the far side of the pens, the greenhouse's end door opened directly onto the garden. I propped that door open each morning, giving the birds access to the outside during the day.

Fig. 2.2 Two pens, each 8 by 8, give access to an outdoor feeding/exercise yard. Note that each pen has its own door–the flock can be separated into two groups if needed–and that either door swings into blocking position to keep the birds out of the greenhouse crop beds. The plywood lids cover vermicomposting bins.

Fig. 2.3 All the comforts of home: Each pen is furnished with laying nests, dustbox, waterer (sometimes kept outside), hanging feeder, and one of the most important "accessories" of all–deep litter. Not shown are the 2-by-4 roosts I install at night, fitting them onto the covers of the dustbox, and the framing of the pens.

I installed a door on each pen, so I could maintain two subflocks in them if needed. Each door was designed to swing into blocking position, on the near side of the pens, to keep the birds out of the greenhouse's growing beds.

I surrounded the garden with electric net fencing to protect the birds and confine them to their work. When I first put them in the greenhouse pens, they feasted on the green forage while tilling in the cover crops I had grown. And to be sure, the last green shred of the cover crops disappeared, leaving that intolerable patch of bare unprotected dirt. Ah, but I was ready for the next phase—turning the space in effect into a giant compost heap, using several loads of round-bale hay from a nearby farm, spoiled for use as feed but just fine for my needs.

I would have preferred taking the necessary mulching material off my own pasture with the scythe, but it had been a dry season, and there wasn't nearly enough to be had. Any number of other organic debris types—autumn leaves, spent crop residues, feed mill residues like corncobs and husks—would have served as well.

The deep organic duff absorbed the poops, preventing their loss to the rains as runoff pollution while retaining their mineral fertility and boosting the diversity and health of the garden's soil food web. The soil never froze under the heavy mulch, so earthworms and slugs remained active, furnishing superb live foods for the busy chickens. They feasted as well on worms from my vermicomposting bins while working the worm bin bedding into the mulch, and enjoyed small but essential "salads" from the greenhouse, either green forage cut from the growing beds or trays of greened sprouts. In late winter the seeds in the round hay bales (they're the reason you should *never* use hay as a garden mulch, right?) began sprouting, and the birds cleaned them up for more free food.

Using this strategy, my flock stayed healthy and contented all winter long. Poultry are cold-hardy, so they enjoyed the sunshine, fresh air, and exercise in their winter yard in all but the nastiest weather. When the weather turned foul, they retreated inside—as they did when night fell, when I would shut the greenhouse doors. My layers maintained both quality and quantity of egg production at levels never matched in previous winters, and I had no problems with frozen combs.

I had no way of measuring the differences, but theoretically the body heat of the flock—maybe 400 pounds or so of live bird at times (the first winter I used the mulched yard, for example, I had two African geese, half a dozen heavy-breed ducks, and four dozen dual-purpose chickens)—moderated the temperature extremes inside the greenhouse at night. Further, the carbon dioxide in the birds' exhalations could well have boosted the growth of the winter crops, which metabolize CO_2. (If the idea of factoring in chicken breath seems far-fetched to you, note that gardeners in the Netherlands, known for expert greenhouse growing, spend good Dutch money to buy bottled CO_2 to pump into their greenhouses and boost crop growth. Why they don't just get some chickens is beyond me.)

By spring the flock had assisted the breakdown of

Fig. 2.4 **Feeding my chickens earthworms from the worm bins and fresh green forage grown in the greenhouse ensures that they are constantly turning the deep mulch over their winter yard.**

the hay into something between a finished compost and a mulch, the soil underneath becoming increasingly friable as the cover crop roots decayed, and with not a weed in sight. I moved the flock onto the pasture, laid out the garden beds, and the new growing season was instantly up and running at full stride.

How better than in that mulched winter yard could I have enlisted my flock to address so many needs at once? How more completely could my concerns—for soil fertility; feeding, contentment, and health of the flock; and prevention of spillover pollution into the wider environment—have become "one great subject"?

3 | YOUR BASIC BIRD

This chapter is an introduction to the domesticated chicken. It is not a zoological treatise: Each topic here about derivation, anatomy, and behaviors of *Gallus gallus domesticus* ties in, later in the book, to practical strategies to take full advantage of our chickens' natural inclinations and talents.

There is disagreement about the domestication of the backyard chicken. It is descended largely from *Gallus gallus*, the wild Red Junglefowl of Southeast Asia, shown in figures 3.1 and 3.2; though it is possible that the domesticated version was the result of a cross between *Gallus gallus* and *G. sonneratii*, the Grey Junglefowl. Domestication may have occurred as early as ten thousand years ago in modern Thailand or nearby regions of Southeast Asia.

For our purposes the precise time and place of origin, and details of ancestry, are less important than a couple of other interesting things. Chickens are birds of the order Galliformes, along with wild relatives like partridge, grouse, and prairie hen. Interestingly, with the exception of domestic waterfowl and pigeons, other domesticated fowl are all galliforms as well—turkeys, guineas, pheasants, quail, and peafowl.

Fig. 3.1 Red Junglefowl cock. PHOTO: ISTOCKPHOTO.COM

Fig. 3.2 Red Junglefowl hen. PHOTO COURTESY OF DR. L. K. YAP

Galliforms are primarily ground-dwelling species. Though they all have the ability to fly, they do so for only short distances—to escape predators, get over barriers, or roost for the night in trees. The most significant thing for us about their ground-dwelling nature is the fact that food—a diversity of both plant and animal materials in all species—is something to be found primarily *on the ground*. If we give our chickens—and turkeys and guineas—the opportunity, they will do what self-respecting galliforms are used to doing: hustle up a lot of their grub on their own. As well, some of these species—chickens prominent among them—*scratch* for food such as worms and grubs beneath the surface of the ground. We will see how, as smart flocksters, we can "capture that technology" for our own purposes.

Though domesticated chickens and Red Junglefowl are so closely related that they easily hybridize along the edges between their native habitat and human settlements, there have been a couple of key genetic mutations since *Gallus gallus domesticus* split off from its *G. gallus* ancestor.

Like all wild birds, the Red Junglefowl hen laid eggs only during the breeding season—egg laying was exclusively about reproducing the species. Selective breeding following domestication seems to have altered a hormone receptor that relates reproduction to change in day length. From that point egg laying in chickens was no longer tied to a specific breeding season but continued through much or all of the year. Other mutations seem to have increased capacity for appetite, growth, and body size. The result of these genetic changes was a fowl capable of greater production of eggs and flesh than required for survival by their cousins in the wild.[1]

Nomenclature

A *hen* is a mature female chicken, one year or older; a younger female is a *pullet*. A male up to a year old is a *cockerel*; an older one is a *cock*. The distinctions are more about size than sexual maturity: Cockerels begin mounting the females in the flock as early as three months of age, even earlier for particularly precocious breeds; and onset of lay for pullets averages twenty-two weeks or so—whereas hens and cocks reach their adult weights at about one year.

A *chick* is usually understood to mean a young chicken, still in the down stage (not yet feathered). But then what is a young chicken from the just-feathered stage to adulthood? If we specify gender, we can use *pullet* or *cockerel*. Technical terms in the butcher's trade might be *poussin* (from the French word for "chick"), a tender young chicken no older than twenty-eight days at slaughter, or *spring chicken*, a young table chicken a little less than 2 pounds. But I really do not know a good, generic term for a chicken at this stage of life. In this book I am most likely to use the awkward term *young growing birds*.

I sometimes use *chook* to mean "chicken." The term originated in Australia and New Zealand, but is being used increasingly among North American flocksters as well, usually I think with an implied sense of affection.

And what is a *flockster*? That is my own word, to define someone who is enthusiastic about keeping chickens or other poultry as enjoyable, respected, and productive partners.

Finally, what is a *rooster*? A prudish euphemism. The proper word for a male chicken is and always has been *cock*. At one time it was the only word used—in the King James translation of the Bible, it was the "cock" that crowed three times, not the "rooster." The cock has for a long time, and in many cultures and languages, been a symbol of resurrection, of the sun, and of the male sexual member—doubtless because he stands up proudly at dawn to greet the sun as it ushers in the renewal of the day. In old Anglo-Saxon, the word for the male chicken and for the penis were one and the same—a no-longer-tolerable usage in the priggish Victorian era. By the nineteenth century in

America and Australia, the male chicken was most typically a "rooster"—an especially silly euphemism, if one had to be found, since female chickens "roost" the same as males. Note that the euphemizing compulsion became even more extreme—the culinary terms *leg*, *breast*, and *thigh* were renamed *drumstick*, *white meat*, and *dark meat*. The leg was even at times referred to as the *second wing*! Ah well, at least we still describe someone who is brashly self-assertive as "cocky."[2]

The Working Model Chicken

Good management requires an understanding of the chicken's basic working systems. In fundamental ways they are similar to our own—the digestive processes for converting food to energy and tissue, for example; or reproduction based on specialized male and female cells, each carrying half of the DNA code. We need to pay closest attention to the points at which we differ—a digestive system that does not include chewing, for example, and reproduction based on that biological marvel, the egg.

Dimorphism

Notice in figures 3.1 and 3.2 that the Red Junglefowl is strongly *dimorphic*—that is, the cock and the hen are distinct in many ways. The cock's size is noticeably larger—and his comb and wattles are much larger—than those of the hen. The cock is much more extravagantly colored, and bedecked with large showy feathers around the neck, over the saddle, and arched out from the tail—as befits his courtship displays. The hen has much tighter feathering, in subdued colors—which better camouflages her for protection from predators while she incubates her eggs and rears her young.

This dimorphism is retained by domesticated chickens as well. In the majority of particolored breeds, the male will have the more showy coloring. Even in solid-colored varieties, in which cock and hen share the same basic color, the cock is likely to exhibit a somewhat deeper tint, with more gloss. The cock's hackle, saddle, and tail feathers will be more prominent, with a higher sheen. Hackle, saddle, and covert feathers will be pointed and narrower in the cock—equivalents on the hen are broader, with round ends. Such differences in feather shape are obvious by the third set of feathers. The rare exception is hen-feathered breeds, such as Sebright bantams. However, the cock always has the larger comb and more fleshy wattles and attains larger adult size. Such dimorphic traits make sexing of chickens easy, once the secondary sexual characteristics manifest, and the size difference may be relevant in culling strategies as well.

Anatomy

The illustrations in this section show the key features of the external anatomy, including the spur and the larger comb and wattles of the cock, and the internal organs, both digestive organs (the same in both genders) and the reproductive organs of the hen.

Comb and Wattles

The comb, as noted above, is larger in the male and serves to attract females. Functionally, it helps radiate heat out of the chicken's body as blood circulates through this exposed tissue—such dissipation is second only to panting in helping the bird deal with hot weather. Thus breeds with large single combs might be preferred in hotter climes. Such combs are more subject to frostbite in winter, however. Far northern flocksters might prefer breeds with minimalist combs for this reason.

The comb can be an indicator of health—changes in color or texture could signal a problem. Combs of pullets turn a brighter red when they are about to start laying.

Selective breeding has resulted in a diversity of comb styles: single, rose, pea, cushion, strawberry, buttercup, and V comb, as illustrated in figure 4.1.

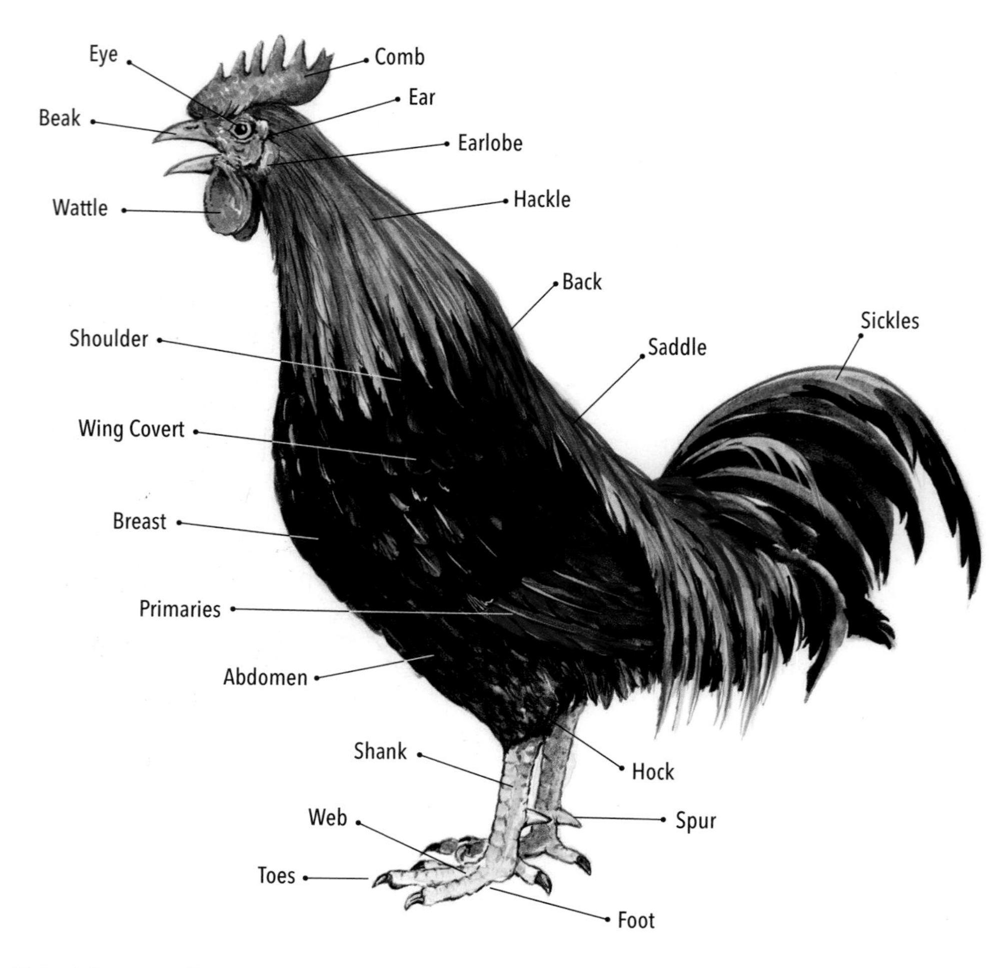

Fig. 3.3 The chicken's basic parts list.

Shanks and Feet

The color of the shank varies by breed. Shanks of ancestral Red and Grey Junglefowl are "clean"—lack feathers—but a mutation emerged among certain Asian breeds that resulted in feathering of the shank as well. When feather-legged breeds arrived in Europe in the 1800s, breeders enhanced the trait.

Growing from the back of the shank of the cock, above the back toe, is the horny spur—shorter and more blunt in a farm breed cock, longer and quite sharp in a game cock such as the Old English Game—and capable of killing a rival. My Old English Game hens occasionally grow spurs as well—much shorter than the cock's but needle-sharp.

Most breeds of chickens have four toes, but a minority such as Dorkings, Faverolles, and Silkies have five.

Feathers

One of the distinguishing features of birds as a class among other vertebrates is feathers or plumage. Feathers constitute the most complex of all *integumentary systems* among vertebrates—the covering of the body, including the skin and its associated appendages, such as hair or feathers or scales. Feathers help protect, waterproof, and insulate the bird—and serve also as sensory apparatus. Feather color and pattern can either enhance the bird visu-

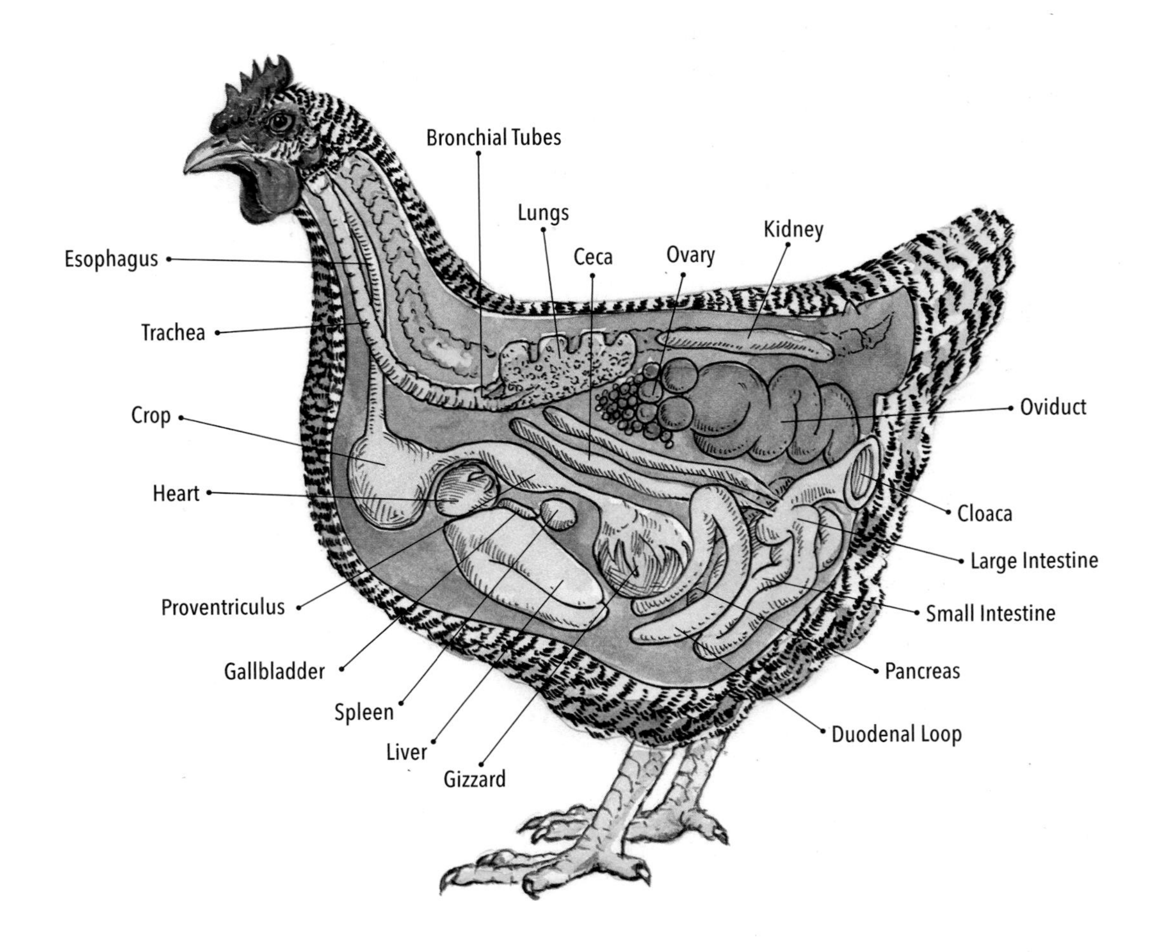

Fig. 3.4 Digestive and reproductive system of a chicken.

ally—as for mating displays—or help it blend into its background to escape the notice of predators.

It is essential that the feathers remain in good condition to fulfill their functions, so chickens spend a lot of time *preening* their feathers: Their tails, which support the large tail feathers, also contain the uropygial gland, or preening gland. The bird expresses preening oil from the gland and uses it to dress the feathers, cleaning and conditioning them in the process and waterproofing them with the oil—partially in the case of chickens, almost completely in the case of waterfowl.

Even given the best of care, however, feathers become worn and broken and lose function. A periodic *molt* is necessary—shedding all the feathers and growing new ones. A chicken does not molt in its first year, since its feathers are still so new. Thereafter, however, it molts once a year, in fall or early winter. Be prepared for this annual ritual—the birds can look pretty appalling missing half their old feathers and with new ones coming in like porcupine quills. Do remember as well that feathers are almost pure protein, so replacing them all requires a *lot* of resource. Despite the mutation discussed above that led to egg laying by domesticated hens year-round, it is not surprising that they lay fewer eggs when providing for this critical requirement.

Perhaps the most important thing to appreciate

Molting

An annual phenomenon among your flock is *molting*—replacement by the chicken of every one of its feathers—which may occur anytime from late summer to early winter. Molting does not occur simultaneously in the whole flock—individuals molt on their own schedules. In my experience, the flock average for molting varies widely from one year to the next, for reasons unknown.

Over the course of several weeks—up to three months or more—the chicken sheds its feathers and grows new ones. It is never completely bare of feathers but can look pretty odd when most of the feathers are new and haven't fully opened—like porcupine quills. The molting sequence—the pattern in which feathers are shed and replaced—starts with the head and ends with the tail, with some areas undergoing molt simultaneously.

Molting is strongly related to egg production. While molting, hens lay far fewer eggs and may cease production entirely. Top-performing layers tend to molt later and complete the molt faster than less productive hens.

If your chickens are well fed and free of stress, don't worry about them as they begin losing feathers at a rapid rate in the fall.

about feathers is just how tough they make our birds. They protect so well from the weather that many keepers of turkeys allow them to roost outside even through the harshest parts of winter. Chickens are not quite that hardy, but their feathers insulate them so well in winter that we're probably better advised to keep their housing open to plenty of fresh air rather than add heat to it. On the other hand, in summer the feathers inhibit escape of heat from their bodies, so shade is essential to prevent overheating.

Digestive System

Figure 3.4 shows the main features of the chicken's digestive system. Food is picked up into the *mouth* by the *beak*. The bird has no teeth—hence anything in short supply is said to be "as scarce as hen's teeth"—and does not chew its food. The mouth adds saliva, containing digestive enzymes, as it passes the food on into the *esophagus* or *gullet*, which deposits it into a flexible pouch-like enlargement of the esophagus called the *crop*. The most important function of the crop is storage. If the chicken finds a treasure trove of food, more than can be accommodated in the digestive organs at the moment, she can take advantage of the windfall by stuffing quite a bit of excess into her crop—like a chipmunk stuffs its cheek pouches. Pick up one of your hens after she's been feeding, and you can feel the distended crop, slightly to her right at the base of the neck—the size of a golf ball or larger. In addition to storing the food for later digestion, the crop bathes it with fluids, which soften it and ready it for digestion. Hard foods such as seeds might remain in the crop twelve hours or so.

When the digestive system farther down is ready, foods in the crop are drawn into the first of two parts of the chicken's "stomach," the *proventriculus*, which mixes them with hydrochloric acid and various enzymes that start the process of digestion. But the food still hasn't been chewed, right? That's the next part of the process: The mix of softened foods and digestive juices enters the *gizzard* or *ventriculus*, the organ in avian species that takes care of both the grinding and mashing of foods we accomplish with teeth and tongue and the early stages of digestion that take place in our stomach. The gizzard is made of two sets of powerful muscles surrounding a pouch with a thick, tough lining that holds the food. One opening into the gizzard from the proventriculus brings in the food; another passes it on into the small intestine for continued digestion and absorption of nutrients.

As for the "chewing" of the food in the gizzard, it depends on the presence of *grit* inside the pouch, bits of stone the bird swallows to serve as grinders. Birds eating a natural diet with a lot of whole seeds and fibrous green plants need plenty of grit—more than confined birds eating commercial feeds. The pieces of grit are retained in the gizzard until completely worn away, so the chicken needs to continually renew its supply. If foraging outside, she will likely pick up all the grit she needs on her own. However, it is easy to supplement with purchased granite grit, which is cheap, by way of insurance.

The soupy puree from the gizzard enters the *small intestine*, which adds digestive enzymes from the *pancreas* for protein digestion and *bile* from the *gallbladder*, synthesized in the *liver* and essential for the digestion of fats and absorption of fat-soluble vitamins A, D, E, and K. By the time this liquefied food has passed the length of the small intestine, most of its nutrients have been absorbed through the intestinal walls and into the bloodstream.

The residual fecal material fills the *ceca* (singular *cecum*), where some of its water is reabsorbed, and where fermentation breaks down remaining fibrous material, a process that produces several fatty acids and eight B vitamins. Some, though not all, of these nutrients are absorbed.

In the *large intestine* or *colon*, more water is absorbed, concentrating the fecal matter, which then moves into the *cloaca*, a stretchy pouch that serves as the final staging area for the poops. Note that the eliminatory functions that are separated into bladder and bowel in our case are combined into one in the cloaca. In other words, chickens do not urinate. Their kidneys do extract nitrogenous metabolic waste as ours do, but they excrete it as uric acid, in the form of a coating of white crystals over the surface of the feces, which are expelled through the *anus* by contraction of the cloaca.

And really, that's just the beginning of the story. A good part of the mineral content taken in with the food is not absorbed and gets passed on into the soil's miraculous digestive processes, a gift to its fertility. As well, the poops carry inconceivable numbers of microbes essential to digestion in the chicken's gut. Do they survive expulsion and enter the soil food web as living members of that complex community? I'm intrigued with that question, but my research to date has failed to answer it.

Reading the Poops

How does a mother judge the state of her child's internal health? One way is to see what his bowel movements look like. In the same way, you can judge whether your chooks' digestion is efficient and healthy with a form of divination I call "reading the poops."

A dropping from a chicken who is eating a diversity of natural foods is gray—or greenish if the bird has been eating a lot of fresh plants—and firmly shaped, with a coating of white uric acid crystals from the kidneys. In the henhouse you would notice something else about it: It does not present a foul stench to a nose at standing level.

A poop from a bird who has been eating commercial feeds, with limited access to fresh greenery and other natural foods, is loose and splattered, black or darker in color, with a foul odor even at standing height.

There is one caveat to the above reading of the poops: The ceca empty their contents directly to the outside two or three times daily. Cecal droppings are looser, mustard to dark brown in color, and smell worse than normal droppings. So do not be concerned if you see such droppings here and there, so long as most of the birds are producing poop that looks like the healthy one described above.

Reproductive System

Birds suffer high rates of predation, a threat to their kind they could not survive without an extremely effective adaptation: the egg. Egg laying and the hatching of young in sizable groups—young who mature at an astoundingly rapid rate—allow for reproductive rates far beyond those of mammals, ensuring that enough offspring live long enough to reproduce despite the relentless onslaught of predators. This high reproductive rate is one of the things that make the raising of chickens so easy and so practical.

Another practical aspect of reproduction via eggs: An embryonic mammal receives nutrients for growth on an as-needed basis via the womb. That is not an option for the chicken embryo, who must make its entire growth—from fertilized speck on the yolk to fully formed chick ready to hatch—entirely out of nutrients stored in the egg. The egg *has* to be a powerhouse of nutrition, and that is a major reason it has been so revered as a food.

Reproduction on the cock's side centers on two *testes*—held deep in the body cavity, as opposed to externally as in male mammals—which produce sperm in massive numbers. (You can see the testicles in a young male in figure 28.28.) Two ducts called *vas deferens* carry sperm from the testicles to two elongated bumps on the back wall of the cloaca called *papillae.* When the cock copulates with the hen, he ejaculates sperm from the papillae into the hen's cloaca, from which they are moved by muscular action up into the *oviduct.*

Reproduction on the hen's side, shown in figure 3.4, begins in the single *ovary*—as in most birds, a second ovary exists but is vestigial—containing thousands of reproductive cells, some of which develop into a cluster of tiny *yolks* or *ova*, which grow as yolk material is added. *Ovulation* is the release of the most mature yolk into the second part of the hen's reproductive system, the *oviduct*, a convoluted tube about 25 inches long. If the hen has been mated by a cock, his sperm will fertilize the ovum in a brief stay in the first part of the oviduct, the *infundibulum*. The yolk next moves into the longest part of the oviduct, the *magnum*, where over the next three hours the *albumen* or egg white is added. The inner and outer *shell membranes* are added in the next section, a constricted part of the oviduct called the *isthmus.* In the next section, the *shell gland* or *uterus*, the shell is laid on, largely as a deposit of *calcium carbonate.* The hen actually mobilizes almost half the calcium to make the shell out of her bones, so a large intake of calcium in her diet is essential to support both shell strength and bone density. After about twenty hours, the finished egg moves into the last segment of the oviduct, the *vagina*, a muscular sheath whose contractions force the egg out when it is time to lay. Before expulsion, the vagina coats the egg with the *bloom* or *cuticle*, which both eases passages and upon drying helps protect the porous egg shell from invasion by bacteria.

When thinking about the laying of the egg, your reaction might well be one of consternation: *Oh yuck—you mean the egg is laid through the poop-hole? You'd think nature would have come up with a better design than that!* Actually, she did. The contraction of the vagina to expel the egg coincides with a prolapse of the uterus that completely occludes the cloaca—there is no contact with the interior surfaces of the cloaca as the egg is expelled directly to the outside.

Of interest is the presence of *sperm host glands* at the juncture between the vagina and the shell gland. These glands serve as storage sites from which sperm are moved by the oviduct up to the infundibulum. Storage of sperm in the glands enables continuing fertilization of ova long after the hen's last mating with a cock—unlike mammalian sperm, the cock's sperm can survive at the hen's body temperature for two weeks or even longer. The function of the sperm host glands isn't just a fun fact—it will have practical implications later on in chapter 26.

Note that the laying cycle is not likely to be keyed exactly to a twenty-four-hour day, even though it is strongly influenced by photoperiod. Since it takes twenty-five or twenty-six hours for a new egg to form, the hen—who starts the cycle in the morning—will lay her egg later each day. She will not lay much later than three o'clock in the afternoon, however; so when the laying of the egg occurs this late in the day, she ceases production for a day or so to reset the clock and begin laying once again in the morning.

The Inner Chicken

An understanding of natural chicken behaviors will help you better care for your birds. The more you spend time observing your birds, the more you will understand the nuances of their social organization, what makes them tense, and what makes them content—even the language they speak. You will end up a more skillful flockster—and will have a lot more fun.

Flocking

Most galliforms are *flocking* species—a major survival strategy is living together in closely bonded groups. The flock has a hierarchical structure, with a place in the pecking order for each flock member. Cooperation and mutual protection are key parts of flocking behavior. You will notice that a hen who finds a trove of good things to eat will not only start gobbling them up herself but will call others in the flock to come and get in on the feast as well. One flock member who sees a predator will make an alarm cry, triggering a rush by everyone for cover.

Understanding flocking will prepare you for behaviors that might otherwise be unsettling. Hens who seem to be fighting may well simply be having a "discussion" about relative position in the flock's hierarchy, not a cause for concern. To be sure, introducing new numbers to the flock results in more intense discussions with potential for actual injuries, so requires careful monitoring and occasional intervention (more on this in chapter 20). On the other hand, flock members have an instinct to zero in on one who is sick or injured—at the extreme, a literal pecking her to death. While gruesome, in a natural setting this behavior is a survival strategy—elimination of the weak sister could make the flock as a whole less vulnerable to predators. As a flockster, do everything you can to prevent stress in the flock, which can bring on elimination behaviors. Isolation of a temporarily disadvantaged member can help give it the opportunity to recover and then rejoin the flock.

Language

Communication within the flock involves a complex set of unique vocalizations, which in the broadest sense we could consider a "language." Communication begins even before hatch: The pipping of a chick in the egg alerts the mother hen that it is beginning to break out of its shell. Among adults, there are unique vocalizations to mean "food here"; to express contentment, disturbance, fear, or frustration; or to issue threats. Alarm calls can signal whether the predator is on the ground or in the sky. Crowing is an especially complex vocalization, varying with the position of a cock in the dominance–submission hierarchy, or whether he is trying to attract the attention of a hen. A cock in your flock may engage in crowing duels with one in a neighbor's flock hundreds of yards away.[3]

Spend time with your birds and pay close attention, and you will learn to recognize the chicken talk used for specific situations and needs. A frantic note from a brooder chick communicates distress to the flockster, instantly distinguishable from its usual clear peep. A mother hen on pasture has a call to let her chicks know she has found food, another to signal "follow me," and another to help a chick find its way back to her if it has wandered away. You will learn the call of a cock when "tidbitting" his hens, described below—you may even notice that the intensity of his food calls varies with the palatability of the tidbit he is offering (a nutrient-dense cricket being signaled more excitedly than a berry, for example). Hens foraging contentedly together engage in a quiet conversational clucking. And some hens use a sharp, distinctive cry to signal to the world their extraordinary accomplishment of having just laid an egg.

Sexual Behaviors

Wild Red Junglefowl cocks have several mating behaviors we may see in our own flocks. Their brassy *cock-a-doodle-doo* call is issued both as invitation to hens as prospective mates and as warning to rival cocks in the area. Should another cock not be deterred, a serious fight may ensue to establish access to the hens. Given their long, sharp spurs as weapons, the fight can result in the death of the loser. Recognize that aggression in pursuit of dominance in mating success is normal behavior. In a settled flock with plenty of space, cocks will typically work out dominance–submission places in the hierarchy, with little subsequent conflict. Occasionally the competition gets deadly serious, however. (See chapter 20 for more on this issue.)

Once he's taken care of rivals, the cock courts the female by "dancing" in a circle around her, his stretched wings stiff and pointed to the ground. If the hen is impressed, she allows herself to be mounted for copulation. Look for this behavior on the part of cocks in your flock as well. (See the sidebar "Sex in the Barnyard.")

Sex in the Barnyard

One behavior on the part of the cock you might look for in particular is "dancing." Temple Grandin, well known for her work with domesticated animals, has remarked on this behavior: The cock performs a strutting display—circling the hen, wings spread stiffly toward the ground and quivering—which persuades her to squat and erect the shoulders of her wings, welcoming his advances. When the hen's cooperation is encouraged by dancing, mating is not violent and the hen is not injured, even when the cock's spurs are long and sharp. Grandin believes that, having ignored this dancing behavior when breeding our chickens, we have bred modern cocks who have "forgotten" how to dance, with the result that mating is carried on with more violence, sometimes injuring the hen.[4]

I have begun looking for dancing on the part of the cocks in my flocks, and favor for use as breeders those cocks who dance for the ladies.

Further light on such mysteries was offered by an elderly neighbor who kept chickens for many years. Sitting by his chicken pen, he would observe the flock's behavior four or five hours at a time. I think he knew more about chicken mating behavior than any expert in any ag college in the country.

"So if you got two roosters," he told me, "the top guy is gonna have his pick of the hens, and he'll have his own special group that are his. He'll look out for 'em, and he'll tread 'em. The other rooster can tread the other hens, but the top guy will keep him away from the hens he's picked. Well, sir, after about four hours there'll be a change—suddenly the top guy will be treading a different group of hens!"

If my friend was accurate in his description of natural flock behavior, think of the implications: The dominant male gets his pick of the hens—that is, priority when it comes to passing on his genes. But the flock has the instinctual wisdom to know that the subordinate cock also has his role to play in ensuring genetic diversity—in keeping some wild cards in the hand—and affords him the opportunity as well to pass his genes on to the future flock.

The junglefowl cock—along with other galliform males—also engages in a behavior called "tidbitting": If he finds an especially choice bit of food, such as an insect, he bobs his head, makes a special call, and picks up and drops the tidbit repeatedly, in order to attract the hen to come receive the special treat. He may even pick up the tidbit and offer it to the hen with his beak. Though tidbitting is doubtless a bid for romance, I like to think as well that the cock has the instinctual wisdom to know that the hen needs the protein boost in order to make nutrient-rich eggs to produce sound offspring. In any case, tidbitting is about mating and species survival, and is an endearing behavior you will enjoy seeing in your flock as well.

As for the hen, both among junglefowl and among domestic chickens, she takes exclusive responsibility for making a hidden nest, assembling a clutch of eggs, incubating them, and nurturing her young. The instinct to brood may be so powerful that the broody hen can even be used to hatch eggs of other fowl species and nurture the young, or to adopt a clutch of chicks not her own. (More information on working with broody hens is in chapter 27.)

Other Thoughts on Behavior

Remember what was said about the role of predation in shaping gallinaceous fowl. Chickens are extremely sensitive to anything they read as a predator in the environment. Avoid crowding them or making sudden moves within the confines of the henhouse. Avoid carrying large flat objects such as an empty feed bag or large scrap of plywood—they see it as a flying predator and freak out. Because of their instinct to be out of reach of the predator while asleep, chickens will feel a lot more secure with roosts on which to spend the night.

Tidbitting behavior could be your key to making friends with your birds. Especially in the case of a skittish cock inclined to be too assertive in defense of his ladies, you can offer special treats—crushed hard-boiled egg or grubs from garden beds, for example—in order to encourage more friendly relations. It really works.

Keying on natural behaviors is a lot smarter than trying to figure out everything on your own. The first year we had chickens, my daughter Heather—who usually took care of them at the time—was away for the weekend, and I was doing the chores. Each evening when I went out to shut up the coop, I found that some chicks, just days out of the shell, had somehow escaped the fence and, with the coming of evening, had hidden under the coop, cheeping miserably. And each evening I crawled under the coop to rescue them, belly-down in the dried litter that sifted down through cracks in its floor. When Heather returned I asked her, "Say, have you had any problem with those chicks getting out of the fence and being stranded when the chickens go in for the night?"

"Oh sure," she replied matter-of-factly. "I just get the mama hen and hold her by the edge of the coop outside. She clucks and the chicks come to her, and it's easy to pick them up and get them inside."

Never assume you've seen it all when it comes to unique behaviors in your flock. We had an Old English Game cock named Charlie Brown who couldn't mate the hens because his feet were badly deformed. However, he would dance most handsomely for the ladies and would get quite excited when a more fortunate cock mounted a hen, standing close to the action and eagerly clucking advice. Goldilocks, a Golden Sebright hen, exhibited an inexplicable but endearing behavior during her long life here (about seven years). For about three weeks each spring, every time I entered the henhouse she would immediately swoop up and land on my shoulder—and stay contentedly perched there as long as it took for the chore that had brought me in.

A friend passed on my favorite chicken behavior story: His family flock ranged freely during the day and returned to roost in the henhouse in the evening. One windy night while watching television with his family, he heard a persistent tapping at the front door. Opening the door revealed one of his hens looking up at him expectantly. Following her out to the henhouse, he found that its door had blown shut, with the rest of the flock clustered outside. In a crisis, the flock had analyzed the nature of the problem, formed a committee to decide on the best course of action, and commissioned an emissary to carry an appeal for assistance.

I hope this chapter has given you some factual information about chickens that will guide your management practices. But remember, most of what you really need to know you will learn by spending time with your flock and observing their behaviors.

4 | PLANNING THE FLOCK

There is no ideal flock in the abstract—your ideal flock is the one that best achieves your purposes, within the limitations of your own situation. Thinking through your goals and management choices is likely to make for a smoother start with fewer surprises. Here are some questions to address.

"Is It Necessary to Have a Rooster . . . ?"

One of the most common questions I get from folks contemplating starting a small flock is: "Do I have to keep a rooster with my hens for them to lay eggs?" The answer is simple biology: Just as a woman ovulates on a regular cycle whether or not she has a mate, a hen ovulates (makes an egg) in *preparation* for reproduction, whether or not there is a cock around to mate her and fertilize her eggs.

Thus it is not necessary to have a cock in your flock of laying hens. If, for example, you live cheek by jowl with neighbors who might object to the crowing of a cock, you can keep a flock of much quieter hens only. Of course, the eggs will not be fertile absent the attentions of a male; however, the hens will lay just as many eggs. And they will form their own hierarchical social structure without a cock.

Our own preference is to include one or more cocks in our flock. As with most avian species, the boys are the more spectacular of the genders, and we enjoy their flash of color and long graceful sickle feathers. We also like the way the cock completes the natural social structure of the flock, how solicitous he is of the welfare of his ladies, and even his clarion call.

If you plan to have a small flock, say up to a couple of dozen hens, including only a single cock should be sufficient if you just want him there to complete the flock. If you want to be sure all the eggs are fertile, one cock can service from eight to twelve hens. If you do include more than one cock in your flock, see chapter 20 for cautions about dealing with aggression.

Flock Size

Proper size for your flock depends first on your production goals and your level of experience. For egg production, as a rough average you can anticipate two eggs per day for every three layers in your flock. Of course, production is cyclical, so inevitably you will have to lean more toward having a surplus of eggs in spring and summer, when rate of lay is highest; or a deficit in winter, when layers decrease production significantly. Numbers needed to produce dressed poultry should be pretty straightforward if you raise batches of meat birds to put in the freezer—how many do you want to eat, and on what schedule?—but the question gets more complicated if your table fowl will be produced by necessary culling of an ever-changing, mixed-ages flock. In any case, if you are a complete beginner, it's probably good advice to start with a smaller flock and work up as you gain experience.

A good place to start is the number of eggs your family eats on average. Imagine a family of four who are content with an average of an egg a day. Given the rough three-to-two ratio above, six good layers would fill the need for eggs for such a family. If they want to eat on average two eggs a day per person—and chances are good they will, once they start eating the best eggs anywhere—then they would want a flock of a dozen hens.

That's a good start on calculating flock size, but remember that production will drop a good deal in winter, and decide whether you want to increase flock size to compensate (and preserve some of the summer surplus with one of the methods in chapter 29). Remember also that some breeds will not meet the three-to-two production level. You might well choose a less productive breed such as Dorking or Brahma because you love their look or temperament, or one that lays eggs with pastel or deep brown shells such as Ameraucana[1] or Marans, but in that case you should add more hens to the flock. And be assured that you will want to share the world's best eggs with friends and relatives—I would factor another dozen eggs a week into your calculation for that purpose alone.

Now, what about meat birds? As noted, if you want to raise batches of meat hybrids, the math is obvious: How many do you want to put on your family's table, and on what schedule? But let's consider the question assuming that your table fowl will come exclusively from culling your first group of started chicks, and that you want to raise two dozen layers. A good plan might then be to start with fifty chicks "straight run"—that is, in the natural ratio of male and female, which for chickens as for humans is roughly half and half. I'm going to tell you in the next chapter why I always order chicks straight run, and never as sexed pullets only. In this case, starting with the fifty chicks, you'll have about two dozen cockerels to cull for the table—almost enough to have chicken from your own backyard every other week. If you want to eat more chicken but still end up with two dozen started pullets, increase the number of chicks to seventy-five, or a hundred. But in this case, order the first fifty straight run, and the remaining ones as all-cockerels. (There should never be a problem with ordering sexed cockerels only. Again, see the discussion about ordering chicks from hatcheries in the next chapter.)

Of course, the more chickens you want to cull as table birds, the more sense it would make to start with the initial fifty, then start subsequent batches of all-cockerel chicks later in the season.

Some questions about flock size are less obvious. If you plan to pasture your flock, remember that the pasture will not support as many birds in the drier part of summer when the grass isn't growing as fast. If you want to supply as much of your flock's feed as possible out of home resources, recognize that the limits on the amounts of foods you can make available may set limits on flock size—you can come closer to being self-sufficient for feed with a smaller flock than a larger one.

We raise our flocks on a little more than 3 acres, about an acre of which is in pasture over which I range our flocks using electronet fencing. I usually range our waterfowl over several large areas of lawn as well. Flock size fluctuates, but we maintain on average a layer flock of about two dozen, raise two or three times that number as table birds each growing season, and raise between a dozen and two dozen waterfowl for slaughter each fall. With these numbers, we are able to rotate our flocks over available pasture (or lawn) about three times in the growing season.

If you want to produce eggs or broilers for local markets, by all means start small and expand as you discover the carrying capacity of the ground on which you range them.

Profile of Your Flock

Check your local zoning regulations regarding keeping of poultry where you live. Far too many localities forbid the keeping of poultry. Many more place certain restrictions, especially regarding number of birds, whether keeping cocks is allowed, and setback distances between flock housing and property lines.

If local ordinances are not to your liking, you may be able to change them. When you look for allies, remember this phrase: *Out of the mouths of babes* . . . Ellen and I participated in an effort in our county to liberalize its zoning ordinances about keeping livestock. The debate before the board of supervisors seemed a close thing—until two young brothers, desperate to keep their pet goats, made their appeal. After the younger boy, eleven years old, told the supervisors earnestly: "My goat doesn't smell bad, and she's never bitten anybody"—it was scarcely a surprise when they liberalized the county's codes regarding family livestock.

Another approach where restrictive ordinances are concerned is to keep your flock under the radar. I've heard from many, many flocksters who quietly keep a small laying flock despite the fact their locality technically prohibits doing so. In such cases it is especially important to encourage a benevolent attitude toward your project among your neighbors—if they don't complain, it is rare that the powers that be will send in the posse.

There are times when the regulations can actually work to your benefit. If a neighbor unreasonably objects to your keeping a flock under any conditions, you may find that local ordinances guarantee your doing so as a right, as my daughter Heather discovered. (See the sidebar "Heather's 'City Cousins' Flock.")

Whatever the regulations, relations with neighbors are among the most important keys to happy poultry husbandry; do everything you can to make sure your flock does not create problems for them. You will not make friends if your birds are free-ranging your neighbor's garden and prize rose beds. I've already mentioned omitting a crowing cock from the flock, but you may need to forgo geese and guineas as well—they love to vocalize. Do notice I didn't say, "And make sure your poultry operation doesn't

Heather's "City Cousins" Flock

Don't assume that you're prohibited from keeping poultry where you live—check the ordinances. When my daughter Heather—our resident flockster during the several years she lived with Ellen and me—moved with her mother to Greenville, North Carolina, she missed her chickens. So on her own initiative, this thirteen-year-old marched herself down to city hall and asked at the help desk whether ordinances allowed for keeping poultry inside city limits. The clerk, never having encountered the question before, had no clue. But she looked up the ordinance, reporting—to her surprise—that Heather would be permitted to keep up to four domestic fowl, no specifications regarding species or gender.

Despite the latitude about species and gender, I advised Heather that—given the close presence of neighbors on either side of her postage-stamp lot—a low profile was best: no turkeys, no cocks. My father and I made a cage—something like a rabbit hutch, wire on light wooden framing, mounted at chest height on four legs. We placed it in an open-front garage, where it received sunlight but was protected from wind and rain.

We populated the cage with four bantam hens I had started for her. In addition to feeding them purchased layer feed, Heather made sure they got fresh greens every day as well, whether trimmings from the kitchen or grass cut from the lawn.

Those petite hens kept Heather and her mother supplied with all the eggs they needed in the remaining years before Heather went off to college.

smell"—of *course* it's not going to smell, because you're going to practice the sensible manure management you'll learn about in chapter 7.

Sharing with neighbors a dozen of the best eggs they've ever tasted will go a long way toward making sure they're on your side where the flock is concerned. Even more effective is initiating their children into the joys of poultry husbandry—feeding the chickens, gathering the eggs, or watching chicks grow in the brooder or with a mother hen.

Choice of Species

Though *backyard flock* likely makes us think first of chickens, there are other species of domestic poultry that might fit your own interests, limitations, needs, and goals. Where their management differs from that of chickens, this book will also refer to the care of turkeys, guineas, ducks, and geese and will discuss them in more detail in chapter 24.

Pheasants, peafowl, or pigeons might suit you as well, but I have no experience with them and cannot advise about their care.

By the way, as you think through the question of mixing it up where species are concerned, you might like an idea of the average life spans of common domestic fowl:

Chicken	7–8 years
Goose	Up to 80, with a possible record of 104, but an average of 20–22
Duck	10 or less
Muscovy	8–10
Turkey	10

Standard or Bantam?

Most flocksters interested in a productive homestead or farm flock choose the standard or large breeds—they produce larger eggs and a lot more meat. But the bantam or miniature breeds might be the best fit in some situations, especially where space is extremely limited. One of my correspondents, for example, reported using her bantam flock for insect patrol in her garden: While standard-sized chickens would trash the garden, the diminutive bantams were effective bug eaters while being "quite gentle" on the growing crops.

The diminutive bantam chickens seem to have originated on the isle of Java. Brought to the West, they were used to develop the first bantam breeds there: Nankins, Rosecombs, and a little later Sebrights—known as "true bantams" because they have no large-breed equivalents. Fanciers became enamored with the idea of miniature chickens, however, and began crossing the bantam gene into standard breeds. Crossing the hybrid offspring back to the standard fowl, and selecting always for small size, yielded miniature versions of the standard breeds, one-fifth to one-fourth standard size but with other traits of the standard breeds intact.

Bantams lay small eggs—though they are actually larger in proportion to body size than those of standard breeds—but they can supply all the eggs the family needs all the same. A flock of bantams, like standard breeds, will need to be culled on an ongoing basis. And although they are small, they make tasty additions to meals.

Factors Influencing Breed Choice

A *breed* is a related group of chickens all close to the same size, conformation, and carriage. Originally, unique breeds tended to emerge in particular localities, where farmers were selecting for the same production traits in the same climate and farming systems—and swapping the results of their breeding efforts. These days there is a bewildering array of breeds and varieties to choose from, but many variations are only skin deep, having emerged from exhibition breeding for feather style and color and comb shape, rather than production traits.

Breed choice has a great deal to do with goals for your flock—what you want to get out of the project. I assume that the reader of this book is looking for a breed that is naturally sturdy and robust, productive, and economical to keep. Such qualities are to my mind more easily found among the traditional breeds rather than the latest "superhybrids" (such as Cornish Cross among meat hybrids; and among layer hybrids, crosses onto White Leghorn, and various sex-link hybrids with such names as Red Star and Black Star). The following are some of the traits to consider as you contemplate breed choice for a homestead or farm flock.

Broodiness

Broodiness refers to the set of instinctive behaviors for hatching a clutch of eggs and nurturing chicks. Most modern breeds have "forgotten" how to brood, while hens of some of the traditional farm breeds and most of the historic breeds have a greater inclination to be mothers. Broodiness is a boon if you want to hatch your own stock using natural mothers, but an annoyance if you don't, or if you want to maximize egg production—a hen lays no eggs while incubating and raising her young. If you want both good egg production and the opportunity to hatch new stock with natural mothers, keeping a subflock of broody hens (such as Old English Game or Malay or, among bantam breeds, Nankin or Silkie) is a good compromise. See chapter 27 for more information on working with broody hens.

Comb and Feather Style

Among the many variations in chicken breeds, a couple that stand out are differences in comb structure and plumage—not only of color and pattern,

Comb Styles

These drawings illustrate most of the comb styles you will see among the many breeds of chickens. Most of the breeds mentioned in this book are categorized by comb style in the following list. Note, however, that a number of breeds have been bred with more than one comb style. For example, you may see Leghorns, Anconas, Minorcas, and Dorkings with either single comb or spiked rose comb.

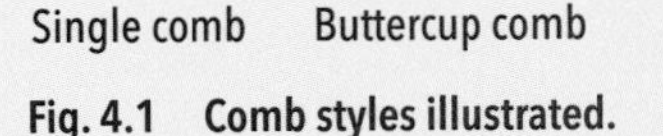

Fig. 4.1 Comb styles illustrated.

Single comb: Australorp, Cochin, Delaware, Faverolle, Java, Jersey Giant, New Hampshire, Old English Game, Orpington, Plymouth Rock, Rhode Island Red, and Sussex
Rose comb: Wyandotte
Spiked rose comb: Dominique, Hamburg
Pea comb: Ameraucana, Araucana, Asil, Brahma, Buckeye, Cornish, Shamo
Cushion comb: Chantecler
Buttercup comb: Sicilian Buttercup
V-comb: Crèvecoeur, Houdan, La Flèche
Strawberry comb: Malay

but of texture and of feather placement on the body. Such variations may be today largely aesthetic points on which exhibition breeders compete, but they can have practical implications the homesteader might consider. For example, the most common comb, the *single* comb (like a vertical serrated fleshy blade attached to the top of the head), is more subject to frostbite in severe winters—thus, far northern flocksters might choose breeds with more compact *rose* combs or *pea* combs, or even the minimalist *cushion* comb of the Chantecler, bred for cold Canadian winters. On the other hand, large single combs help dissipate heat better in hot weather, so they might be a better choice for more southerly flocksters.

Placement of feathers can affect self-reliance. Feathers on the legs, when bred to extravagance, can inhibit efficient scratching, reducing intake of self-foraged foods. Some feather-legged breeds such as Brahmas, however, retain their ability to forage—indeed, the Faverolle, a utilitarian feather-legged breed developed in France, will lay and grow much better if given free access to the outdoors but will languish and lay poorly in confinement. The extravagant headdress on crested breeds such as the Polish, Crèvecoeur, and Houdan limits vision, making them more vulnerable to predation.

Even feather color can be a practical issue. All-white breeds stand out like neon to the eyes of predators so would be a poor choice for flocksters who want to free-range their birds. The barred gray pattern is said to be the best camouflage, followed by the black-breasted red pattern. (See "Breeds We Have Raised," page 41.)

Eggshell Color

Most hens lay either white eggs or brown eggs. That the latter are nutritionally superior to the former is one of those facts that "everybody knows" but which is in fact not a fact. The misunderstanding probably arises from the *fact* that supermarket eggs (almost exclusively produced by white-egg layers) are vastly inferior to farm eggs (often produced using brown-egg breeds). However, there is no *inherent* difference: If layers of white and brown eggs are managed and fed the same, there is no difference nutritionally between their eggs. If you have any preference as to eggshell color, you can make your choice free of nutritional guilt.

If you get carried away with a preference for brown eggs, you might want to try one of the breeds known for extremely dark, almost chocolate-brown eggs—more commonly the Marans and Welsummer, more rarely the Barnevelder and Penedesenca. I raised Cuckoo Marans for years. Our Marans hens never laid chocolate-brown eggs, but they were extravagantly speckled—a nice look as well, as you see in figure 4.2.

Incidentally, I said *most hens* lay white or brown eggs. Actually, one widely available breed, the Ameraucana, lays eggs that are tinted pastel—greens, blues, occasionally even pink. The colors are never as emphatic as the pictures in the catalogs touting

Fig. 4.2 Cuckoo Marans eggs.

"Easter Egg Chickens!" would have you believe. Still, the novelty of the rainbow effect appeals to some—especially children. We usually keep a few Ameraucanas for that reason.

Do note that most breeds that lay eggs with greater color appeal tend not to lay as well as more traditional farm breeds—the visual special effects come at some cost to production. But that is one of the joys of keeping your own chickens—*you* get to decide what is just the right mix of eye appeal and daily payoff of eggs.

Temperament

Some breeds (the lighter-bodied layer breeds as a group, such as Leghorns and Hamburgs) are more flighty or excessively nervous; others, more docile and laid back (such as the Cochin, Buff Orpington, and Dorking). Cocks of some breeds, such as Rhode Island Red, are said to be especially aggressive. You might therefore consider reputed temperament before settling on a breed. Be aware, however, that there is probably no more "your mileage may vary" a trait than temperament; your management style and treatment of your birds have a lot more to do with temperament than innate disposition. If you are calm and respectful around them, there is much more chance they will be so in return.

Winter Production

Egg laying declines dramatically in winter—indeed, some breeds cease production altogether. Our Cuckoo Marans hens, for example, laid only an egg per week or so in winter, while the Old English Games quit laying entirely. (Remember that it is not natural for them to lay eggs in the winter at all, so we flocksters should be grateful for what we do get.)

Some breeds hold production better in winter than others—though as in most things chicken, the source of your particular *strain* (based on traits selected for by a specific breeder) may have more influence on rate of lay in winter than *breed*. For planning purposes, the Henderson chart (www.ithaca.edu/staff/jhenderson/chooks/chooks.html) tries to summarize what to expect from different breeds regarding winter laying. (Look for the snowflake icon in the egg column.) Breeds we have included in our flocks specifically to keep egg production up in the winter include Partridge Chantecler, Delaware, New Hampshire, Rhode Island Red, and Wyandotte.

Choosing Your Breed(s)

There is no "right" or "better" breed. While the fit of a given breed with your goals and needs will largely determine your choice, inevitably intangible factors will influence you as well—how much you like the look of a given breed, or its temperament. Your project will be the more successful the more you enjoy it, so choose the breed that speaks to you. If you respond to a beautiful, more mellow breed, such as Brahma, in preference to a breed that is twice as productive of eggs, such as the more flighty Leghorn, there is nothing illogical about your choice.[2]

Note that, in the age of the computer, many hatcheries let you order chicks individually by breed. It is easy, therefore, to put together an order for a kaleidoscopic array of breeds. And remember, if you're not especially pleased with the breed you start with, it will be easy to switch to another when it's time to rotate stock.

A couple of breed organizations are especially useful for putting you in touch with breeders of poultry stock. The *Society for the Preservation of Poultry Antiquities* is dedicated to perpetuating and improving rare breeds of poultry. One way it does so is through an extremely well-organized Breeders Directory that helps members locate rare stock from fellow breeders.

The *American Livestock Breeds Conservancy* is dedicated to the proposition that traditional breeds will be conserved not through heroic efforts to keep them alive essentially as zoo specimens or genetic libraries, but through growing them for their *economic* (production) traits. Only when homesteaders and

Some Useful Traditional Breeds for a Small-Scale Flock

There are hundreds of breeds of chickens from which to choose. The following list attempts to highlight those that have been most prominent as sturdy, productive breeds on homesteads and small farms, categorized by type. Inevitably categorizing certain breeds—Ameraucana and Australorp, for example—as either "layer" or "dual purpose" is somewhat arbitrary. This list is by no means definitive—I have omitted the more ornamental breeds, such as Buttercup—and you may well discover breeds that strike your fancy not listed here. But checking out these breeds should be a good beginning toward choosing a breed or breeds with which to start a small-scale home or market flock.

Note that I have included the category of "game" breeds—that is, breeds that in some times and places have been used in the cock-fighting pit. I include them not to encourage the "sport" of fighting cocks—indeed, I strongly oppose it—but because the game breeds have made important contributions to the development of modern breeds and should be conserved; because some breeders like raising them for their beauty and spirit, even if they are less productive of eggs and not as fast growing as farm breeds; and because the hens are almost certain to be superb mothers.

LAYER BREEDS

- Ameraucana
- Ancona
- Barnevelder
- Hamburg
- Leghorn (White Leghorn most common, many other varieties)
- Minorca
- Welsumer or Welsummer

MEAT BREEDS

- Brahma
- Cornish or Indian Game
- Naked Neck or Transylvanian Naked Neck or Turken

DUAL-PURPOSE BREEDS

- Australorp
- Buckeye
- Chantecler
- Dominique
- Dorking
- Faverolle
- Houdan
- Jersey Giant
- Marans (Cuckoo Marans most common)
- New Hampshire
- Orpington
- Plymouth Rock (including White, Barred, and Partridge varieties)
- Rhode Island Red
- Sussex
- Wyandotte (a number of color varieties)

GAME AND OTHER BROODY BREEDS

- Aseel or Asil
- Cochin
- Malay
- Modern Game
- Nankin (bantam)
- Old English Game
- Shamo
- Silkie (bantam)

small farmers appreciate the mealtime virtues of traditional and historic breeds will they seriously commit to their conservation. The ALBC maintains a list of breeds (of all farm livestock, not just poultry) in danger of extinction.

One of the most important questions is whether your emphasis will be on egg production only or putting dressed poultry on the table—or both. But remember that you'll want to integrate your flock with the total work of small-scale or homestead food production—what this book is all about—so I recommend one of the sturdier, more self-reliant breeds typical of small farms in earlier eras.

In the following overview of breed types, I will ignore the fancy types—of interest mostly as ornament or in the show ring—and focus on breeds likely to be of more use in small-scale food production projects.[3]

Breeds Worth Conserving

You can keep a small-scale home or market flock while helping with the important work of conservation breeding if you choose from among breeds in the American Livestock Breeds Conservancy's Conservation Priority List.[4]

CHICKENS

CRITICAL:
- Campine
- Chantecler
- Crèvecoeur
- Holland
- Modern Game
- Nankin
- Redcap
- Russian Orloff
- Spanish
- Sultan
- Sumatra
- Yokohama

THREATENED:
- Andalusian
- Buckeye
- Buttercup
- Cubalaya
- Delaware
- Dorking
- Faverolles
- Java
- Lakenvelder
- Langshan
- Malay

WATCH:
- Ancona
- Aseel
- Brahma
- Catalana
- Cochin
- Cornish
- Dominique
- Hamburg
- Houdan
- Jersey Giant
- La Flèche
- Minorca
- New Hampshire
- Old English Game
- Phoenix
- Polish
- Rhode Island White
- Sebright
- Shamo

RECOVERING:
- Australorp
- Leghorn–non-industrial
- Orpington
- Plymouth Rock
- Rhode Island Red–non-industrial
- Sussex
- Wyandotte

STUDY:
- Araucana
- Iowa Blue
- Lamona
- Manx Rumpy (aka Persian Rumpless)
- Naked Neck (aka Turken)

Continued on the following page

Continued from the previous page

DUCKS

CRITICAL:	THREATENED:	WATCH:	RECOVERING:	STUDY:
Ancona	Buff or Buff Orpington	Campbell	None	Australian Spotted
Aylesbury	Cayuga	Rouen		Dutch Hookbill
Magpie		Runner or Indian Runner		
Saxony		Swedish		
Silver Appleyard				
Welsh Harlequin				

GEESE

CRITICAL:	THREATENED:	WATCH:	RECOVERING:	STUDY:
American Buff	Sebastopol	African	None	Gray
Cotton Patch		Chinese		Steinbecher
Pilgrim		Toulouse		
Pomeranian				
Roman				
Shetland				

TURKEYS

CRITICAL:	THREATENED:	WATCH:	RECOVERING:	STUDY:
Beltsville Small White	Narragansett	Black	None	Broad Breasted Bronze
Chocolate	White Holland	Bourbon Red		Naturally mating nonstandard varieties of turkeys
Jersey Buff		Royal Palm		
Lavender/Lilac		Slate		
Midget White		Standard Bronze		

Laying Breeds

Most of the traditional breeds developed for high egg production are small, putting more resource into laying than making large frames and muscle mass. They lay white eggs; come into lay early (about five months); tend to be flighty around people; and almost never go broody. Ancona, Hamburg, and Minorca are among the easier to find of the traditional laying breeds. The queen of this group is the Leghorn. Innumerable white strains of this breed have been developed for use in the egg industry, but traditional strains—active and hardy in addition to being prolific layers—are readily available. Leghorns also come in many colors other than white—Light Brown Leghorns, both male and female, are among the most beautiful chickens that ever graced a homestead—and with either single or rose combs.

When selecting a layer breed, remember that the production figures you see are widely variable breed averages, suitable as a rough guide only. It is said that an Australorp hen holds the world laying record—364 eggs in 365 days! But the average flock of Australorps will not lay as many eggs as the average Leghorn flock—not by a long shot.

Meat Breeds

A few traditional breeds were developed primarily as table fowl: Jersey Giants (White and Black) in America; Cornish in England; and in Asia, Brahma, Cochin, and Shamo. These days, though, such breeds have been supplanted by fast-growing hybrids, foremost among them the Cornish Cross, currently the foundation of the broiler industry (*broiler* is another word for a young meat chicken), grown worldwide at an estimated thirty-two to forty-two billion per year. A fundamental question is whether *meat chicken* at your house will mean fast-growing broiler strains, grown in batches to fill the freezer, or the product of routine culling of a dual-purpose flock, discussed next.

Dual-Purpose Breeds

The most typical farm chickens in the past were breeds now called dual purpose. While such breeds will not lay as well as the egg-specialist breeds, nor grow as fast

Fig. 4.3 Trio of Cuckoo Marans. Note the auto-sexing color–the cock is noticeably darker than the hens. This barred pattern is said to be the best camouflage for avoiding predation.

Fig. 4.4 This Silver Spangled Hamburg hen is an example of the smaller-bodied traditional egg breeds.

as meat hybrids, they serve as the best compromise of the two: They are reliable layers but grow fast enough to a generous size to make good table fowl. They tend to be more tranquil than the flightier egg-specialist breeds—a bit easier on the nerves. Most breeds in this group lay brown eggs; in some, occasional hens may retain the broody trait and make good mothers. These breeds are likely to be your best bet for a small-scale flock, and there are many to choose from: Buckeye, Delaware, Dominique, New Hampshire, Plymouth Rock, Rhode Island Red, Wyandotte, and more from America; Chantecler from Canada; Faverolle, Houdan, Orpington, and Sussex from Europe; Australorp from Australia. Many of the traditional dual-purpose breeds are seriously in need of conservation breeding.

Breeds We Have Raised

The following are some of the breeds we have worked with over the course of almost three decades to make up our typical mixed flock of adult layers, young growing replacement birds being continually culled as table fowl, and a working subflock of broody hens:

- **New Hampshire:** Our first flock of chickens were New Hampshires, developed out of Rhode Island Reds by Andrew Christie in

If You Want to Grow a Meat-Bird Hybrid . . .

The Cornish Cross—developed out of crosses between traditional Cornish, massive and broad-breasted but slow growing, and White Plymouth Rock for faster growth—has become the foundation of the commercial broiler industry and is often the meat bird of choice in local pastured broiler markets as well. Even many homestead flocksters choose this hybrid for its astoundingly fast rate of growth—to as much as 5 pounds dressed weight in eight weeks, some strains reaching slaughter size in as little as forty-four days. But that seemingly miraculous growth comes at a cost: Development of muscle tissue outstrips all other systems, and the Cornish Cross suffers leg ailments, low vitality, and heart problems. It is as well exceptionally lazy as a forager, not to say stupid, taking only limited advantage of self-feeding opportunities on pasture. My chicken buddy Mike Focazio told me he saw one of his Cornish Cross broilers tentatively pick up an earthworm—and then spit it out!

In recent years hybrid meat strains coming out of the *Label Rouge* system of range-produced

Fig. 4.5 Cornish Cross broiler.

Fig. 4.6 Traditional Dark Cornish, broad-breasted and massive. Today's strains of commercial Cornish Cross started as crosses between White Cornish and White Plymouth Rock.

Fig. 4.7 "Freedom Ranger" broilers, typical of hybrids bred for the French *Label Rouge* system of certified free-range broiler production.

poultry in France have become increasingly available in this country—and to my mind they are a much better homestead choice if you want to raise a meat hybrid for your freezer. These broiler strains are indeed all *hybrids*, rather than genetically stable *breeds*, and are often colored rather than white. The one you may hear the most buzz about is the "Freedom Ranger," which might be a good one to try.[5] They require a longer grow-out—though only an additional ten days to two weeks—but are more hardy and resilient than the Cornish Cross, and much better suited to taking advantage of natural forages if allowed to range. I have found them greatly superior in flavor to the Cornish Cross as well.[6]

I raised Cornish Cross for a number of years to produce plump roasters for the table. But in early summer one year, when I was raising a large batch on our pasture, we had a sudden, unseasonable heat spike. When I arrived on the pasture, I found many of my Cornish Cross, right at slaughter size, prostrate with heat exhaustion—I lost twenty-two of them all told. They had sat on their butts in the shade of their shelter and *died* rather than walking 10 feet for a drink of water! I turned 180 degrees to another section of pasture on which I was ranging a group of New Hampshires, the same age as the Cornish Cross to the day: They were scooting about like little water bugs, running across their entire enclosure when they wanted a drink of water. It was hard to imagine them the same *species* as their sadly compromised cousins nearby. From that time I never raised another batch of Cornish Cross.

the 1920s. Christie bred for a superior meat bird, ensuring their hardiness by keeping his breeders in pasture shelters through his harsh New England winters. His success is attested to by the fact that the New Hampshire was at one time the dominant broiler in the poultry industry; numerous broiler crosses were made with the New Hampshire as half of the mix. We took care to locate the Newcomer strain, developed by Clarence Newcomer in the 1940s for better egg production. (I'm not sure it's even possible to find the original Christie strain.) Our New Hampshires were vigorous and hardy, laid a lot of large brown eggs, and made plump table birds. Some of the hens would go broody and were excellent mothers. Though I have not raised them for several years, I have just started a new batch of seventy-five in the brooder and intend to make them a mainstay of our flock in the future.

- **Rhode Island Reds:** Both male and female rich, lustrous red, darker than the New Hampshire. Among the best layers of all brown-egg breeds, and an excellent dual-purpose breed. This breed has a reputation for being more aggressive, especially the cocks. Both New Hampshire and Rhode Island Reds lay well in winter.
- **Barred Plymouth Rock:** A quintessential

Fig. 4.8 Old English Games. Note how similar their black-breasted red coloring is to that of the Red Junglefowl, cock and hen, in figures 3.1 and 3.2. PHOTO (LEFT) BY BONNIE LONG

rock-solid American farm breed, with sharply defined parallel bars of dark gray and lighter color (not quite white). It is said that this barred gray pattern is the best camouflaged of all color patterns against the eye of the predator. Among the best of dual-purpose breeds.

- **Cuckoo Marans:** This dual-purpose breed (much the same look as the Barred Plymouth Rock, though not as sharply etched) was developed about a hundred years ago in the town of Marans, on the western coast of France—known for extremely dark brown eggs (see figure 4.2). We found it an excellent table fowl, though its winter egg production was poor. Some of our Marans hens were excellent mothers.
- **Welsummer:** Another dark-egg breed is the Welsummer, bred in the village of Welsum in the Netherlands in the early years of the twentieth century. In our experience it's not as large nor as vigorous as the Marans. The color pattern is black-breasted red—the one most typically seen in "cock at dawn" illustrations and photographs. (The male has flaming red hackle and saddle feathers, with a black breast and tail. See figure 4.8 as an example of this pattern in both cock and hen.)
- **Delaware:** One of the crosses of the New Hampshire (with Barred Plymouth Rocks), in its heyday a foundation of the broiler industry. A handsome bird, mostly white, marked with black barring in the hackle and tail feathers (see figure 4.9). Though the Delaware was bred to be a meaty chicken and a good layer of large eggs, it is hard these days to find stock that lives up to its full potential. (For more about Delawares, see chapter 25, which includes a picture of both cock and hen.)
- **Partridge Chantecler:** Bred in Canada in the early twentieth century as an exceptionally cold-hardy dual-purpose fowl (due in part to its minimalist comb and wattles, which are less subject to frostbite). Though they're not champion layers, we keep a few in the flock because they hold their egg production well in winter. Most of our Chantecler hens have gone broody and been competent mothers.
- **Light Brahma and Buff Cochin:** Both these Asian breeds are large and make good table fowl, though they are slow to mature. Neither is a champion layer, though both tend to go broody and make excellent mothers. I no longer keep either, since I prefer clean-legged chickens (both these breeds are fully feather-footed), but if you're looking for a breed that is both beautiful and among the most docile of all chickens, either could be a fine choice for you.
- **Buff Orpington:** A heavy general-purpose fowl, another extremely mellow breed for those seeking docile backyard companions. Though average in egg production (of larger-than-average eggs), they hold their production well in winter. Our Orpingtons have usually been excellent mothers.
- **Silver Spangled Hamburg:** An ancient breed originating in Holland—the only egg specialist we have raised in the past (though recently we started some Light Brown and some Dark Leghorns to help keep egg production up).

Fig. 4.9 Delaware and Black Australorp hens. PHOTO BY BONNIE LONG

Though too small to be considered a meat breed, they are visually striking and prolific layers (the eggs are small). A breed once termed "thrifty," both because they are economical eaters and because they forage well.

- **Australorp:** Striking coal-black bird with bright red single comb, developed out of the Orpington in Australia in the early twentieth century (see figure 4.9). Though a bit on the small size for a dual-purpose breed, its egg production is excellent. We have a friend whose Australorp hen successfully hatched a clutch of eggs she laid in a bucket of rusty nails!
- **Wyandottes:** Another quintessentially American dual-purpose farm breed, which originated in New York State in the 1870s. We've kept a number of the color variations available, especially the Silver Laced. Consider Wyandottes if a docile breed with a rose comb (less subject to frostbite) and good winter production appeals to you.
- **Dorking:** If you want a bit of history in your backyard, consider keeping Dorkings, brought to England by the Romans with Julius Caesar over two thousand years ago. The varieties of this breed we know today were refined in England. Dorkings are short-legged, and males have quite large single combs. (There are rose-combed varieties as well.) They are scarcely champion layers, though friends of ours had a Dorking hen who at six years old laid more eggs than anyone else in the flock. They were valued in England in earlier times as an excellent meat breed. Dorkings are among a few breeds distinguished by having five toes (Faverolles and Houdans are others).
- **Old English Game:** If I knew that starting tomorrow I could purchase neither feeds nor chicks from outside sources and could choose only one breed of chicken for my flock, I would take the Old English Game. Like Dorkings, the OEG also has a long lineage, going back a thousand years. Though much used during that time as "game" chickens for cockfighting, they have as well been valued as a utilitarian farm breed. While hopelessly unproductive by modern standards for rate of growth and egg production, they can feed themselves if given enough ground on which to range. Though small, they are surprisingly plump under their feathers, and rich in flavor—at one time in England they were considered the standard against which all other meat fowl were judged. The hens are among the best chicken mothers on the planet and are the backbone of my working broody subflock. Despite their aggressiveness toward one another, I have always found the cocks friendly toward both my visitors and me. (The one exception was an OEG cock I brought in from elsewhere who, I suspect, had not been respectfully treated.) Old English Games have contributed their vigor, hardiness, and longevity to innumerable modern breeds.

5 | STARTING THE FLOCK

Which came first—the chicken or the egg? Most beginning flocksters prefer to start with the chicken—either just-hatched chicks, started birds just out of the brooder up to onset of lay, or adult birds—though a few go-getters might prefer to start by hatching eggs in an incubator. I have never used an artificial incubator, but my friend Don Schrider shares his experience with us in the sidebar "Hatching Chicks in an Incubator" (chapter 27).

Sources of Stock

There are many sources of stock for your flock, each with its own advantages and disadvantages. Whichever you choose, try to find stock that's true to its origins, with emphasis on the breed's economic (production) qualities. For example, almost any hatchery will offer "New Hampshire Reds," but most of that stock should more properly be labeled "Production Reds," for they are seldom true to Andrew Christie's rugged original, nor to the Newcomer strain, selected in the 1940s for better egg production. When I bought my first New Hampshire chicks, I followed a tip from a friend: The hatchery he suggested offered run-of-the-mill "New Hampshires" in its catalog, but one of the hatchery's employees was still maintaining a breeding flock of Newcomer strain, and I was able to special-order.

Local Sources

Check the classified ads in local newspapers, check bulletin boards at your post office or farm co-op, and ask friends and relatives in order to locate flock owners near you with good stock they're willing to sell. County fairs often have poultry on display—talk with the exhibitors. Find out from leaders of the local 4-H club who has poultry projects. Vendors at farmer's markets who sell eggs might sell live birds as well or might be in a network of local small producers who will.

All such purchases via direct contact have the advantage that you can see the birds themselves, and ask their owners about their flock goals and management. Inspect the birds by hand to be sure they are well fleshed; and check under the feathers for lice and mites (tiny creepie-crawlies), especially on the skin around the vent. Even if you are not (yet) an expert, it will be obvious whether the birds have the bright eye of health, or are dull-eyed and listless.

Most farm co-ops offer poultry stock, typically just-hatched chicks in the spring and ready-to-lay pullets later in the season. These are likely to be truly mass-market birds, sometimes of mediocre quality, and you can be sure the staff knows nothing about their breeding. Selection is usually extremely limited, most typically highly hybridized layers rather than traditional breeds, and often available only debeaked.

My First Batch of Chicks

Long before Ellen and I married and moved to Boxwood, I lived for a while out in the sticks by myself, mostly off the land. One day on a rare town trip I heard an advertisement for "twenty-five free chicks" at the local farm co-op. When I stopped to check it out the offer really was: Buy a bag of chick feed and get the chicks free. When I said to the clerk that seemed like quite a deal, he laughed and said, "They're excess layer cockerels—won't be much meat on their bones!" But the kid in me recognized a fun project—I paid for the feed and grabbed my box of free chicks.

I had zero experience brooding chicks on my own, but it turned out to be easy enough: I just kept them in a closet, with a lightbulb for warmth. When they were feathered, I set them up in a spacious part of a barn on the place, over a deep mulch of waste hay. They were fine companionship to a lonely hermit.

As they grew it became obvious the clerk was right—they weren't especially meaty. But all the same I began putting them on the table, a welcome alternative to the groundhogs from the surrounding fields and bluegills from a nearby pond that had been gracing my table. (Hmmm, I even remember eating a possum who proved a bit too slow.)

Not long afterward an acquaintance paid me an unexpected visit. We took a walk through lovely hay fields, shimmering in the breeze the way they do when just ready to cut. I asked her if she'd like to stay for supper. I guess I'd been off to myself way too long: When she accepted, I nonchalantly picked up my hatchet, walked out to the barn with my friend, and proceeded to make dinner. From scratch.

When I saw her next, she admitted diffidently that she'd never accepted a dinner invitation that turned out quite like mine.

Breeders

I referred in the previous chapter to two organizations dedicated to conservation of traditional breeds, the Society for the Preservation of Poultry Antiquities (SPPA) and American Livestock Breeds Conservancy (ALBC). Membership in either will put you in touch with breeders of most breeds you are likely to be interested in.

Be especially careful buying stock from someone breeding for competitive exhibition. To be sure, some breeders who show their poultry are dedicated to preservation of their production traits and would be fine sources of stock. For example, I got my starter stock for Old English Games from a fellow member of SPPA dedicated to conservation of Old English as a utility farm fowl (rather than as a fighting breed)—selecting for a more plump, rounded body style (not the leaner body preferred for the fighting pit); for enhanced egg production; and for retention of the broody instinct, for which this breed is noted.

For many breeders, however, the birds they show are essentially works of art sketched in DNA, *not* productive participants in the home economy. I have met show breeders who have no compunction about mixing in "a little of this and a little of that" in order to fine-tune the plumage or conformation of their birds, heedless of any loss of production capabilities or possibility they are creating a genetic cesspool. It shouldn't be a surprise when such breeders produce the proverbial grand champion show hen who produces only a couple of dozen eggs per year.

Hatcheries

It is possible to order just-hatched chicks through the mail from one of many commercial hatcheries around the country. I like to order from smaller, regional, family-operated hatcheries—especially those committed to conservation breeding—which often give more personalized service. On the other hand, the huge mega-hatcheries typically feature larger selections of breeds, more available ship dates, and greater opportunities to order small numbers—even individual chicks—of different breeds. Hatcheries I've ordered from tailor their offerings toward the productive home flock (rather than exhibition), and most furnish decent stock and attentive service. When I get consistently poor results with chicks from a particular hatchery, or indifferent or clueless service, I avoid it like the plague thereafter. If you know experienced flocksters, heed their advice about hatcheries that are reliable, or not.

Typically, the required minimum number of chicks in a hatchery order is twenty-five (for maintenance of body heat in transit).

Straight Run or Sexed?

If you order chicks from a hatchery, you will frequently have the option of choosing "sexed" or "straight run." Sexed chicks are separated by professional sexers trained to detect minute, breed-specific differences between newly hatched cockerels and pullets. You can order all pullets if you want layers only in your flock—or all cockerels, which are cheaper and reach larger butchering size, if you want to raise a batch for the freezer. A straight run (not sexed) batch should be the natural hatch rate—roughly 50 percent each males and females—though flocksters who receive batches heavily weighted toward cockerels swear somebody is stacking the deck.

If you do order straight run chicks, remember that they will *have* to be culled sometime before maturity—a flock with large numbers of excess males is not a workable proposition. Don't be naive about this fact of life—include it in your plans for the flock before it's a last-minute crisis. Don't think it will be easy to give away your excess cockerels—"free to good home" roosters are as much in demand as ants at a picnic. And unless you know someone who does custom butchering, don't assume that "of course" you'll be able to find that service for a fee when the time comes. You wouldn't believe the number of times I've received panicked calls from such naive beginner flocksters. I always assure them I'm happy to put in the grunt labor of butchering to furnish my own table, but doing it for someone else would be drudgery—and I don't do drudgery at any price. (Then I invite them to bring their birds and join me next time I slaughter.)

Though of course the option of placing all-pullets orders is convenient, flocksters inclined to consider the question more deeply may conclude that the choice of straight run or sexed chicks is more complicated than it first appears. Since the majority of hatchery orders are for pullet chicks only, it becomes impossible for hatcheries to sell the unwanted cockerel chicks—however many "cockerel specials" they offer. It is simply a fact of life in the business, therefore, that excess cockerels are killed, by the hundreds of thousands—by conveying them alive into a high-speed chopper; with "controlled atmosphere killing" (using carbon dioxide, argon, or nitrogen); sometimes by simply dumping them into a barrel and leaving them to suffocate. The reader may well choose otherwise, but my choice—since learning that my pullets-only orders *necessitate* the treatment of living creatures like so much disposable garbage—has been to make straight run orders exclusively.[1]

Debeaking

Many hatcheries offer the option of debeaking—chopping off half the upper beak of the newly hatched chick. Debeaking is routine in the poultry industry as a "necessary" alteration to prevent cannibalism: relentless pecking at one another, to the point of death, both in the brooder and among mature layers. But such behavior emerges *only* in situations of extreme stress—thus debeaking is an admission up front that the management system is a profoundly stressful one. The reader who agrees that our duty, to the contrary, is to give our birds a stress-free life can be assured that a flock managed as recommended in this book will *never* require debeaking. Let's recognize debeaking for what it is: mutilation, plain and simple—mutilation that may well result not only in chronic pain the rest of the bird's life, but in decreased ability to forage, and an inability (in the absence of a fine point on the beak) to properly preen the feathers and rid the skin of external parasites.

You may read—in literature targeted not at the industry but at you, the backyard, small-scale flockster—why you might want to debeak and how to do it. My question to anyone with that advice is, "What is it about your management that causes crazed behavior among your birds?"

Vaccination

Many hatcheries offer vaccination of the chicks you order as an option. If so, it will likely be for Marek's disease; occasionally a hatchery will also offer vaccination for coccidiosis. The additional charge is minimal, and some flocksters choose to have their chicks vaccinated at the hatchery. Vaccines for other diseases—laryngotracheitis, Newcastle, bronchitis, avian encephalomyelitis, fowl pox, infectious bursal disease—may be purchased and administered by the flockster.

Marek's is a viral disease that kills more chickens than any other disease. Coccidiosis is caused by any of nine species of protozoans and can cause weakening of growing birds, slow growth, and death. That sounds pretty scary, and the temptation is to choose vaccination by way of insurance. However, I have never had chicks vaccinated—for any disease, ever. It's interesting to note that both the Marek's virus and cocci protozoans are virtually universal—that is, to be found anywhere chickens are raised. Since my flocks have never had a problem with either, it is reasonable to assume that neither is the *cause* of the associated diseases (in the same way that we can say that viral organisms do not *cause* the common cold, since they are universally present in the environment). Healthy chicks gradually exposed to cocci develop an immunity to them (just as healthy children develop immunities when challenged by exposure to normally present pathogens). The Marek's virus is called out of dormancy by stress. In contrast to the enormous highly confined (which is to say, highly stressed) flocks of the poultry industry, small home flocks—receiving normal exposure to universally ambient pathogens and managed to minimize stress—do not require vaccinations to thrive. Since there is as well a small chance of harmful reactions to vaccines, I avoid them.

Don't just take my word for that: Ideal Hatchery is one of the two largest suppliers of chicks for backyard flocks in the United States and obviously earns more on orders that add vaccination than on those that don't. Yet Ideal says unambiguously on its website: "Ideal Poultry *does not recommend [vaccination] for small flocks*" (their emphasis). (They make the same statement about debeaking.)[2]

Enrollment of the hatchery you order from in the National Poultry Improvement Plan (NPIP) might be of more importance to you than vaccination. Breeder flocks in the program are certified to be free of pullorum and typhoid, diseases that were once widespread. Some flocks may be certified free of mycoplasmosis as well. Do note, however, that smaller breeders may choose not to get involved in maintaining the paper trail required for NPIP certification, while still taking care to ensure that their stock is disease free.[3]

Starting Chicks in a Brooder

Many flocksters, even those with no experience whatever, start their flocks with just-hatched chicks in a homemade brooder—a nursery for baby chickens. Doing so is not at all difficult, so long as you remember they are dependent on you for their every need: *You are Mama.* If you frequently monitor your babies, and heed what their behavior tells you, all will be well.

The Chicks Are in the Mail

People are often surprised that live chicks can be sent through the mail (see figure 5.1). The key to the mystery is nature's provision for the fact that the earliest chicks in a natural hatch may be out of the shell hours or even a whole day before the last ones hatch: Just before hatch, the chick absorbs the last of the yolk material and thus has in reserve sufficient internal resources to remain comfortably on hold without feed and water while waiting for its slower siblings to hatch—easily two full days, even up to three. It is during this on-hold period that chicks can be shipped through the mail.

Do note that the chicks must not be fed or watered before shipping—they remain in their suspended state only during the no-intake period. Once they begin feeding and drinking, their metabolism shifts to a more active phase, and they must have feed and water regularly.

Note the ship date for your order on the calendar so you will be home to receive it. Shipping usually takes two days, but orders sometimes arrive the day following shipment. Advise your postmaster or letter carrier that you are expecting the shipment. You might want to have the postmaster call so you can pick the chicks up at the post office yourself, rather than waiting for your carrier's delivery. Open the box to check the chicks in the presence of the postal employee—most hatcheries insure their orders but require substantiation of a claim by postal personnel. Sometimes one or two chicks die in transit—and many hatcheries include one or more extra chicks in case of such losses. I have had losses beyond this level only a couple of times in dozens of shipments over the years.

Fig. 5.1 Just-hatched chicks have the remaining contents from the egg yolk on reserve, so they can be sent through the mail before they begin to eat and drink.

And why do chicks die in transit? Because shipping them through the mail is stressful on them, even in the best of cases—let's be clear about that. However common—or necessary—the supplying of chicks via the mail, the compassionate flockster should reflect on the cost to the chicks, and on possible alternatives. Local sources, for example, might provide a more humane alternative, if you find a source you can trust. If there is a local hatchery near you, support them with your purchases if you can. Unfortunately, commercial hatcheries have been subject to the same relentless centralizing as other aspects of contemporary agriculture. In my own state of Virginia, there is a single small custom hatchery. I have purchased chicks from them in the past, but only when making trips their way in connection with other business. It is not practical at this time to make a three-hour round trip to get their chicks, extremely limited in selection in any case.

Fortunately, an organization called Animal Welfare Approved assists farmers who want to set up small regional hatcheries, with the specific goal of reducing the current level of postal transport of chicks. Perhaps in the future, we will as a result have greater access to chick stock from hatcheries close by.

In the meantime, there is no better way to get new chick stock without having them sent through the mail than hatching and rearing with *real chicken*

Animal Welfare Approved

A program of the Animal Welfare Institute (AWI), Animal Welfare Approved (AWA) certifies humane treatment of farm animals on pasture- or range-based farms. Clearly they put their money where their mouth is, having given $15,000 in grants for incubators, in order to encourage more local production of chick stock. In the coming five to six years, they plan further grants to increase significantly the number of regional hatcheries serving local stock needs. Both these measures will reduce the shipment of live chicks, which is highly stressful on them in the best of circumstances.

Animal Welfare Institute, unlike extremist animal-rights groups, does not oppose the raising and slaughter of domestic animals as food. However, it advocates for humane treatment of all animals, both those in our care and wild species, and has worked for the adoption of the majority of federal laws to protect animals—the Animal Welfare Act, the Endangered Species Act, the Marine Mammal Protection Act, and the Humane Slaughter Act.

While I don't agree with AWI/AWA on every point, the standards that AWA has developed are excellent guides for anyone who wants to raise livestock species with respect—whether backyard flockster or farmer. For example, AWA favors a return to traditional dual-purpose breeds. While that is an idea not likely to fly in the current climate of super-specialization, in smaller, more distributed, naturally based paradigms, all dual-purpose chicks hatched would be equally valued: females for future egg production, males as meat birds.

Learn more about the work of Animal Welfare Approved at www.animalwelfareapproved.org. Their standards for humane raising of poultry—currently for chickens, turkeys, ducks, and geese—are linked from www.animalwelfare approved.org/standards.

Learn more about the parent organization, Animal Welfare Institute, at www.awionline.org.

mothers. Using broody hens as our source of replacement stock is such a win–win proposition, it has its own extensive discussion in chapter 27.

Setting Up the Brooder

The down of newly hatched chicks does not insulate as well as the feathers of adult chickens, so they are vulnerable to rapid loss of body heat. An artificial brooder is *any* arrangement that substitutes for a mother hen's maintenance of warmth for the chicks and protection from wetting and from harsh drafts. You can purchase a brooder, like the one shown in figure 5.2, complete with heat source, feeding troughs, wire floor, and clean-out tray. Or you can save your money and house your babies in a brooder as simple as a cardboard appliance carton—we brooded our first batch of chicks in the carton from a new refrigerator. Many flocksters temporarily partition off a corner of a toolshed or stall in the barn, such as those in figures 5.3 and 5.4. In our grandmothers' time many a batch of chicks was started in a box behind the woodstove.

Some flocksters who are concerned about piling up (crowding into a corner so tightly that chicks suffocate) avoid corners by setting up the brooder as a long strip of cardboard about 2 feet wide, fastened into a circle. One of my correspondents uses 14-inch metal roof flashing, set up in a circle of any needed size, with the ends fastened together with metal siding screws when in use, and rolled up and stored when not in use. I have never used this option, since I've never had a problem with chicks piling up, usually a result of some stress such as being frightened. Small batches of chicks should be free of such stresses in a well-managed brooder. The brooder I usually use for colder parts of the season is four pieces of scrap particleboard, topped with a large piece of scrap cardboard, screwed together using corner cleats when in service, and disassembled and hung flat on the poultry house wall when not. In warmer parts of the season, I simply shut off the smallest section of my poultry house with a wire partition and a wire-on-frame door.

The brooder must have a heat source. Most brooders for small home flocks rely on 250-watt heat lamps, available at the local farm supply, *securely affixed at least 18 inches from any flammable surface.* Some folks prefer a red lamp; others use a clear one. Again, if the chicks are not unduly stressed, I've never seen much difference between the two. Another option is a hanging warmer with an electric heating element on a rheostat control, available from larger hatcheries and poultry supply houses. The brooder should be well ventilated but should permit no direct drafts onto the chicks. Since warm air rises, constant airflow can be assured by having small openings at floor level and a sizable vent at the top of the brooder.

A more elaborate setup, especially for larger groups of chicks, features a hover like the one in figure 5.5: an insulated structure of metal or plywood suspended a few inches above floor level that contains the heat source. The feed and water are offered outside the hover. Chicks self-regulate body temperature by leaving to eat and drink, then retreating to the higher warmth under the hover as needed. Such a setup can encourage faster feathering, since the chicks spend more time in the cooler temperatures outside the hover.

Predators must be excluded, of course. Most of the more obvious ones—fox, raccoon, possum—will probably be excluded already by the housing in which you set up your brooder. A couple of predators of chicks are especially difficult to exclude from the housing and so require special vigilance. Rats are no threat to an adult chicken but will eat helpless chicks, sometimes pulling them down their burrow holes. Snakes have a taste for chicks as well, so make sure they have no place to hide. (More on neutralizing these threats in chapters 6 and 21.) Don't forget your pets as potential predators: Make sure the family dog or cat cannot get at the chicks, or frighten them by trying.

Fig. 5.2 This purchased brooder for small batches of chicks includes its own heat source, feeders that reduce waste, and a clean-out tray.

Fig. 5.3 This brooder is a corner of a horse stall, set off with straw bales, topped with a heavy wire mesh gate. Heat is supplied by a 250-watt lamp.

Fig. 5.4 This brooder is an end of a toolshed, temporarily closed off with plywood and furnished with an adjustable hanging waterer, waste-inhibiting feeder, grit, and rheostat-controlled heating element.

Introducing the Chicks to the Brooder

Make sure the brooder is completely ready by the time your chicks arrive: The chicks, as said, have had a stressful trip, so you want to get them into the more compatible conditions of your brooder without delay. Turn on the heat source a couple of hours in advance of expected delivery, so the temperature is already nice and cozy before you put in the chicks.

The floor of the brooder should have a layer, several inches deep, of absorbent, high-carbon litter. I prefer coarse, kiln-dried pine shavings. Straw is fine too but should be replaced as it becomes damp, to prevent growth of molds. I don't like fine shavings or sawdust, which tend to pack down—I prefer the higher oxygen level in a coarse medium, which helps

Fig. 5.5 A homemade plywood hover in a brooder, with short feet to raise it a few inches above floor level. It's shown here with its top open to reveal the interior, which contains the heat lamps. Feed and water are outside the hover, and chicks self-regulate for temperature.

Fig. 5.6 Be sure to give the chicks a drink of water right away, to help them over the stress of their trip through the mail.

inhibit anaerobic pathogens. Shredded paper and cardboard work well if they are available, but the floor should *not* be a slick surface such as sheets of newspaper or cardboard, which can cause the chicks to slip and injure their joints—even more likely for ducklings and goslings. Not only is a thick litter in the brooder the best provision for management of droppings, but it gives the chicks something to scratch in and entertain themselves with, preventing the stress of boredom.

The first priority for the chicks is giving them a drink of water. Don't wait for them to find the waterer on their own: As seen in figure 5.6, take them individually from the shipping box, gently dip their beaks in the water for a few seconds until you see them swallowing, then release them onto the brooder floor. Be sure the water is lukewarm, not cold.

Some flocksters offer electrolytes—available from hatcheries and poultry supply houses—in the first drink. My "country boy" version of this restorative boost is the addition of some honey (maybe a quarter to a half cup in half a gallon of water), a couple of tablespoons of raw apple cider vinegar, and garlic—a couple of fresh cloves squeezed with a garlic press. I keep this solution on offer the entire first day, then switch to plain water.

Fig. 5.7 These chicks are huddling together to keep warm. Time to dial up the heating element.

Managing the Brooder

Remember that you are acting as a second-best substitute for a mother hen. Be as vigilant to their needs as she would be—frequent monitoring of brooder chicks is the key to success. For example, the standard recommendation is to keep brooder temperature at 95 degrees the first week, then decrease by 5 degrees each week. But I have never found it necessary to use a thermometer in the brooder. Simply observe the chicks' behavior and apply the Goldilocks principle: If they huddle together under the heat source, as in figure 5.7, it is too cool; if they are hanging out around the edges, away from the heat source, it is too warm. Adjust the temperature, raising or lowering the heat lamp (or turning on or off a second lamp), or dialing a different setting on the electric brooder heater. Scooting about over the litter like water bugs on a pond indicates that the temperature is "just right." (Of course, like all babies, chicks need a lot of sleep, so don't be perturbed to see immobile chicks beak-down in the litter.)

Watering

One thing to monitor especially closely is watering. Do *not* use an open waterer—that is, any waterer into which a chick can clamber and get wet or drown. Even in moderate temperatures, a soaked chick can rapidly chill and die. The best waterers for small brooder batches are the vacuum-seal type, featuring a reservoir that is filled, then screwed onto a base and inverted. The base has a lip between its edge and the body of the waterer, which provides easy access for drinking but is not wide enough for the chicks to get in and wade. A hole in the base allows water to run out of the reservoir until the hole is covered by the rising water level—the resulting vacuum prevents further flow until the hole is again partially opened by the dropping of the water level as the chicks drink. Three designs are shown in figures 5.2, 5.4, and 5.8.

The litter should never be soaking wet, a condition conducive to disease organisms. If water splashed by the chicks does wet the litter, either remove it or spread it so it will dry.

Typically there is little problem with wetting of the litter when you're brooding chicks, guinea keets, or turkey poults. Waterfowl hatchlings, however, are exuberantly messy with their water, so wet litter is inevitable when brooding them. Frequent removal of wet litter is necessary. A labor- and litter-saving alternative is to set the waterer on a wire-on-wood frame over a catch basin, like that in figure 5.8—most of the splashed water will end up in the basin, emptied as needed. There will still be some wetting of the litter, so monitoring its condition frequently is necessary.

Remember that ducklings and goslings in the downy phase can, like chicks, quickly chill and die if they get soaked. Later on you will want to provide water for them to bathe in, but in the brooder use a waterer that does not allow them to wade.

Some flocksters like to keep a light on brooder chicks even at night, to encourage more feeding and faster growth. I figure that chickens naturally fit the same circadian rhythms I do, so I turn off the brooder light at night and let the chicks sleep. Note, however, that I use a heating element to keep them warm. If you use a heat lamp to warm the brooder, don't be concerned about the full-time presence of light.

Fig. 5.8 Three-gallon waterer on half-inch wire mesh, over a catch basin. In the brooder I either dig the basin into the earth under the litter, or set up cinder-block steps to give ducklings and goslings easy access. PHOTO BY BONNIE LONG

Sterility?

The distinction between *sanitation*—essential for preventing disease and distress—and *sterility* is critical. The waterer, for example, should be cleaned frequently—but boiling it or soaking in bleach is unnecessary. The most important distinction between the two relates to proper litter management. I've already mentioned the need to make sure the litter doesn't get wet from the waterer. But it will become damp as well as it absorbs the moist droppings of the chicks—who produce a considerable amount of them. Your nose will be your best clue that the litter has become overloaded—when you get that first whiff of ammonia, top off the litter with a nice thick layer of new shavings or whatever other clean, absorbent litter you are using. (And after that, add the fresh high-carbon material a little *before* the "whiff" point: Ammonia can actually do some respiratory damage at levels too low for us to detect by its distinctive smell.)

If you are doing successive broods of chicks in a brooder, do not be misled by advice you will likely see to "clean out and sterilize" between batches—advice that comes out of an obsession we've developed about the threat of "germs," and the delusion that we can defeat them with heroic feats of cleanliness and the use of sterilizing chemicals. So long as the litter is maintained in good condition using the management described above, *it is actually good practice to leave it in place to ripen between batches*—that is, to become more alive with beneficial microbes.

You might want to skip forward to chapter 7, "Manure Management in the Poultry House," for a fuller discussion of the benefits of the deep-litter system. In my practice, *manure management with deep litter begins in the brooder.* I brood chicks on deep litter that the adult flock has been working "forever," topped off with a 2-inch layer of kiln-dried pine shavings. Beneficial microbes are already at work in the established litter and proliferate into the chick litter on top. Are there pathogens present as well? Probably. There are almost certainly the cocci that cause coccidiosis—as said, almost universally present wherever chickens are raised. The exposure to modest numbers of cocci in the litter challenges the chicks' immune systems, and they develop natural resistance to them.

My own experience is borne out by many of my poultry-keeping correspondents who raise successive batches of broilers to serve local markets. They report that subsequent broods of chicks actually do *better* than the first batch of the season, as the litter ripens. Indeed, reports from those who make a practice of leaving the litter in place have been so unanimously positive, I'm inclined to think that the usual "sterilize between batches" is mere superstition. (See the sidebar on deep litter in the brooder.)

Cannibalism

In the literature you may see scary references to *cannibalism*—chicks in the brooder may constantly peck at one another's feathers and toes. Once raw wounds develop, everybody starts zeroing in on them—and things get ugly, fast. It is often recommended to use an infrared lamp to prevent this gruesome abuse. The most extreme prevention is debeaking, discussed previously. But cannibalism emerges only among chicks under enormous stress, as in massive industrial brooder operations. Assuming that the chicks' basic needs are met—proper temperature and ventilation, easy access to the waterers and feeders, and sufficient protein in their feed—the only inducement to cannibalism would be overcrowding or boredom. Give them plenty of room to run around in, and a litter that they can scratch and have fun in, and you are unlikely ever to have a problem with cannibalism.

Feeding Brooder Chicks

Chicks are notorious wasters of feed, so choose (or make) a feeder with holes for the chicks' heads or a lipped edge to inhibit scratching in the feeder or billing out of preferred bits, which scatters the finer portions of the feed into the litter. Some flocksters prefer a hanging feeder, which can be raised as the chicks grow.

Though I know of some producers for local broiler markets who practice otherwise, I think the best advice for the beginner is to feed brooder chicks free-choice—that is, keep feed in the feeder at all times. Be aware, though, that the finer portions of the feed sift to the bottom—they should not be left to go stale. From time to time, pour these "fines" into a shallow container, mix with a bit of milk, whey, broth, or water, and let the chicks clean it up. (And make sure they do so quickly—if it sits around too long, it will become moldy.)

As for feed formulations, see the discussions in chapter 16, if you plan to buy your feeds; and chapter 17, if you would like to make your own. It is worth remembering that modern superhybrids—meat

"Mature" Deep Litter in the Brooder

Contra the stern advice to "sterilize the brooder" between batches, here is a real-life report from a producing member of American Pastured Poultry Producers Association, who finds many benefits to leaving the litter in the brooder through successive batches of chicks to mellow (build up and become more biologically active):

> Our deep litter brooder generates a surprising amount of gentle heat. The setup is far from fancy, just a circular cardboard wall made from old corrugated boxes set on a concrete garage floor bedded with commercial pine shaving bedding, over which we hang one or two 250-watt heat lamps for extra warmth. (We had a big metal electric brooder hood in it early on, but we like the heat lamps much better—easier to monitor the chicks and very easy to adjust the heat immediately by raising and lowering.)
>
> Our brooder has had chicks or poults in it almost continuously since a week before Easter this year, and we just keep adding clean shavings as needed. Broiler chicks spend two weeks in it before moving outside. The pack is now almost a foot deep.
>
> Once the pack got perhaps six inches deep (quite a while back since it settles a good bit) you could feel warmth radiating up from it when you held your hand over it.
>
> And not only does the bedding pack provide heat, but it also seems to be generating food and forage training for the chicks. A month or so back we started to notice the chicks scratching more than previous batches had. We thought we must have gotten a particularly active batch (we pick up broilers from our local hatchery every few weeks so the chicks should be about the same); but closer examination revealed masses of wiggly brown critters anywhere the pack was a bit moist (not housefly larvae, I know what those nasties look like). No idea what they are but the babies love them. They now are actually digging down four or five inches (yes, week-old, "lazy" broiler chicks!) and creating great craters. And after we moved the last two batches outside I think they do forage a wee bit more on the pasture than earlier batches did.
>
> Deep decomposing bedding is also supposed to help prevent and combat disease problems. Does it? In the months since the pack started heating up we have not seen any diarrhea or pasted butts, and have lost just one poult to a respiratory illness and one chick that failed to grow and eventually starved at about 10 days old (not sure what might have caused that, but I suspect a defect in the chick rather than disease—the fifty other chicks in the same batch grew normally). Out of twenty-five

poults and 250–300 broiler chicks that seems a quite respectable showing. Has the deep bedding helped? Who knows, but it certainly doesn't seem to hurt.

It doesn't smell bad either, in case anyone was wondering. Just a mild earthy-chickeny smell.

—Jean Nick, Happy Farm, Kintnersville, Pennsylvania

If the above is not scientific enough for you, here are quotations summarizing research on deep litter at the Ohio Experiment Station in the 1940s.[4]

> Sanitation in brooder houses has been largely restricted to the everlasting use of the scoop shove, fork, broom, and spray pump. What's new is the discovery of how to let nature's sanitary processes do a better job using built-up litter . . .
>
> The prevention or control of coccidiosis by starting day-old chicks on old built-up litter could have been prophesied years ago. It has long been recognized that chicks exposed to small dosages of coccidia at an early age developed a resistance which gave protection against heavier dosages to which they are often exposed from 4 to 12 weeks of age. Built-up litter has thus proved the most practical and effective means by which this resistance can be established.
>
> A second reason why built-up litter could have been expected to limit coccidiosis is the fact that nearly all, if not all, living organisms including bacteria, protozoa, etc., have their parasites. Old built-up litter would seem to offer a favorable medium and conditions for the functioning of the parasites and enemies of coccidia and perhaps other diseases, too.
>
> The first experimental evidence with reference to the use of built-up litter as a sanitary procedure was secured by the Ohio Station in 1946 when it was first used in the brooder house. During the three years previous when the floor litter was removed and renewed at frequent intervals, the average mortality of 10 broods, or a total of 18,000 chicks, was 19 percent. During the succeeding three years with the use of built-up litter, the average mortality of 11 broods, or a total of 10,000 chicks, was 7 percent. Seldom did a brood escape an attack of coccidiosis before the use of built-up litter. Afterward there was no noticeable trouble from coccidiosis in 11 consecutive broods started and raised on the same old built-up floor litter. Old built-up litter is floor litter which has been used by two or more previous broods of chicks.

strains ready to slaughter at seven or eight weeks, or layers who begin laying at sixteen or seventeen weeks—have higher protein requirements than traditional farm breeds.

Like adult chickens, chicks need grit in their gizzards to grind their feeds—the only difference being grit size (say, the size of radish seeds). Choose the "chick size" of granite grit if you buy it; or find a deposit of coarse sand on your place. Sprinkle the feed lightly with grit each time you feed. After the first couple of days, furnish the grit in a separate container, free-choice. They will know how much of it to eat.

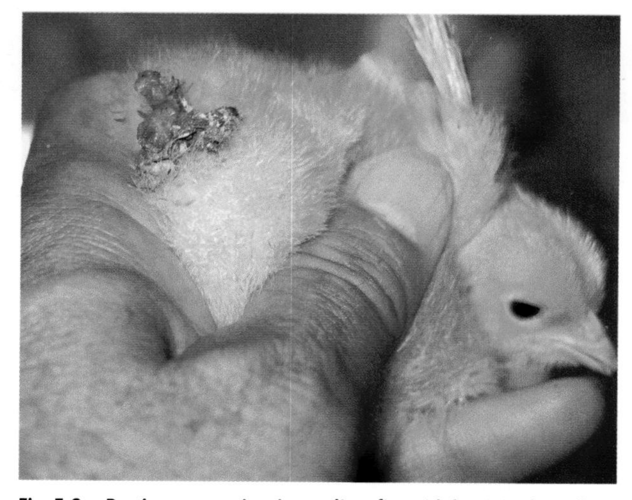

Fig. 5.9 Pasting up requires immediate first aid–but it is also telling us something is wrong, either with management of the brooder or with feeding.

Pasting Up

Remember what was said in chapter 3 about "reading the poops" as a guide to your birds' well-being? Here in the brooder is your first chance to practice this form of divination. A condition called *pasting up* or *pasty butt* is sometimes seen in chicks in the brooder: The expelled fecal matter is not a neat little dropping that gets incorporated into the litter, but a viscous mess that sticks to the down around the vent, becoming larger as the chick continues to poop. As it dries, it tends to occlude the vent and if left unattended can even cause the death of the chick—simply because it cannot poop.

If you find that your chicks are pasting up, take action immediately. First aid consists of holding the chick gently in one hand while you pull the accumulated feces from around the vent. It may help to soften the deposit first with warm water. You may pull out some of the down it's stuck to, but that is no problem. Indeed, you might want to pull out a bit more, to clear enough down to reduce the chance expelled poop will stick. Feeding a little raw cornmeal or fine oatmeal can help clear up the condition. But the most important thing you should do is heed what pasty butt is telling you: "Something is *wrong*!" Then make changes to make things *right*—rather than repetitively treat the symptom.

A good starting point might be to note that I have *never* had a single case of pasting up in a chick on pasture with a mother hen—obviously the condition is not an inevitable part of a chick's early growth. The hen makes sure her chicks stay warm enough, gathering them under her breast and wings for a warming session whenever needed, and helps them find high-quality natural feeds such as green plants, wild seeds, insects, slugs, and earthworms from day one. Imitate her good work.

Being chronically too chilled can cause chicks to paste up, so make sure the brooder is warm enough. But in my experience the major cause of pasty butt is mediocre feeds. If you do not have an alternative to purchased feeds for your chicks, then follow the mother hen's lead and provide them in addition the widest range of natural feeds you can—figure 5.10 shows some of the possibilities. In my experience the beneficial effects of such feeds offset the negative effects of poor feed, and pasting up is rarely a problem.

Mixed Brooder Batches

The conventional wisdom is that all chicks in the brooder should be of the same age—and certainly the same species—lest older brooder mates, or those of faster-growing species, bully the younger and smaller members. Like so many cases where small home

Fig. 5.10 I like to offer live feeds to brooder chicks from day one. Here, three-week-old Freedom Ranger chicks enjoy dandelions, crucifers, and grass clumps, with soil attached–plus crushed hard-boiled egg and soldier grubs. Note that, with these foods available, they are ignoring the feed in the feeder, prepared with such effort and expense by yours truly.

flocks are concerned, there is some room for bending the rules. I have occasionally brooded different ages and species in the same brooder without significant problems. Indeed, one year I raised in the same brooder a total of one hundred hatchlings—made up of two separate batches, hatched one week apart, each batch consisting of ducklings, goslings, *and* chicks. Though the older goslings did get a bit bossy toward the end of the brooder phase, no harm was done, and nobody seemed seriously stressed. (I never combined mixed batches more than a week apart in age, and wouldn't expect that I could much exceed that interval without problems.)

Graduation from the Brooder

How long do chicks need to remain in the brooder? The answer depends on the point in the season, and how you manage them afterward. You can think of three weeks as an approximate minimum, and they should not need longer than five, even in the early part of the season when night temperatures can be pretty cold. If you want to make the extra effort, when the weather is good you can give the chicks day outings in a sunny part of the pasture, inside a circle of 1-inch poultry wire, and return them to the security and warmth of the brooder at night.

The chicks should be completely past the downy

stage when moved from the brooder—that is, completely covered with feathers. Be aware, however, that at this stage, even though visibly they are fully feathered, feathering is still rather sparse, and not as insulating as it will become later. As well, they are inexperienced and do not understand how deadly even a light rain can be, so attentive monitoring at this stage is especially important. We once had a new batch of layer pullets on the pasture just out of the brooder. They were happy to be in the big wide world and left their open shelter to explore in the early morning, getting soaked in a drizzling rain. When I went out to feed, I found comatose chicks scattered about the pasture. Grabbing them up and heaping them in my shirttail, I rushed them into the poultry house and under the brooder lamps. Most recovered after they had warmed up, though some died from the exposure. Even running about in dewy grass can be too wetting, and too chilling. These days I keep the shelter door closed until I get out onto the pasture and can supervise.

Where you move the chicks when they're ready to leave the brooder is up to you. Many seasoned flocksters advise that you should keep them in a group separate from the older established flock. My own experience is that, despite some initial hazing from the older birds, feathered chicks out of the brooder have no great problem integrating with the larger flock. For details on integrating young growing birds with adults, see chapter 20, especially regarding key feeding issues should you choose this option.

Do You Need to Identify Your Flock?

Decide as you get started whether you need to permanently identify birds in your flock. Some tracking of the flock is necessary for effective management as it ages and changes composition through the years. Most keepers of the homestead flock, however, will not need to keep elaborate records and may never need to identify individual birds. If you are not too particular about breeds, tracking your flock might be as simple as switching breeds when you replace your layer flock. ("Let's see now, the Wyandottes and Barred Rocks are the old girls, and the New Hampshires and Australorps are the young 'uns.")

For more elaborate record keeping requiring detailed identification, see "Identification of Breeders" in chapter 25.

This chapter has focused on starting young chicks, a period in which they are vulnerable and require a lot of careful attention. Once they are well started, they are robust, and their care gets easier. Requirements for housing and water are easily met, and the most natural means for managing their manure also happens to be the simplest. They will be happiest of all if given plenty of room to range.

PART TWO
Basic Care

6 | HOUSING

Gallus gallus—the Red Junglefowl, at least in good part the ancestor of domestic chickens—did not have housing; and I've heard from intrepid flocksters who follow suit, allowing their chickens to live outdoors, fending for themselves and roosting in trees at night. That is the cheapest housing option, but it certainly opens the field to the neighboring predators. Equally problematic, chickens have a tendency to revert to the feral state if not kept in close association with their keepers. A tree-roosting flock would be too wild for my taste. It is *housing* as much as any other factor that keeps our birds *Gallus gallus* ***domesticus***.

Chickens' notions about housing conditions parallel our own. In summer we want shelter from the broiling sun. In winter we want to be out of the wind. We can bundle up against cold temperatures per se, but a sharp wind in the winter cuts to the bone. The chickens are bundled up in a layer of feathers, among the best of all natural insulation, so they're ready for the cold—but windchill is as much a challenge to staying warm for them as for us. And think how miserable you feel when getting soaked in a windy rain, even in summer. Despite their feathers, getting wet while exposed to wind and low temperatures can be lethal.

But it is good to remember tree roosting when thinking about housing. The fact that some flocks roost in trees—in some cases by preference—reminds us that our birds are pretty tough critters. When thinking about housing, it's important to banish the concept of pampering from the beginning.

One of the most frequent questions I encounter about chickens is, "How do you heat their house in the winter?" In my mid-Atlantic climate (Zone 6b, where winter temperatures may get down into the subteens), I add *no* heat in the winter housing. And I have not installed any artificial insulation, nor doubled the walls of the henhouse.

Flocksters considerably to the north of me should need no artificial heat or insulation either (though I can't advise those keeping fowl up in Alaska, or northern Canada). Indeed, *adding either would probably be more detrimental than beneficial*: A warm, airtight house will be a damp house—both the manure and the chickens' exhalations accumulate moisture in the coop—and dampness encourages both disease pathogens and molds.

Henhouse Design

This chapter will not give you a blueprint for building a coop but will instead help you consider the options. There are as many design possibilities as there are flocks.[1] Materials left over from other projects, begging to be put to use, will help determine design—as will management style, flock size, climate, the nature of local predation, and more. Native materials and primitive techniques—adobe, wattle and daub, cordwood masonry—may fit your needs, and will certainly be cheaper than building exclusively from purchased dimensioned lumber. Cast-off truck

Did He Say *Ventilation*?

Don Schrider is a champion breeder of Light Brown and Dark Leghorns; he has written for such poultry publications as *Backyard Poultry* and headed up ALBC's groundbreaking Buckeye breeding project. Here he shares his thoughts on what I mean by *plenty of ventilation*:

> I house about three dozen show-quality Light Brown and Dark Leghorn chickens. I make my pens from frame-and-mesh, bolt-together steel panels used to make dog kennels. I use tarps (secured with plastic cable ties) and half a roof (corrugated flexible roofing panels—brand name Ondura—attached with pipe clamps) to block the wind and keep rain and snow off the birds. The roosts are set up high, about four feet, so that the birds use their wings—if I had Orpingtons or another heavy breed I would have lower roosts so the birds could climb up. Feed and nestboxes are under the roof and protected by the tarps. Bedding is straw, and it should be thought of as a compost area. I feed whole corn and toss it into the bedding in the afternoon daily so the birds turn the bedding, get exercise, and get a warming snack before bedtime. You can't overestimate the value of clean, fresh air and exercise for the health of chickens and to keep them warm by causing their bodies to work.
>
> Frostbite is minimal because no moisture can build up as it does in a closed pen. Combine a lack of moisture in the winter air with exercise, a late-day heat-generating meal, and a circulatory system at full function—and you have birds that are less likely to frostbite.
>
> This system is superior to closed houses at least as far north as Pennsylvania.
>
> —Don Schrider

Fig. 6.1 Don's open-air pens in summer. Note shade cloth to provide additional protection from the sun, and the electric net fencing to deter predators. PHOTO COURTESY OF DON SCHRIDER

Fig. 6.2 When the snow flies . . . PHOTO COURTESY OF DON SCHRIDER

Fig. 6.3 . . . the pocket created by the tarp and the roof protects the birds from snow, rain, and harsh wind. PHOTO COURTESY OF DON SCHRIDER

Fig. 6.4 Sheltered interior space includes roosts, nestboxes, feeder, and straw litter. PHOTO COURTESY OF DON SCHRIDER

caps and even abandoned recreational vehicles have been converted to housing. You're limited only by your imagination. *So long as the housing for your flock provides shelter from wind and rain, adequate sunlight, and protection from predators, all else is secondary.*

How much space to allow depends on how the house will function. If it is simply sleeping quarters for a flock that spends most of its time outdoors, a small space can accommodate a lot of birds. If the flock is confined to the building (for example, in winter), be generous with space—give them vastly more than what is allowed for industrial layer flocks. I recommend a minimum of 3 square feet per adult chicken, ideally up to 5 square feet. (More on this later.)

Make *flexibility* a key part of housing design. Unless you *know* your flock is never going to expand, allow extra space from the beginning—or design for ease of adding on to the coop. I like plenty of interior partitions and doors, made of poultry wire on light wood framing. Partitions make possible separation of groups within the flock—younger birds or breeders from layers, waterfowl or guineas from chickens, brooding hens from everyone else—or turning one section into a brooder for chicks. When not needed, the doors between partitions can be removed from their hinges and hung on a wall.

Remember your own convenience. If the coop is small, hinged access to the nestboxes from the outside beats crawling into a cramped interior to collect eggs. Cleaning out an interior too low to stand up in comfortably is a strain. Electricity in the coop is convenient if you want to make the investment. I don't want running water inside the poultry housing, but I do like having a hydrant just outside the door.

If beginners tend to overestimate their flock's need for artificial heat in winter, they may underestimate the need for shelter from the hot sun in summer. Siting the coop in the shade of trees is a good idea. If their house is the birds' only access to shade, make sure it is open and well ventilated. (See below.)

If feed is stored in the coop, make sure it is absolutely rodent-proof. Metal trash cans provide secure feed storage. Stout plywood may do the job as well, though rodents sometimes chew into wooden containers. Reinforcement with metal roof flashing or quarter-inch hardware cloth should solve the problem.

Fig. 6.5 Our original chicken house, now subdivided into separate pens to manage matings in the breeding season.

Ventilation and Sunlight

We don't want to live in "dark, damp, and stuffy"—we want to bring the sun and fresh air inside our house. Ventilation and sunlight in the poultry house are just as important.

A complete exchange of air in the coop four times a day should be considered the absolute minimum—to keep the interior from getting too damp (conducive to growth of molds and pathogens, and to respiratory ailments); and to get rid of gases such as ammonia and carbon dioxide while maximizing oxygen. But when it comes to ventilation, don't think small. In an earlier era, open-front poultry houses had a strong following as far north as New England.[2] In these poultry houses the entire front was left open, right through the winter, with only a covering of poultry wire to protect from predators. Keeping in mind the necessity to shelter the birds from direct wind and from rain, it is pretty much the case that, even in winter, the coop *cannot have too much ventilation.*

Since so many flocksters are not prepared to

Fig. 6.6 Our main poultry house, the Chicken Hilton. Inner mesh doors provide predator protection and confine the flock when needed, but for maximum ventilation the outer doors stay open full-time, even in winter, except when there is a driving snow.

understand just how literally to take that last statement, I've added a sidebar showing you the housing my friend Don Schrider, not far from where I live in northern Virginia, Zone 6b, uses for his champion Dark and Light Brown Leghorns—year-round.

Remember the flock's need for sunlight as well. Bring plenty of sunlight into the interior using wire mesh over all exterior openings. It's easy to install solid doors that open outward, and additional inner doors of wire mesh on light frames that can keep the flock inside while maximizing sunlight and airflow. Whether or not windows have hinged or sliding sash, permanently install wire mesh in the openings. When the sash is open, ventilation increases, but predators are excluded.

Poultry Housing at Boxwood

Our original chicken house, figure 6.5, was on our place when we moved in close to three decades ago—8 feet by 18, with a wooden floor. Over the years we have changed the window layout, and replaced the siding and metal roofing. We have also added a lot of interior partitions and now use it mainly as the Breeders Annex, for isolating pairs and groups of breeders in the breeding season. (More on this in chapter 26.) The two windows are glazed with double-walled polycarbonate left over from construction of our greenhouse, and remain open almost all the time. Half-inch hardware cloth is permanently installed in the window frames to exclude predators. There is a single doorway, with two hinged doors:

The Chicken Hilton

This is a sketch of the layout of our main poultry house, a pole barn construction with plenty of headroom, 24 feet long and 13 feet wide. It has an earth floor, covered deeply with oak leaf litter, with a barrier of 24-inch roof flashing dug in around the perimeter. For additional ventilation a ridge vent runs the entire length of the roof's peak.

Interior partitions (A, B, C, D, and E): The interior was initially divided into three separate sections (A, B, and C) using chicken wire on 2-by-4 framing. Partition framing includes 40-inch-wide doorways, into which I easily placed lightweight doors, also wire on frame. Just as easily, I can remove the hinges and hang the partition doors on a wall.

I subsequently installed additional wire-on-frame partitions to subdivide sections A and B, to yield sections D and E. A and B are now 8 by 8, and D and E are 5 by 8. Removable partition doors into D and E are 36 inches wide.

Doors: We placed three doorways in front, 32 inches wide by 72 inches high, one in the center of each interior section. I made for each doorway an outer door that swings outward and can be latched open. The outer doors are almost

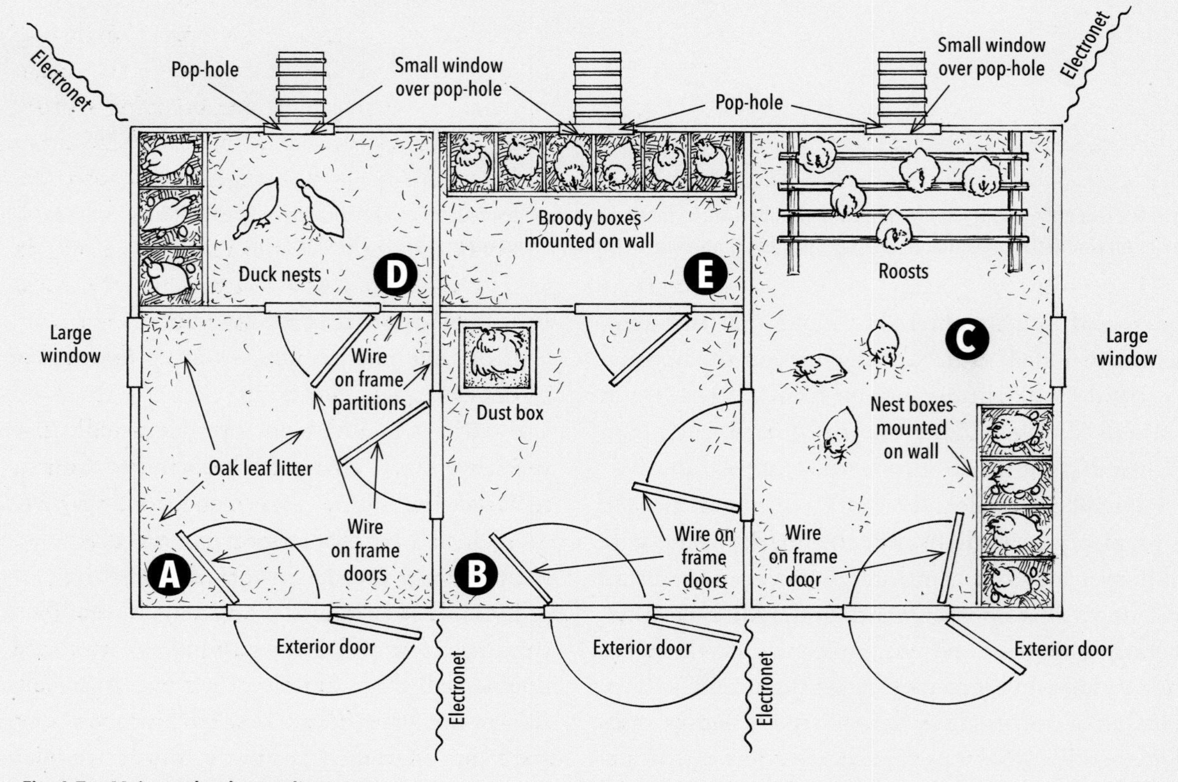

Fig. 6.7 Main poultry house diagram.

always open, even in winter, except when there is driving rain or snow, or when temperatures in the teens coincide with heavy winds. The inner doors, half-inch hardware cloth on wooden framing, swing inward to give the flock access to the outdoors; or they can be latched closed in order to keep the birds inside (and keep predators out) while allowing maximum ventilation.

Pop-holes: At the rear of each section is a pop-hole—a chicken-sized door (24 inches wide and 18 high) that usually remains shut but can be opened to give the flock access to the outdoors through the rear.

Electronet placement: Note that I can anchor the ends of electronet fencing at any point along the outside of the house, allowing separate flocks access to separate pasture enclosures. For example, I could install one electronet such that on one end it encloses the door of section A, surrounds a 3,000-square-foot pasture enclosure, and terminates on the far left corner of the building. At the same time I could set up a second electronet to make a separate enclosure of similar size, with access via the door of section B, with its far end connected to the first net.

Windows: There is a 24-by-42-inch window set into each end wall. Smaller windows, 20 by 24, are set in the middle of each section in the rear, over the pop-holes. All window openings have half-inch hardware cloth permanently installed. Together with the wire mesh doors in front, the side and rear windows provide maximum ventilation in every direction.

In the winter I block all five windows with plywood, turning the interior into a pocket into which the wind cannot blow. At the same time, there is plenty of airflow through the mesh doors in front and out the ridge vent up top.

Roosts: A ladder-style set of roosts fills the rear of one of the sections. It isn't attached but instead simply leans against the wall, so that it's easy to swing out of the way for adding or removing litter.

Nests: Nestboxes for the hens are mounted on a wall, well above floor level, with a landing perch in front that the hens can use to approach the nests. All the nests were designed to be fitted with doors when I need to use them as trap nests. When there is no need to nest trap, I put the doors into storage.

Ducks have a nest at floor level for laying outside the breeding season. In late winter, however, as we approach the breeding season, I keep the ducks and geese outdoors full-time, with a pasture shelter and sturdy, spacious nest units.

Broody boxes: I mounted a set of six broody boxes on a wall at a comfortable working height. I can thus give privacy to hens who are incubating eggs, while losing no floor space for the rest of the flock.

Dustboxes: The deep litter is usually dry and fine enough for the chickens to use for dust bathing. But just in case I provide a 24-by-24-by-16-inch dustbox, which can be placed in one section or another as needed.

Flexibility: You are unlikely to make a duplicate of my Chicken Hilton. But study the layout with care, and strive for the same flexibility, the same broad range of options, that underlies the design. Here are some examples

Continued on following page

Continued from previous page

of the management flexibility made possible by this design.

At the beginning of winter I might have both partition doors in place between sections A and B, and between B and C. Ducks and geese spend the night in section A and during the day have access through that section's front door to graze a patch of cold hardy rye and wheat enclosed by an electric net fence. A 140-gallon tank keeps their bathing and watering outside. Meanwhile I have a group of excess cockerels awaiting slaughter in section B. Their front door gives them access to a heavily mulched winter feeding yard in a separate electronet enclosure. (I set up a temporary roost for their use.) The layer flock is in section C. Since the other two front doors are inside electronet, I need the door into C for my own use. However, during the day I open the pop-hole out the back of C, giving the layers access to yet a third electronetted area out back, another heavily mulched winter yard.

Late winter comes, and I start keeping the ducks and geese outside—that's a simple matter of shutting the front door to A, and parking my A-frame shelter for their use on their grazing area, which is already protected by electronet. By this time I've slaughtered the last of the cockerels, so I remove the partition door between B and C to give the layer flock more space.

I receive a big batch of chicks right at the end of winter and set up a brooder temporarily in section A. (It disassembles and stores flat on the wall the rest of the year.) And it's not long before my mother hens start going broody, and I move them into the broody boxes and set them on eggs I've collected from my breeders for hatching.

You might need a coop much smaller than my Chicken Hilton. If you design for flexibility, however, it will allow for changes in management and for meeting different needs as you go through the seasons.

The solid outer door swings outward and can be latched open; the interior door (wire mesh on wood frame) swings inward and can either be latched open (allowing the flock access to the outside) or closed (to keep them inside while maximizing sunlight and ventilation into the interior). Placement of door and windows furnishes maximum cross-ventilation, while the corrugations in the metal roofing allow flow of air up through the living space and to the outside.

We built our main poultry house ourselves, shown in figure 6.6, designed for a larger flock and to serve many functions. The "Chicken Hilton" is a 13-by-24-foot structure, divided into three equal sections in the long dimension by wire partitions; two of those sections are further subdivided. Mesh-on-frame doors block access among the resulting five sections. Or they can be removed and hung out of the way on the walls, giving maximum flexibility of use—the layers in one section, waterfowl in another, a third converted temporarily to a brooder for a group of chicks, and more.

Each of the main sections has its own doorway in the front—again, with solid wood doors opening out, mesh-on-frame doors opening inward—and a small rear window. Beneath each rear window is a chicken-sized pop-hole hinged into the wall—when open, it provides an alternative access to the outside. Each end of the house has a larger window. All windows have half-inch wire mesh permanently fastened in the frames.

In the winter I install fitted plywood shutters in all the window openings to block the wind, but the three solid outer doors in front remain latched open in all but the harshest weather. During the last winter, I shut the outer doors overnight fewer than a dozen times—when there was blowing rain or snow, or when temperatures in the low teens coincided with heavy winds. The deeper recesses of the interior—where the roosts are located—form a pocket impervious to wind, even as maximum ventilation is provided by the open doors and a ridge vent that runs the entire length of the roof. Like Don Schrider, I have almost no problems with frostbitten combs since I started keeping the henhouse more open in winter—in contrast with earlier years when I kept it closed up as tightly as I could at night.

The big henhouse is indeed designed for maximum flexibility. At the time of this writing, midwinter, I have the flock subdivided into three groups, each with its own separate access to an outdoor yard surrounded by electric net fencing: ducks in one end; cull cockerels awaiting their turn at the table in the middle; and the "keeper" flock—the layers—in the other end. The ducks go out onto a plot of rye, wheat, and barley grasses for winter grazing; the yards for the other two groups are heavily mulched as described in chapter 2. The ducks and the cull cockerels have access to their yards via the left and middle doors, I come and go through the right door, and the layers get into their yard through the pop-hole in the back of the right section. As you can see in figure 6.7, I have not exhausted the possibilities—I could hang the doors that further subdivide the left and center sections as well if needed.

We retained an earth floor in the Chicken Hilton, with a perimeter barrier of metal roof flashing—half-inch hardware cloth would serve as well—dug in to a depth of 18 inches to deter digging predators.

The standard response to our main poultry house, with a profile higher than any sane flockster would choose, is: "That looks like a horse barn!" But the large amount of headroom increases air circulation—and offers intriguing future possibilities for multi-species housing.

My main regret about this building is having opted for pole barn construction, based on white oak and red oak poles, which I treated myself using sodium tetraborate. (I will not use lumber pressure-treated with heavy metals.) The deep litter in the house is highly bioactive and will eventually defeat my attempts to preserve the posts with the boron. I wish I had spent the additional money up front for a concrete-block foundation—solid against both predators and rot.

Multi-species Housing

It is possible to house more than one livestock species in the same housing. I referred previously to the large amount of headroom in our main poultry house. I've often fantasized about turning that overhead space into a pigeon loft, releasing the pigeons to fly free during the day, foraging their own food at no cost to me, while providing them a protected space to roost at night and to rear their squabs to be harvested for the table. The chickens, at floor level, would provide manure management by scratching the pigeons' droppings into the litter. A friend warns of what he sees as a fatal flaw in my scheme: On his farm, not far from where we live, the pigeons would be easy pickings for hawks who learn that they'll be flying out in the mornings. However, I expect that this is a "your mileage may vary" situation—in earlier times those who maintained dovecotes or pigeon lofts did indeed allow their occupants to fly free and gather their own wild foods—and that some adventurous flocksters trying that option might be rewarded with additional payoff from their chicken house.

Joel Salatin's Raken House (housing *ra*-bbits and chic-*kens*) is an excellent example of the benefits of multi-species housing. The breeding rabbits live in wire cages suspended at waist height, over a large flock of layer hens, over 12 inches of wood chips as a deep litter. The chickens turn the droppings and urine of the rabbits into the litter, speeding its conversion to compost for the farm's fields. Offal from butchering the rabbits is thrown to the chickens as a valuable protein feed.

Some flocksters set up a stall in the barn for their chooks' quarters and give them free rein to clean up

Fig. 6.8 Year-round, portable housing for a flock of up to fifty hens and guineas, made with a stout wooden frame, PVC pipe, poultry wire, and heavy canvas tarp. PHOTO BY JON WILSON

Fig. 6.9 Caroline Cooper of British Columbia assembles this hoophouse for her flock in winter, setting it up near her house to make caring for them easier, and takes it apart for storage the rest of the year. It is made from 2-by-6 lumber as foundation, with attached 20-foot-long PVC pipes and plastic vapor barrier sheeting. PHOTO BY CAROLINE COOPER

spilled feed, denying it as a resource for rodents, and to scratch through the manure of horses and ruminants, ridding them of parasites and their eggs.

Our poultry flock is always a mix of fowl species, which from time to time offers challenges. If not carefully managed, waterfowl may soak the litter. (See the next two chapters, on deep litter and watering, for management strategies.) In late winter, when the testosterone starts rising in preparation for the breeding season, cock guineas may harass the chicken cocks relentlessly—I have gone so far as to hobble them to give the poor cocks some peace![3] The important question of blackhead is discussed in the section on turkeys, in case you are considering a mix of chickens and turkeys. But for the most part, keeping a mix of fowl species in the same housing has not been a problem for us.

Other Housing Options

There are no limits when it comes to housing options. Flocksters of my acquaintance have made chicken coops from every imaginable converted outbuilding; cordwood masonry, daub and wattle, strawbale, and adobe construction; discarded recreational vehicles, cars, and school buses; pickup truck caps. They have mounted them on poles and bermed them into the earth to keep them cooler in the summer and warmer in the winter. Hoophouses have become a popular choice for lightweight structures that are easily assembled, most often for winter use, and just as easily disassembled for storage. Figures 6.8 and 6.9 show a couple of examples.

As long as you take into account ventilation, predators, space, and sunlight, you can let your imagination run wild. Indulge your whimsy. Have fun.

Accessories

Just as we furnish our own houses to round out their comfort, the chickens' quarters will need a few "furnishings." The best-furnished chicken house is the one that provides both maximum comfort and sense of security to the birds, and maximum convenience and efficiency for us.

Roosts

Chickens have a strong instinct to roost: to get up off the ground, out of reach of the predator, for their night's sleep. Roosts (perches) should be provided for all chickens—and guineas and turkeys—with the exception of the fast-growing broiler hybrids that are too heavy and clumsy to utilize them. While chickens denied a roost will settle down for the night on the litter, they will be more content and feel more secure with a place to perch at night.

There is nothing complicated about providing roosts. The cheapest, easiest option is to cut some sturdy saplings around 2 inches in diameter—leave them rough—and attach to the walls in the sleeping area. Plastic or metal pipes are not good choices—they are too smooth for grasping securely, and metal gets too cold in winter. A simple "ladder" of 2-by-4s—or 2-by-2s for smaller numbers of birds—is easy to make. Remember to round off the edges so the chickens can grasp them more easily and comfortably. I simply lean mine against the wall without attaching, as you see in figure 6.10, so it is easy to swing out of the way for cleaning.

Fig. 6.10 Accessories in my henhouse include nestboxes mounted on the wall, hanging feeder, and roosts–plus one of the most important "accessories" of all, deep litter.

Allow 8 to 10 inches of roost space per adult bird, 18 inches between a roost and a wall, and 12 inches both vertically and horizontally if the roosts are in a ladder configuration.

Nestboxes

A hen's instinct is to hide her eggs from the predator. Nest design should satisfy this instinct—the nest should convince her that she has found a hidden, secure place in which to lay. The top and sides should provide a sense of isolation, and keep the interior fairly dark. Line the nest well with straw or wood shavings, which will not only keep the eggs clean but also help cushion them and prevent breakage. Fasten a retaining strip of some sort, about 4 inches high, along the bottom edge of the front of the nest—otherwise they'll scratch out the nesting material and you'll have to renew it more often.

Provide plenty of nests in relation to the number of layers—if urgent hens crowd a nest, there is increased chance that the jostling will crack an egg, an invitation to egg eating. (See below.) One nest for each four to six hens should be sufficient. Make nests 12 to 14 inches in each dimension. Even when provided with plenty of individual nests, though, hens will sometimes perversely crowd together into one of them in order to lay.

I have always provided individual nests as described above. Some flocksters prefer a communal nestbox—a single, larger interior space that can be accessed by a number of hens at once. In either case a top over the nests, slanted at a 45-degree angle, is a good idea.

You can buy manufactured nests like the one in figure 6.10, but I don't recommend them—both because of their expense and their flimsy construction in comparison to home versions you can build or rig yourself. See appendix A for instructions for homemade nestboxes. Those particular instructions are actually for making trap nests, useful for testing layer performance and for selecting hens as breeders. However, you can simply leave off the doors and their tracking strips to make ordinary nestboxes.

Flocksters I've spoken to have used 5-gallon buckets mounted on the wall (maybe with a hole cut in the back for gathering eggs), cast-off plastic milk crates, wooden boxes, or a few straw bales to define a private nook in a dark corner.

I make my nests of plywood, mounting them on a wall to save floor space. Setting the nests above floor level also helps prevent egg eating. In my experience this unwelcome behavior may start with the cock of the flock, who gives an exploratory peck to an egg he finds in a nest at floor level. If a crack in the shell results, other flock members peck curiously as well. It's not long before the birds discover there's something good to eat inside, and egg eating becomes a nasty habit that spreads in the flock. Mounting nests on the wall helps prevent that first exploratory peck. Landing perches affixed in front of wall-mounted nest entrances make it easier for hens to enter the nests.

Fig. 6.11 In case my chickens don't have sufficient access to dust bathing elsewhere, I provide a dustbox as a backup.

I don't like solid bottoms in nests. The lining material such as straw eventually gets dirty and disintegrates, and a solid bottom makes cleaning out more of a chore. Instead, I make the bottom of the nest of quarter-inch hardware cloth. As it disintegrates, the lining sifts down through the wire mesh, making the nest somewhat self-cleaning as I renew the straw on top from time to time.

Chickens sometimes roost in the nests, or on top of them; and—since they do much of their pooping at night—the nests get soiled, or the tops become caked with a layer of dried droppings. Younger birds are more likely to roost in the nest, since the older hens—higher in the social hierarchy—out-compete them for preferred real estate on the roosts. Providing more than adequate roost space can help. A slanted top over the nest will prevent chickens' roosting there. If a slanted top has not been built on the nest, it can be added, in the form of scrap plywood or even a piece of stiff cardboard, tacked into place over the nest.

It may be necessary to block access to the nest itself at night, again with a piece of plywood or cardboard. In my experience if I block access each night

Fig. 6.12 Don Schrider ensures constant access to dust bathing outdoors with this section of scrap plastic culvert, which keeps the covered clay soil dry even when the weather is not. PHOTO COURTESY OF DON SCHRIDER

for a week or so, the birds get used to sleeping on the roosts, and I can discontinue blocking the nests with no further incursions into them at night.

Collecting eggs as frequently as practicable is a good idea. With frequent egg collection, the eggs will stay cleaner in wet weather, with less time for hens to track mud onto eggs already in the nest; and there will be fewer eggs in the nest at a given time, with less chance that jostling by hens will crack one as they come and go.

If a hen is on the nest, she may make a fuss when you try to check under her for eggs. Generally she will make no serious objection to an exploratory hand underneath—even a peck at that hand will rarely be more than a token. Occasionally, though, a hen will put serious intent into the peck—somewhat painful but bearable. I have lost blood to guinea hens on the nest. Such hens quickly establish respect.

Dustboxes

A major chicken ritual is dust bathing: Chooks find a dusty spot somewhere and thrash about in it, fluffing the dust up under the feathers and onto the skin, in the process ridding themselves of external parasites. Outside, chickens hollow out a depression in dry ground and dust-bathe in it. Inside, they use the driest parts of the litter for bathing. But in case they have no other access to dust bathing—during a rainy spell, or when the litter is not fine enough—it's a good idea to furnish them a dustbox.

You can bang together a dustbox out of scrap materials on hand. Do remember that the chickens will scatter the dusting material out of the box if it's too shallow—a deeper box will help retain the contents, as will adding a lip around the edges. My dustboxes are simple edge-nailed plywood boxes, 24 inches square on the bottom, 16 inches deep, with 2-inch plywood strips around the top edges as a lip. Instructions for building my dustbox are in appendix B.

Fill with 4 inches or so of any dusty, nontoxic material, renewing it as needed. Sphagnum peat moss is excellent, though it has to be purchased and is a non-renewable resource. Clay soil, dried and sifted, works well. Wood ash is good, though I do not add it at more than maybe 2 parts out of 6 or 7. A few handfuls of garden lime, diatomaceous earth, or pure sulfur powder make the mix even more effective against exoparasites. *Wear a good dust mask when handling dusting materials.*

In lieu of building a dustbox, my friend Don Schrider covers a bit of ground outside with a piece of scrap plastic culvert, which keeps it dry and dusty even in rainy weather, as you see in figure 6.12.

Dropping Boards?

Poultry do at least half, or more, of their pooping at night. So even if we release our flock to the outside during the day, we all have to deal with manure management in the coop. A common recommendation from traditional poultry husbandry guides is the use of *dropping boards* under the roosts: The droppings fall on the boards and are scraped off by the flockster and moved to the compost heap. An alternative approach is the use of a *dropping pit*—a framed pit underneath the roosts, topped with wire mesh. The droppings fall through the mesh, which prevents the birds' getting into them. Periodically the accumulated droppings are removed and composted.

While sources I respect continue to advocate these poop practices, my own reaction to dropping boards and dropping pits is sheer incredulity. Scraping the boards, or mucking out the droppings from a pit, would be unpleasant and weary work. The basic premise of both designs, as far as I can tell, is that the flock's droppings are vile, nasty, and threatening; the birds should be kept separate from them at all costs. Fortunately, there is a far better approach to manure management in the poultry house: *deep litter over an earth floor*, in which the birds do the work themselves, turning something that is repellent and a possible substrate for disease—the poops—into a substrate for health. That subject is so important, we will consider it separately, in the following chapter on manure management in the poultry house.

Preview that chapter on another important question as well before finalizing your henhouse design. If you have an existing building to convert to poultry housing, by all means avoid the effort and expense of building new. But if you are going to build from scratch, my strong recommendation is to *leave an earth floor in the coop*. Not only will you save the expense of framing and installing a floor, but you will be ready to create the conditions for best manure management.

Rodents

With a perimeter barrier in place, and wire mesh secured in window frames and on inner doors latched shut at night, you should have few worries about predators entering the coop. An exception is rodents, who are a lot more difficult to exclude than, say, raccoons. Mice are no threat as predators, though they can be a serious threat to your feed supply. Rats can be even more voracious eaters of your expensive feed but are also a threat to chicks, whom they may drag down their burrow holes.

My best advice is to focus on *preventing* a rodent infestation of the poultry house, not carrying on a war of attrition after they move in. Two key practices in the poultry house should help keep rodents at a minimum. First, I try to give my rodent friends *nowhere to*

Fig. 6.13 Covering the entire chicken run with a mixed organic mulch turns it into a giant chicken-powered compost heap, providing both responsible manure management and entertainment for the flock. PHOTO BY BONNIE LONG

hide inside the poultry house. I try to keep all storage space such as shelves as open as possible, so a mouse or rat nest is immediately apparent. I make sure nothing is lying around at floor level other than the deep litter—constantly turned over by the chooks, leaving it unusable as tenancy by rodents. If the birds do occasionally scratch up a mouse nest, the mouse doesn't have a chance—one of the hens snatches it up, then plays a desperate game of keep-away with her sisters while trying frantically to gobble it down. As for baby mice—"pinkies"—they're *caviar.*

The second strategy for rodent prevention is: *Don't feed them.* For more rodent prevention strategies, see chapter 15.

Access to the Outdoors

The shelter you provide your flock is a house, not a prison. Give your birds access to the sunshine, fresh air, exercise, and natural foods to be had outside. Best of all—for maximum health, contentment, and foraging opportunities—is getting the flock out on *pasture.* This is such a desirable option it will be discussed in three separate chapters later in the book. But pasturing is not an option for many flocksters, so let's here consider best access to the outdoors for their flocks.

I cannot readily imagine any situation where full-time confinement of chickens in the coop is absolutely necessary. Even in situations where space is extremely limited, it should be possible to construct a double-decker coop and run—living quarters above on stilts, wire-enclosed run in the footprint below. If space is available, of course, provision of a larger run is better.

The problem with a static run—whether large or small, and whatever the size of the flock—is that eventually the chickens, hungry for green, eat every last blade of grass. From that point the run looks like the surface of the moon, dotted with chicken poops. Not only is the result unsightly, but the accumulating manure virtually poisons the soil, serves as breeding ground for flies and vector for disease pathogens, and runs off in the next rain as pollution to groundwater and streams. The responsible flockster will come up with a more wholesome alternative, such as:

- **The composting run:** After you've read the next chapter on manure management in the henhouse using deep litter, note that it is possible to take that concept *outside*. That is, cover the entire enclosure with a deep litter, using whatever organic "wastes" come to hand—such as dry leaves, spoiled hay, crop residues—turning the run in effect into a giant compost heap, as pictured in figure 6.13. The organic debris will absorb the poops, retaining them for soil fertility applications, while preventing their running off as pollutants—a win–win solution in every way. And remember how well it works as a winter exercise/feeding yard.
- **Chicken pie:** Make the coop the center of a foraging-ground "pie," with "slices" defined by electric net or poultry wire fencing, either temporary or permanent. Release the chooks into one slice at a time, rotating through the entire area around the coop. If the flock isn't too large in relation to the total available space, slices grazed earlier will recover completely before the birds are released onto them later in the rotation. Almost inevitably, the area immediately around the entrances to the coop become worn free of cover, so lay down a thick organic mulch in the bare spots as described above.
- **Dueling gardens:** Whatever your space limitations, if you both keep chickens and grow a garden, you already have space for a solution I call dueling gardens: Make the run and the garden each the same size; enclose each with wire or electric net fencing; and set the coop in between them, with separate doors opening onto the two spaces. In the first season, release the birds onto one of the spaces—set up as a composting run as described above—and garden in the other. Alternate the use of the two plots in the following season. The soil in the new garden side will already be enriched with compost. Since it has been continuously under a heavy mulch for a year, you may not even have to till—just lay out the garden beds, and plant.[4]

7 | MANURE MANAGEMENT IN THE POULTRY HOUSE: THE JOYS OF DEEP LITTER

If you are around any livestock operation, regardless of species, and you smell manure—you are smelling mismanagement.

—Joel Salatin

Repugnance for what comes out the far end of an animal is not merely cultural conditioning—our senses are warning us of potential danger: Feces can be a vector for disease. Joel's quote above implicitly advises us to *trust* that repugnance: If it smells bad, it could be dangerous. But it also implies that *there are ways to manage manure so it doesn't stink*, giving us our most important hint that its threat has been neutralized. Properly handled manure, in other words, is not a danger.

Many readers of this book have already experienced the transformation of things yucky into not only something pleasant, but a valuable resource: the alchemy of the compost heap, which starts with manures and rotting vegetation and ends with compost, smelling as sweet as good earth, ready to fertilize the garden. *The compost heap is our model for making the same transformation in the henhouse.*

You assemble a compost heap from nitrogenous materials such as manures and spent crop plants, mixed with carbonaceous ones such as leaves and straw. Coarse materials will eventually compost, but if you make the effort to shred them more finely, the composting process speeds up considerably. Inconceivable numbers of microbes multiply in the pile, using the nitrogen in the manures and fresh green matter as a source of energy to break down the tough, fibrous high-carbon materials into simpler components. The ideal balance of carbon to nitrogen in the mix is 25 or 30 to 1. Too much nitrogen is signaled by the smell of ammonia, meaning that some of the nitrogen—a potential source of soil fertility—is being lost to the atmosphere. (Ammonia is a gas of nitrogen and hydrogen, NH_3.) Moisture in the heap is essential to the microbes driving decomposition, though it must not be soaking wet—a condition that would inhibit decomposers while favoring pathogens. Oxygen is also essential for the decomposers, so you turn the heap over completely at least twice during decomposition, maybe more. Heat is a by-product of the composting process—a well-made compost heap becomes amazingly hot. The end result of this devoted effort is compost, one of the best possible fertility amendments the gardener can find.

It is possible to make the chicken coop in effect a slow-burn compost heap if you *leave the earth itself as the floor, and keep it covered deeply with high-carbon organic litter.* The sorts of decompositional microbes at work in the compost heap—and in the soil food web—migrate out of an earth floor into the deep litter; the slight wicking of moisture out of the earth helps them proliferate and thrive. (If you have an existing building with a wood or concrete floor to use for poultry housing, by all means avoid the effort and expense of building new. You can still use deep litter to keep the henhouse sweet, with a couple of tweaks noted below.)

Oh, and all that laborious shredding and turning of the compost to assist its breakdown? Just leave that to the chooks.

Materials for Deep Litter

The poops laid down by the birds are rich in nitrogen, so naturally—as in the compost heap—we want a lot of carbonaceous material in the litter to balance it. In contrast to the ideal C:N ratio for a compost heap, however, *the higher the carbon content of the deep litter, the better.* That is, the more carbon in the mix, the more manure the litter can absorb before its nitrogen drives the C:N ratio out of balance, resulting in production of ammonia.

The high-carbon material chosen for the deep litter depends on what is cheapest and most readily available to you. It should ideally be somewhat coarse, so the scratching of the chickens fluffs it up and incorporates plenty of oxygen, assisting its breakdown by microbes and discouraging growth of pathogens. I prefer oak leaves, but that's mostly because a close neighbor, who has half a dozen mature white oaks on her place, prefers to get rid of the accumulating leaves in the fall. She even hauls them over and dumps them in a big pile at my place. I say "God bless 'er!"

Kiln-dried wood shavings are excellent, with their extremely high carbon-to-nitrogen ratio (500:1), but are an additional expense if you have to purchase them. For example, I recently bought some shavings for $5 per 2½-cubic-foot bale (expands to 8 cubic feet) to use as brooder bedding. Buying enough to deep-bed the entire henhouse would be expensive indeed. Wood chips might serve—they too are extremely high in carbon and last a long time before they have to be replaced. Joel Salatin uses them as the litter in his Raken House—he cleans out only once a year, when even this coarse woody material has been reduced to compost by the microbes and the constant working of the chickens. Sawdust is satisfactory, though it doesn't fluff up as much as other materials. Whether using sawdust, wood shavings, or wood chips, be sure to use either kiln-dried or aged material—"green" woody materials may support the growth of molds, whose spores could be bad for your birds' respiratory systems—and yours.

Note that old hay and certain crop residues such as soybean vines are *not* appropriate as litter materials—with a significant nitrogen content of their own, they do not effectively balance the nitrogen in the poultry droppings and quickly heat up.

What about straw? Many flocksters avoid the use of straw because, especially in the presence of the slight dampness of an earth floor, it can support the growth of *Aspergillus* molds, whose spores can cause respiratory problems. I have corresponded with flocksters, however, who report that they use straw over an earth floor without problems. Though I have in the past avoided straw litter, I am now experimenting with it as an addition to litter with a much higher proportion of oak leaves—so far with no mold problems. Note that there is no problem using straw as the litter over a wooden floor—the drier conditions in such a litter prevent growth of *Aspergillus.*

Nearby processing of agricultural crops may furnish other litter materials. Milling of corn, cane, buckwheat, or peanuts, for example, may generate corncobs, chopped corn or cane stalks, or hulls that are available cheaply enough to be used as deep litter.

Alchemy

Over many years showing countless visitors through my poultry house, I have found that—if my visitor has previously been in a chicken house—at some point she will stop talking, sniff the air with a puzzled look, and ask, "Why doesn't it *stink* in here?" When that happens, I know I'm on the right track with manure management.

But the transformation of "nasty" to "pleasant" is just part of the magic. Remember the comparison of the deep litter to an active compost heap—the process in deep litter is driven by the same busy, happy gang of microbes. And among the metabolites of the microbes—by-products of their life processes—are vitamins K and B_{12}, in addition to other immune-enhancing compounds. The chickens actually ingest these beneficial substances as they find interesting things to eat in the litter. Don't ask me what they're eating, but chickens on a mature deep litter do little other than scratch and peck. This is alchemy indeed: *What started as repugnant and a potential vector for disease has been transformed into a substrate for health.*

Should you think I'm spinning fairy tales, know that scientific experiments have borne out the benefits of a bioactive deep litter. In 1949 a couple of researchers at the Ohio Experiment Station published research on deep litter. I urge you to read the full report, but to summarize: One experiment compared two groups of growing pullets, both on old built-up deep litter, one group receiving a complete ration, the other fed a severely deficient diet. Mortality and weight gain in the two groups were virtually identical. In another experiment comparing pullets fed a severely deficient diet, groups on old, thoroughly bioactive litter suffered far lower mortality (7 as opposed to 23 percent) and achieved much higher weight gain (at twelve weeks, 2.34 compared to 1.64 pounds) than those on fresh litter. Both these and further experiments demonstrated: "Obviously, the old built-up litter adequately supplemented the incomplete ration."[1]

The Food and Agriculture Organization of the United Nations confirms these observations: "Microorganisms thrive on the manure in the litter and break it down. This microflora produces growth factors, notably vitamin B_{12}, and antibiotic substances which help control the level of pathogenic bacteria. Consequently, the growth rate and health are often superior in poultry raised on deep litter."[2]

Deep-Litter Management

Factor in the use of deep litter when designing housing for your flock—deeper litter absorbs more manure and supports more microbes, so allow plenty of space for it. Aim for a depth of 12 inches if possible. Happily, in winter you can factor in as well the role of that thick layer of organic duff in insulating the coop from the frozen ground outside—and the heat generated in an active deep litter. The temperature is nothing like that of a well-constructed compost heap; but the warmth rising out of the pack moderates air temperature in the winter house. Caroline Cooper of British Columbia, Canada—whose winter hoophouse is shown in figure 6.9—sees temperatures of –13 degrees Fahrenheit for two weeks at a stretch in a typical winter but finds that the bedding, 12 to 18 inches deep, is warm to the touch a few inches below the surface.

The great thing about deep litter is that the birds do most of the work. But there are a few things requiring input and monitoring on your part as well.

Stocking Density

Joel Salatin makes this observation about stocking density on a deep litter: If you allow 5 square feet per adult chicken, the birds' constant scratching will turn into the litter all the manure laid down, even in high-poop areas such as those under the roosts. At 4 square feet, there will be some capping of manure under the roosts—formation of a crusty layer impervious to the hens' scratching. At 3 square feet, there will be extensive capping. If there is capping of the manure in your coop, turn it over with a spading fork from time to time, and the chickens will break it up from the cap's underside.[3]

Let It Mellow

You will see advice that the coop should periodically be thoroughly cleaned out. But as the Ohio experiments demonstrated, it is not fresh new litter that supports the health of the flock, but "old built-up"—that is, highly biologically active—litter. Thus an important implication: *Never clean out the litter completely.* Once beneficial levels of microbial activity are established, don't get rid of them by a de rigueur "thorough clean-out." Over time, the buildup of the litter—or the need for compost for the garden—requires removing part of the litter. Leave as much as you can in place, however, to retain the benefits of the already active microbes and to "inoculate" the fresh material you add.

The Whiff Test

The caveat to the above rule against cleaning out too much of the litter is that inevitably the addition of nitrogen by the incoming poops will overwhelm the carbon in the mix—resulting in the generation of ammonia. Be alert to that first characteristic whiff: It is telling you that an imbalance must be corrected—both because nitrogen for soil fertility is being lost to the atmosphere, and because ammonia damages the chickens' delicate respiratory tissues. Reestablishing the necessary balance is simply a matter of generously topping off with your high-carbon litter material of choice.

Do note that ammonia's deleterious effects begin *below* the concentration our nose can detect (25 to 30 ppm). With experience, you will learn to read the developing condition of the litter, so you can add fresh carbonaceous material *before* it starts generating ammonia.

Avoid Wet Litter

If you water inside, *avoid wet litter.* A soaked litter is anaerobic—deprived of oxygen—and more likely to support growth of pathogens. Wet litter also generates ammonia far more readily than drier litter.

Remember from chapter 6 that a lot of airflow through the coop helps keep the litter from getting too damp. Wet litter is more likely around the waterer, so check conditions there often; scatter any wet litter out over the total litter surface, where the chickens' scratching will help dry it. Waterfowl are especially likely to wet the litter. (See the next chapter for ways to cope with waterfowl and watering.)

Remember as well, however, that the busy critters in the litter need water for their work—monitor the litter to ensure that it is not powder-dry. Caroline Cooper reports that the winter air in British Columbia is extremely dry, so from time to time her husband, Shaen, carefully adds water to the litter to keep it active. If I have a waterer inside the chicken house, I frequently empty the small amount of water in its lip directly into the litter when rinsing it out.

Using the Compost

The deep-litter approach to manure management enlists the flock in the great work of soil fertility. Over time—figure at least a year—the litter will be reduced by the action of chicken and microbe to a finished compost. Sniff a handful: Like any fine compost, it will smell of earth with not the slightest hint of raw manure. In my experience litter at this stage of decomposition is ready to use directly in the garden—it will not burn plants, will not inhibit seed germination, and visibly boosts the growth of crops.

I have found litter from a coop with a wooden floor too raw to apply directly in the garden. Such litter should be further broken down in a conventional compost heap before use in the garden.

Disadvantages of Deep Litter

In close to three decades of relying on deep litter for best manure management, I have encountered only two potential disadvantages. The slight wicking of moisture from the earth into the litter is as said actually a benefit. However, we once had a summer of record-breaking rains, resulting in increased moisture in the soil under the deep litter (remember, we use an earthen floor). The litter was not actually wet as a result but was considerably damper than usual—

Bioactive Litter for a Micro-flock

A tip from my friend Kate Hunter

I'm on my second year of housing my tiny flock of layers in my shed over the winter. I build a temporary 5' x 6' enclosure with garden caging which gives them exactly the same space in the shed as they get in their mobile pen (7.5 square ft/bird). The shed's wood floor is protected from any possible moisture damage, first by a sheet of tar paper, and then a tarp, which I pin up along the inside of the caging. The tarp holds a two-inch layer of garden soil as inoculant—that is, introduction into the litter of the same beneficial microbes in soil and in a compost heap—and then I fill with several inches of mulch. I continue adding mulch for several weeks after the girls have gone in [to an eventual depth of 12 inches]. It absorbs the manure the hens produce without ever turning foul.

The litter makes for great wintertime activity—my girls love scratching in it. When it's frozen, they can't scratch it, so I go in there with a pitchfork every once in a while and turn things over for them. Then they get to scratching again. I also throw some of their feed directly on the mulch from time to time so that they have additional motivation to scratch through it and keep the mulch aerated and loose.

I emptied the tarp as soon as the housing was dismantled and the girls were back outside last spring. I spread the litter as a fertility mulch under the apple tree. It smelled fine. There was no damage to the wood floor either.

—Kate Hunter, homesteader
(livingthefrugallife.blogspot.com)

Fig. 7.1 With a little creativity it's possible to use biologically active deep litter in winter housing for a micro-flock as well. PHOTO COURTESY OF KATE HUNTER

damp enough to encourage the growth of molds. We had a number of eye infections that season, and lost an entire batch of nineteen guinea keets. Once I recognized the problem, I helped decrease the moisture content of the litter by adding a lot of thoroughly dry leaves and kiln-dried shavings.

The other potential disadvantage of deep litter over an earth floor—assuming the henhouse is not on a block perimeter foundation—is the absence of a wood or concrete floor as a barrier against digging predators such as foxes, coyotes, and dogs. My solution was to dig a barrier about 18 inches into the earth—metal roof flashing, but half-inch hardware cloth would work as well—around the entire perimeter of the poultry house. That's a lot of digging (oh, my aching back!), but it prevents a lot of digging (by four-legged neighbors intent on dinner in your chicken house).

A Win-Win Solution

I cannot overemphasize the importance of deep litter in the henhouse for the most natural and therefore the most rational manure management. A deep-litter house is more pleasant for both owner and fowl, with the chooks doing most of the necessary work for us. Microbial action in the litter turns what is potentially disease causing into a substrate for health—indeed, ripe litter as demonstrated in the Ohio studies has positive feeding benefits. Deep litter provides mental health as well—the entertainment of happily scratching an endlessly interesting deep litter, in lieu of the stress of boredom. A deep organic duff insulates the floor of the winter poultry house, while the warmth of its decomposition moderates the chill. Finally, this magic process captures the fertility in the poops for soil building, the key to food self-sufficiency. What better illustration of the *integrating* strategies at the heart of this book?

8 | WATERING

Think of water as the most essential nutrient for your flock, and make sure there is never a lapse in the flock's access to fresh, clean water. Layers drink more than nonlayers—eggs are 65 percent water—and serious water deprivation will greatly reduce production, in the worst case permanently. Fast-growing meat hybrids drink more water than their slower-growing traditional breed cousins.

I do not agree with the advice often seen to "sterilize" waterers with bleach or other chemical cleaners periodically—just scrub them out with a scrub brush, old toothbrush, or handful of coarse dried grass. And do clean frequently—the birds will drink more water if it's clean. Keep in mind what *you* want to see in a glass of water—as much as you can, make sure that's what the birds are seeing in their waterer as well.

If you water inside the coop, remember what was said in the previous chapter about avoiding wet litter. Avoid open waterers—the chooks kick debris into them as they scratch the litter. Remember the vacuum-seal waterers recommended for chicks in a brooder. There are larger versions available for older birds as well, in either molded polyethylene or galvanized steel, from 2 gallons up to 14. All feature a reservoir fitted onto a base with a narrow lip with access for drinking but no room for splashing. Litter gets scratched into these waterers as well, but it is easily rinsed out of the lip, in lieu of dumping out the whole waterer. Setting the waterer on some sort of platform—about the height of the chickens' backs—helps keep the water free of litter.

When placing one of the large vacuum-seal waterers outside, find a patch of level ground to set it on—if it's tilted, the rising water level spills over the edge of the lip before it covers the hole in the base, so a vacuum never forms in the reservoir and all the water runs out. As well, *listen* to it before walking away. The larger waterers of this type feature a cap of heavy molded plastic with a handle, which seals the reservoir at its top by screwing down tightly onto a rubber O-ring. A bit of unnoticed trash on the O-ring will allow passage of air—whose faint whine or whistle alerts you that a vacuum cannot form inside the waterer, and that you must clear the cap's seal of the speck of trash before leaving the waterer.

Automating the Water Supply

Fill-and-tote waterers will be entirely adequate for most small flocks. The larger the flock, however, and the farther out on the pasture, the more inclined you may be to opt for automated watering of some sort. There are several options.

Various automatic watering systems are available for larger poultry houses or exhibition breeders with many individual cages to serve—nipples that flow only when the birds are billing them, small self-fill cups or bowls on individual supply lines—but such systems are not only overkill for most small flocks but may be impossible to set up out on the pasture, which is where your birds deserve to be if possible.

Based on my experience, the one automatic waterer I strongly advise you avoid is a molded plastic trough, 18 or 36 inches long, with a flimsy float mechanism to regulate water supply. It is cheaply made and will not last.

The simplest version of automated watering is the one I use most for my birds on pasture: a molded rubber watering tub, 5 to 7 gallons, with an attached float-operated shutoff valve (similar to the one in a toilet tank) on the end of a supply hose that runs from an outside hydrant. Of course, the result is an open waterer—precisely what I warned against inside the henhouse. The difference here is placement on a grass sod: The water stays much cleaner when the waterer is on grass, and frequently moved. If left in one place, traffic around the waterer degrades the sod—the water always gets dirtier on bare ground.

Fig. 8.1 Simple automated watering based on a float-operated shutoff valve and rubber watering tub. Blocks provide access for smaller birds, while the insert prevents drowning if they fall in.

The setup in figure 8.1 for automating watering requires no plumbing skills. It's also not nearly as finicky as some other options, which may require precise alignment, or clog with sediments or algae. Using Y-connectors readily available for garden hoses, you can serve any number of individual float waterers from a single hydrant. Here are some pointers if you want to give it a try:

- **The valve:** The automatic watering valve I use is not intended for watering poultry—it is designed for attaching to a large stock watering tank. However, it's easy to mount it onto a molded rubber watering or feeding tub, maybe adding a piece of scrap wood between the tub's top edge and the mounting bracket, providing a surface for the mounting screws to bite into. The valve could be rigged onto other vessels as well, so long as they are not too shallow—the float must have enough clearance from the bottom of the vessel to swing into its full-open position.
- **Float design:** It is worthwhile ascertaining before buying a float-operated valve how the float attaches to the body of the valve (see figure 8.2). If the float is a single piece of molded plastic with tabs that snap into indents in the valve body, it is a piece of junk best avoided—the slightest nudge while cleaning or rinsing may unseat the float, leaving it incapable of shutting off the water. Reattaching the float may be difficult without unmounting the entire valve. A much better design features a cotter pin inserted through

Fig. 8.2 Before buying a float-operated valve to make an automated waterer, check how the float seats on the valve. Molded tabs (left) become unseated with even the slightest jostling; a cotter pin (right) locks the float securely in place.

matching holes in the end of the float and its seat on the valve, for a slip-free connection between them.

- **Hoses:** I buy hoses from recreational vehicle suppliers that are manufactured for potable water. Most garden hoses contain lead, which can leach into the water, particularly that initial flow, which has been sitting in the hose awhile. Hoses should be labeled specifically "Not for drinking" or as being "Safe for drinking"—keep looking if the hose you're considering isn't labeled either way.
- **Making it drown-proof:** This float-operated waterer works fine for adults, but young growing chickens may fall into it and drown. If you keep young birds with the adults, therefore, put something into the tank or waterer—a large rock, concrete blocks, or the cutoff bottom half of a plastic bucket, as shown in figure 8.1—onto which chickens can clamber if they fall in.

Watering in Summer

In summer the ill effects of loss of access to water increase, since chickens drink two to four times their normal amount in hot weather. More frequent monitoring is essential to make sure they don't run out.

Chickens will drink more water if it's cooler, so place the waterer in shade if possible. If you use a hose to automate watering, remember how hot the water coming out of a dark-colored garden hose can be. That may be another reason for buying hose labeled for potable water: All such hoses I've seen are white, and in my experience the water in them stays much cooler than in a dark hose.

Watering in summer will be especially challenging if you are raising fast-growing meat hybrids, especially Cornish Cross, who in high heat may sit on their butts and *die* rather than walk 10 feet into the sunshine for a drink of water, as noted in chapter 4.

Watering in Winter

The "gotcha" of water deprivation stress can occur in winter as well, if you forget that the waterer can freeze, denying the birds their needed drink. There are various heating options for preventing freezing—heat tape, heat lamps, birdbath warmers placed in the waterer, a heated base onto which the waterer is placed (if it's metal, not plastic). All such options require electricity, which gets more expensive all the time and is likely to become less reliable as the energy crisis deepens. The simpler option for the frugal homesteader is to either refill the waterer each day or set it in the basement or elsewhere indoors overnight. In my area (Zone 6b), it is rare that daytime temperatures are low enough to cause freezing of the waterer. If that is a problem, two waterers can be used and switched in turn between the people house and the poultry house.

If you're not familiar with a freeze-proof yard hydrant, they make winter watering more convenient.

The *Water* in *Waterfowl*

Watering issues constitute the biggest difference between management of waterfowl and of chickens. Chickens need water to drink; waterfowl ideally need water to splash in. Not only does the volume of water increase dramatically when watering waterfowl, but if not carefully managed, the splash zone can become quite a mess. Watering needs for ducks and geese are the same so will be discussed as one topic.

Bathing

I strongly recommend that you provide an opportunity for your geese and ducks to bathe—not just to splash some water on themselves, but to swim and immerse themselves entirely in the water. Bathing is part of preening the feathers for waterfowl and serves the same function as dust bathing for chickens: Geese and ducks use the water to drown external parasites under the feathers (as chickens use the dust to smother

A Frost-Free Yard Hydrant

Many outside faucets have to be shut off and drained for the winter to keep them from freezing. An exception is this freeze-proof yard hydrant, designed to prevent freezing even when it is being used daily in the winter. The key to prevention of freezing in the hydrant's vertical pipe is a drain port or bleed valve at its base.

The supply line to a valve in the base of the hydrant is installed well below frost line. When the hydrant's handle is up, linkage via a steel rod opens the supply valve and shuts the bleed valve. Returning the handle to the off position closes the supply valve and opens the bleed valve, permitting the water in the pipe to drain out into the unfrozen subsoil. Thus there is never standing water in the pipe, which can freeze, expand, and burst it.

There is one "gotcha" associated with this hydrant—ask me how I know! If you leave any sort of hose connector screwed onto the head of the hydrant, it may hold the column of water in the pipe, and freezing will indeed burst it.

When installing the hydrant, make sure the bleed valve is surrounded by several inches above and below of coarse gravel, with some scrap plastic sheeting over the top of the gravel. With this precaution, soil will never sift down around the opening of the bleed valve and stop it up, permitting water to stand in the pipe and freeze.

Fig. 8.3 When this hydrant is shut off, water drains into the soil below frost line, and none remains standing in the pipe to freeze.

them), so bathing contributes to good health. But they also *love* it—bathing certainly enhances their mental health. Enjoying their exuberance in the water will boost your mental health as well.

If you want to try breeding either geese or ducks, there is a further reason to give them the opportunity to bathe. For waterfowl, mating is easier in the water than on the ground. Indeed, among heavier breeds of both ducks and geese, it may be impossible for the male to mount the female; and successful fertilization will not occur without water to buoy him up and help him perform.

If you have "wild water" on your place—a pond or stream—it should be easy to arrange access to it for your waterfowl. But if you don't, it's not difficult to set up a ducksplash for their use, like the one in figure 8.4. I use a couple of different stock watering tanks, molded rubber or fiberglass—one of 50 gallons, the other more than 100—to provide bathing water for my ducks and geese. I use the same arrangement described above: a

Fig. 8.4 If you don't have a pond or stream, furnish your ducks or geese a bath, such as this stock watering tank on a supply hose with a shutoff valve.

supply hose to a float-operated shutoff valve attached to the tank. If you want to provide your waterfowl a similar ducksplash, setup is basically the same—only the container is larger. Here are a few additional pointers:

- **Cleaning:** The bathing tank should be cleaned frequently—the ducks and geese will drink from it as well—especially in summer. Fortunately, clean-out is easy—I just open the tank's plug and let it drain, swipe the interior with a brush, rinse, screw in the plug, and walk away.
- **Access/egress:** Unless you dig it into the ground, you will need to furnish steps up into the bathing tank for the waterfowl. I make mine by stacking concrete blocks of various sizes. If you build one of wood, make the surface of slats or rough-sawn lumber, or score it well, so that it will never become too slick with water, poop, or algae—waterfowl may injure their legs if they slip on such surfaces. Adults exit the tank without assistance, but ducklings and goslings in the down stage may die of drowning or exhaustion if they don't have something to climb onto. I stack a couple of concrete blocks to provide a platform just below water level—once the young ones clamber onto it, they easily hop over the edge and onto the ground. In a mixed flock, provision of such a platform prevents drowning of young growing chickens as well.
- **Preventing drilling:** Waterfowl, especially ducks, have a habit of drilling in wet soil—and there is bound to be plenty of that around the ducksplash. If tearing up the sod in that area is a problem, set wood frames with attached half-inch hardware cloth around the tank. A simpler option is to set the tank on a piece of

hardware cloth large enough to cover the drilling zone. In this case, however, it is necessary to lift it free of the sod from time to time—if growth of grass through the wire is ignored, it will eventually be impossible to lift the wire.

The Rinsing Waterer

If it isn't possible to provide a full bath for your waterfowl, *at a bare minimum you must provide water deep enough to submerge their entire heads.* Ducks and geese unable to rinse their eyes and nostrils frequently, especially when eating powdery commercial feeds, are more susceptible to eye infections and possibly even choking. The automated watering tub I use for my chickens on pasture (see figure 8.1) provides water deep enough for my waterfowl to rinse their heads.

Waterfowl will splash water over themselves from such a waterer and vigorously work it with their bills, thereby preening their feathers and inhibiting external parasites—though such water grooming is a poor second best to full bathing.

Winter Watering of Waterfowl

Providing water for waterfowl in the winter is more challenging than for chickens. Avoid watering them inside the poultry house if possible—they are incredibly messy with their water and will quickly soak the deep litter, which then is more subject to growth of molds and, because it is more anaerobic, of disease-causing pathogens.

If you must water inside, set a 5- to 7-gallon tub waterer on a wire frame over a large catch basin to reduce wetting of the litter—a setup like that shown in figure 8.5. Again, the setup shown is a poor second best to real bathing—but a real bathing tank inside the poultry house would lead to soaking everything in sight. When I have used it in the past, it was only for a couple of geese and a trio of ducks—it wouldn't be effective with many more than that. Empty the basin as needed using 5-gallon buckets carried outside. A better refinement if you are handy, of course, would be to plumb a drain line from the bottom of the catch basin to the outside.

Fig. 8.5 If you must water ducks and geese inside the poultry house in winter, set the watering tub over a catch basin to minimize wetting of the litter.

Frequently disperse the wet litter around the catch basin onto the drier parts of the litter using a pitchfork. Of course, if you house the winter chicken flock with the waterfowl, they can handle this work for you, especially if encouraged with some scratch grains thrown onto the litter.

My strong recommendation is to keep the watering of the waterfowl outside, even in winter. Your experience may be different if you are far to the north of me (mid-Atlantic, Zone 6b), but I find that even in the dead of winter maintaining the 100-gallon-plus bathing tank for the waterfowl is not a great problem. That volume of water freezes slowly on even the coldest nights. The black color of the molded rubber absorbs solar energy as soon as the sun appears, so it isn't long before the night's ice melts enough that you can break it up and toss it out in sheets.

In winter I don't fill the tank with the hose and float valve, which would freeze—I just top off as needed to replace water lost in the discarded ice. Clean-out in winter is infrequent, since algal growth slows so much in the low temperatures.

9 | PASTURING THE FLOCK

Only those forced by circumstance to confine their flock to a static run should do so. If we base our management on imitation of nature, we will allow our flocks as much space as possible on high-quality, actively growing greensward—to forage the natural feeds most suited to their good health; to benefit from fresh air, exercise, and sunshine; and to satisfy their curiosity and engage in instinctive social behaviors.

Day Ranging

Some flocksters are able to dispense with fences entirely and allow their flocks unrestricted ranging. Such an arrangement maximizes the benefits of being out of confinement and works well where there is no serious threat from predators. Since many predators are active at night, day ranging typically involves letting the chickens range where they will during the day, then religiously shutting them in their coop at night. (However far they range, they will return to the coop as the sun is going down.) Remember, however, that dogs can be daytime predators extraordinaire. (For more on predation challenges, see chapter 21.)

Another limitation on complete free-ranging is the presence of close neighbors—our chickens will not make themselves, or us, any friends if they are eating the lettuces in a neighbor's garden, or scratching up a bed of prize roses.

Finally, the greater the freedom to roam, the greater the chances that hens will follow a natural inclination to hide their eggs, in lieu of using the nestboxes in the henhouse. Since they can be quite clever at doing so, there could be a considerable loss of egg production. (If you do find a nest hidden by a free-ranging hen, don't simply empty it of its eggs—the hen will conclude that a predator has found her nest and will find another nest site, even better hidden. If you leave some plastic

Fig. 9.1 The pastured flock is a happy flock.

The Traveling Layer Flock

A mobile henhouse for the pastured layer flock is a great convenience. Flocksters with market flocks make them from whatever comes to hand, from converted recreational vehicles to school buses, but the most common versions I've seen are mounted on hay wagon chassis or flatbed trailers and pulled with a pickup, ATV, or tractor (see figures 9.2 and 9.3).

A nearby friend, Stephen Day, assisted by his father Richard, recently made an "eggmobile" for a flock of one hundred layers (figures 9.4 and 9.5). Stephen reports:

> We started building the coop with no schematics and only a general idea of how it would come together. Dad and I both have prior experience putting together outbuildings, so we just brainstormed each new problem as we built. The whole idea is based on Joel Salatin's eggmobiles, and my total cash outlay was about $1,500, not including labor time.
>
> We started with an old 16x7 tandem flatbed with an iron gridwork floor that we bought cheap through Craigslist. The grid floor should both keep manure from piling up in the trailer and pull plenty of air up from below and out along the corrugations of the roof panels.
>
> We made six sections of stud wall, two feet on center, bolted them through the floor, and braced everything with 2x4s diagonally. Since the trailer will never be on perfectly level ground, the roof needed no pitch—it's just six sheets of corrugated metal roofing. The walls are plywood with two chicken doors cut into them as well as a human door, two windows, two vents, and two rectangular flaps for gathering eggs from nestboxes without going inside.
>
> The interior has several roosts we made out of saplings, with room to expand to accommodate more birds. There are 25 nestboxes, which I hope will eventually serve 200 laying hens. I'm starting with 100 as an experiment to test the trailer, our predator situation, the egg market, and the chickens' impact on our pasture. We will use a gravity-fed cup watering system and the chickens will be fed primarily by pasture

Fig. 9.2 Eggmobile for a modest layer flock inside an electronet fence.

Fig. 9.3 Eggmobile on a hay wagon chassis.

grazing, supplemented by pellet feed. We will tow the trailer several hundred yards with our pickup every week or two, to keep their pasture area fresh.

Fig. 9.4 Stephen Day's eggmobile for one hundred to two hundred layers on his farm in northern Virginia. PHOTO BY STEPHEN DAY

The Cadillac of eggmobiles is the one built by Tim Koegel of Windy Ridge Farm in southwestern New York State for his market layer flock. An impressive high-tech modification is the addition of solar panels, which can be mounted on either side, depending on the orientation of the shelter in relation to the sun after it is moved (not shown in the picture, unfortunately). The solar panels supply charge to an extensive system of electric nets, as well as interior lighting for work inside during the dark early-morning and evening hours—and to provide supplemental lighting to encourage winter laying. In the winter, with reduced solar input available, Tim connects the shelter and its electric nets to grid power.[1]

Fig. 9.5 Interior of Stephen Day's eggmobile. Note metal grille floor for drop-through droppings and enhanced ventilation. PHOTO BY STEPHEN DAY

Tim's flock uses this shelter year-round, as you can see in figure 9.6. In the winter two 250-watt infrared heat lamps on a thermostatic control prevent freezing of the waterers. When he parks the mobile shelter for the winter, Tim attaches a hoophouse of the same dimension. Each is 10 feet by 20, yielding a combined space of 400 square feet for the flock of 150 layers. Deep litter in the shelter is wood shavings; in the hoop, hay. Tim encourages scratching of the latter by feeding scratch grains there.

Fig. 9.6 Tim Koegel uses his Cadillac eggmobile as housing for his layer flock year-round. He moves it frequently to new pasture in the summer and parks it and attaches a hoophouse in the winter. A scraped yard in winter provides even more space when weather permits. PHOTO COURTESY OF TIM KOEGEL

He pushes back the snow to provide an exercise yard, which the hens enjoy using on sunny days. The dormant sod does not furnish grazing, of course, so Tim sprouts grains for feeding.

or wooden eggs in the nest when removing the eggs, the hen will usually continue laying in it.)

A hen in a day-ranging situation may even decide to set a nest she has hidden—and show up three weeks later with a clutch of chicks in tow, a surprise that may or may not fit your plans.

Electric Net Fencing

The best way to maximize your flock's range, while keeping them where you want them and protecting them from anything on the ground with a nervous system, is *electric net fencing*, or *electronet*. Since electronet is such an ideal solution to the freedom/restriction conundrum for me, I will discuss its use in the following chapter.

Water and Shelter

Be careful not to let watering on pasture be "out of sight, out of mind"—be sure your birds always have plenty of water at all times. The farther a pastured flock is from the water hydrant, the more inclined I am to automate watering. The simplest version of automated watering is a float-operated valve at the end of a hose, attached to some sort of watering tub—I set up such waterers as much as 200 feet out on the pasture.

When you move your flock out onto pasture, remember their need for shelter from rain and for shade, and as a place to spend the night. Getting wet can be life threatening for chickens and guineas if it's windy and cold; and they have a deep instinct to be in a sheltered space at night.

Shade is essential for poultry on pasture. If the area where you have them penned does not have tree shade during all parts of the day, it is essential that a pasture shelter provide this retreat from hot summer sun. We learned this lesson one spring following a sharp temperature spike—when I went out to the pasture, I found my layer flock badly stressed, and several had begun attacking weaker members. One hen in particular walked about in a daze, oblivious to the fact that other flock members were literally pecking her apart. She subsequently died, and we vowed never again to let our flock get that stressed by exposure to the sun.

Waterfowl, once they are well feathered, will ignore the shelter when it rains—they think the bath-from-heaven is terrific. Given the insulation of their heavy plumage, however, waterfowl have an even greater need for shade. They may or may not take refuge in the shelter at night. (Design of pasture shelters is a big topic and will be considered in a subsequent chapter.)

Be especially careful if you have young growing birds on pasture—they have not learned the hazards of getting wet. Make sure young birds not long out of the brooder—still somewhat sparsely feathered—don't get lethally wet in a light rain or heavy dew.

Pasture Management

If the pasture improves rather than deteriorates as a result of keeping our flock on it, we know we're on the right track. Grazing by poultry—like grazing by ruminants—helps prevent the pasture from growing up in brambles and shrubs and following the progression to forest; it also stimulates vigorous new growth of the sward. Ranging chickens eat wild seeds, thus helping to reduce the seed bank for weeds that might otherwise become too aggressive. The manure laid down by the flock fertilizes the sod—a benefit you'll never get from a power mower. We first saw a dramatic illustration of this fertility boost after we started raising our birds on pasture. I had fifty young broilers in a Polyface-style mobile pen, which I moved every day.[2] After a couple of weeks we had a nice spring rain. The band of rich green that emerged in the wake of the pen could have been laid out with a ruler.

It is desirable to have a diverse mix of plant

Fig. 9.7 Lawns can be a resource, not just a dead-end chore.

species in the pasture sward—our goal should not be the monoculture desert seen in a typical suburban lawn. If you doubt the wisdom of encouraging as wide an array of ambient plant species as possible, answer me this: What do the following plants have in common—dandelion, lamb's-quarter, stinging nettle, burdock, and yellow dock? (No points for a dismissive, "They're all just *weeds*, for heaven's sake!") Answer: Each is at least 4 percent, and up to 12 percent, higher in protein than that quintessential high-protein fodder crop, alfalfa. Poultry relish them all. Further, each of these common weeds concentrates a different mix of minerals. So if these plants are part of the pasture mix, our birds will more likely balance their mineral intake as they forage.

While the flock does much of the work of pasture management, there are a few contributions we should make to pasture improvement as well. It is best to rotate the flock over the pasture. Chickens especially wear at the pasture with their constant scratching and should be moved to a new plot before they damage the sod. As well, rotation helps ensure that every part of the sward gets its share of the fertility-enhancing deposition of poops and avoids a possible buildup of pathogens or parasites on the pasture.

You need to do much less mowing on a pasture being grazed by poultry, but it's best to mow occasionally. Grass and other pasture plants are highest in protein and other nutrients when putting on rapid new growth after grazing or mowing. In the spring I like to let my pasture grasses grow as high as possible, then cut with a scythe, and rake up the long-stem grass for use in compost and mulches. A few days to a week later, I move the flock onto the rapidly regrowing grass.

Later in the season, when I would get less return for the labor of scything and raking, I occasionally use a power mower to prevent too heavy a set of weed seeds in the sward. Though nutritious "weeds," as said, are welcome in the mix, some can be pretty aggressive if I don't head them off a bit with preventive mowing.

Grazing by the flock and occasional mowing by the flockster will do much to improve the pasture without additional seeding. However, you can certainly overseed the pasture to improve the mix if you like. After taking the flock off the pasture in the fall, I simply broadcast a mix of pasture perennials likely to do well in my area (timothy, orchard grass, Kentucky bluegrass, various clovers, alfalfa). Alternatively, I might broadcast my seed mix in the late winter and let the freeze–thaw cycle work the seeds into the soil. In either case, germination will not be as good as when drilling the seeds, so I use generous amounts of seed. Check with your extension agent about the best species for seeding pasture in your area.

Many of the pasture species I have been seeding in past years are introduced species. I am currently preparing a part of our pasture for an experimental planting of native grasses exclusively: big bluestem, little bluestem, broomsedge, Indian grass, switchgrass, side oats grama, eastern gama grass, purpletop. And how am I doing the site preparation? Why, with *chicken power*, of course (see chapter 12).

"Pasturing" on the Lawn

If you have no pasture on which to range your birds, be assured that some flocksters pasture their flocks on their lawns. Indeed, one of my correspondents in the American Pastured Poultry Producers Association (APPPA) made her start in pastured poultry on the lawns of her three-quarter-acre house lot and for several years sold three to four hundred broilers a year in local markets.

Most years I rotate my waterfowl flock over four sections of lawn around our house during at least part of the growing season, as shown in figure 9.7. I do far less mowing; the birds fertilize the lawn; there are more trees for shade near the house than on the pasture; and the birds turn that lovely grass into terrific winter feasts and high-quality cooking fats.

10 | MANAGING THE PASTURED FLOCK USING ELECTRONET

Readers who have stuck with me so far will have concluded that my poultry husbandry is as low-tech as I can manage. But there is one technological marvel that is fundamental in the care of my flock: *electric net fencing*, or electronet, shown in figure 10.1. Though electric fencing of any sort is not appropriate in most urban and suburban contexts, for me—on 3 acres in a small rural village—it resolves the restriction/liberation conundrum, allowing me to give my flocks extensive range to forage freely, while keeping them within needed limits. And it is incredibly effective at protecting them against predators on the ground. It does not protect against aerial predation, of course. (For more about dealing with aerial predators, see chapter 21.)

Good electric net fencing is not cheap, but I wouldn't cut corners on quality[1]—money saved if you buy a cheap system vanishes if you lose chickens. If you buy the best equipment from a reliable company such as Premier, the initial investment in a roll or two of netting plus a decent energizer will set you back several hundred bucks. However, that investment buys you a fundamental tool for managing the homestead or farm flock that with good care will last for a long time. It buys you the ideal compromise between maximum health and well-being for your pastured flock and maximum protection from the heavy hitters in the neighborhood.

Since a good electronet system is a considerable investment, research the topic further when planning your system. Before placing an order, talk with the technical support staff at your preferred supplier to make sure your choices are sound.

The larger the area on which you range flocks, the more prohibitive becomes the cost of using electronet to protect them. Are there effective alternatives using single-strand electric fences? I have never experimented with this option myself, though I have used single-strand to deter deer. However, some members of American Pastured Poultry Producers Association (APPPA) who range market flocks over a couple of acres or more use single-strand electric perimeters, some of only two strands, some with up to seven, usually with excellent deterrence of predators, who

Fig. 10.1 Electric net fencing allows birds extensive range–but only where we want them–while preventing most losses to predators.

tend to lead with the nose. It seems a fact of life for most, though, that single-strand fences allow some ranging of chickens outside the perimeter, where they are easy targets for predators. Occasional such losses some members find economically acceptable, given the savings involved and more extensive range for the birds.

Net Design

An electric net fence is a lightweight plastic strand mesh, with interwoven support posts, easily set up and easily moved. The vertical strands, for support only and carrying no charge, are "welded" to the horizontal strands which—with the exception of the bottom one, which lies on the ground—are twisted with several almost hair-thin stainless-steel wires that carry an electric charge. The posts—fixed into the net at intervals of 8 to 12 feet or so and made of plastic, to prevent grounding of the charge—end in steel spikes the user pushes into the earth to erect the fence. More recent designs feature the bottom horizontals much more closely spaced than those higher up, both to contain young growing birds and to make better contact with predators at nose level.

Note that all the horizontals are twisted tightly together at each end of the net so they share a common charge, ensuring that current will always be available in the entire net, even if there is a break in an individual horizontal line. The braided strands on each end terminate up top in a metal clip, which can be fastened to the companion clip on the next net, to yield long perimeters of any size needed.

Electric mesh nets are available for other purposes as well, but those designed for poultry are available in heights of 40, 42, and 48 inches. You might assume "the higher the better," but the higher net is more difficult to gather and bundle when it's time to move the fence. I've always used 42-inch nets with excellent results. Of course chickens can fly that high—but they could fly over a 48-inch net as well. Once they get zapped a couple of times, however, they rarely do.

The standard length of poultry nets is 164 feet, though half nets of 82 feet are available. I keep a few half nets on hand to use where a full net is too long for the perimeter I need to establish.

Setting Up the Fence

A tight fence will make better contact with a curious predator, or a chicken, than a sagging one. Some users prefer posts with double spikes, not only because of the right-angle step on the second spike for pushing it in with the foot (rather than jabbing it in by hand), but because they hold better in wet or loose soil. I've always used the single-spike version, which is lighter and thus easier to move. In compacted or gravelly soil, I drive a hole for the spike with a piece of rebar and a small sledgehammer, then reinforce the fence as described below.

Poultry netting will arrive properly folded and rolled. The end posts will be obvious—they're the ones with all the horizontals twisted together. Holding the entire bundle in one hand, start with one of the end posts and lay out the net flat on the ground, one panel at a time. Now go back around the perimeter, poking the spiked posts in place to stand the fence up. Mow the outside of the perimeter, as close to the fence as you can get. Now remove the fence and toss it on the ground, well inside the perimeter line you've mowed—and cut another swath with the mower, to the inside of the first. Stand the fence back up, in the middle of the mowed swath.

When mowing, set the mower blade low—the shorter the grass, the longer before you have to mow the fence line again. If you prefer using a scythe and can mow a lawn with it, you can make the perimeter with your scythe.

Now tie every corner of the fence—not only right-angle corners, but anywhere the angle of the perimeter changes significantly—to a corner post, preferably with biodegradable twine. These more substantial molded plastic posts, also with steel spikes, are available from your electronet supplier, but should be readily available in the electric fencing section of your farm co-op as well.

The resulting fence may or may not satisfy you, depending on how insistent you are on setting up a sag-free fence. I like a really tight one, so in the middle of each panel—the section of mesh between posts—I place an additional ⅜-inch fiberglass rod, with a screw-on plastic insulator lifting up the top strand, thus adding tension and removing all sag from the fence. Note that I strongly recommend *coated* rods, available from electric fencing suppliers: Those not coated will after some weathering stick fiberglass splinters in the hand.

Plan to *rotate* your fenced areas over the available pasture, in order to prevent excess wearing of the sod, break life cycles of pathogens and parasites, and ensure even application of the flock's manure as a fertility boost to the sward. Frequency of rotation depends chiefly on stocking density and the point in the growing season—lush spring or dry summer—and thus how quickly the pasture recovers.

You can set up each perimeter in turn as an island unto itself, entirely separated from the poultry house—in which case you will need to use a mobile pasture shelter, discussed in the next chapter. Or you can anchor rotation plots onto the poultry house. Ideally, it will have multiple doorways (if only pop-holes of chicken size)—if the chooks always use the same entrance, the area around it turns into the same poop-dotted, unwholesome bare ground discussed earlier. If there is no alternative to continual use of the same entrance, put the bare spot under deep mulch.

Remember when laying out the fence that you do *not* need to make a complete loop—that is, to complete a circle with the two ends of the fence attached to each other. Even if the far end terminates at the wall of a building, for example, contact with the fence at any point will cause a flow of electricity from the energizer through the body of the predator and into the ground.

Note also that it is probably not necessary to include a gate in the layout: At least in the case of the 42-inch fence I use, I can easily swing one leg over to straddle the net, then follow with the other leg. Of course, you want to be *certain* the power is off before doing so.

Moving the Fence

Moving the fence is easy, *if* you do it properly. If you handle the fence carelessly when moving it, it will make you weep with frustration. Most important, avoid the temptation to roll up the net. Begin by laying it out flat. Then, starting at one end, *fold* each panel in half as you draw the end post over to the next post. Repeat until all the panels are neatly folded, like pages of a book, with the posts in a bundle like the book's spine. Now you can pick up the gathered net and lay it out around the new perimeter.

A big challenge when moving the fence is keeping the flock from scattering to the four winds while you do so. Ducks and geese aren't much of a problem—once you have the new fence mostly set up, they are easy to herd into it. But chickens resist herding—save yourself the aggravation of trying. If you set up the new enclosure first thing, you can tempt them into it with their morning feeding.

I have found it a good investment to have extra rolls of netting on hand, so I can set up my new perimeter—usually adjacent to the existing one—while the birds continue foraging in the old one. If they resist moving, I collapse the old fence—that is, pull it ever more closely around the flock—until the birds have no alternative but to enter the new paddock.

If you don't have additional nets, the best strategy is to leave the chickens shut up while you reconfigure the fence in the early morning. If they are in a mobile pasture shelter, move it into the new enclosure before letting them out. If the net is anchored on the poultry house, simply open the appropriate door or pop-hole onto the new enclosure.

Charging the Fence

When you inevitably get a taste of the fence, you will understand why it is so incredibly effective. The jolt you feel, however, will likely be when you're on your feet, with the soles of your boots giving you considerable insulation from the ground. But if you hit it while down on one knee, more solidly grounded, the wizard in the fence will rattle your teeth. *That's* what the predator feels.

Keeping that hottest possible spark in the fence is thus the key to predator deterrence and is a matter of both equipment and management.

Energizer

There are many models of fence charger to choose from, depending on your needs. Remember, though, that electric netting requires a more powerful energizer than single-strand electric fencing. Err on the side of buying a charger with more voltage capacity than you need, rather than too little.

The major division is between chargers that plug into household current and those powered by a battery. I use both. My plug-in model is located in the main poultry house, which is wired. If the fence is anchored on the henhouse, I simply connect the charger's terminal to the end clip on the net. For freestanding fences, I pull current from the plug-in unit, up to 200 feet out on the pasture, using insulated cable on the fencing rods with screw-on insulators, mentioned earlier, to make it more visible and prevent tripping. This is the arrangement I prefer, since the plug-in energizer will charge more nets in a system, and will take more weed load on the fence before losing voltage. I run the cable to a manual shutoff switch so I don't have to walk all the way to the henhouse to cut power to a freestanding net. If there is more than one netted enclosure in the system, their manual switches can be wired either *parallel* (all other nets in the system remain charged when an individual switch is open) or *in series* (nets farther down the line also lose power when a switch is open).

Fig. 10.2 Fence charger with solar panel, backup battery, and controller to manage pulsing the fence and trickle-charging the battery. Note the warning sign, always a good idea, and required by law in some jurisdictions.

Battery units can be small or large, powered by batteries either disposable or rechargeable, from a couple of D-size up to an automobile battery. Depending on its capacity, a battery charger may power multiple enclosures with many rolls of netting.

Some battery chargers accept an add-on solar panel with a controller to maintain charge in the fence while trickle-charging the battery—at night or on heavily overcast days, the controller draws from the battery to charge the fence. For freestanding use in a site too remote to serve conveniently from household current, the solar-powered energizer is hard to beat. I use two of them. (The lead-acid, deep-cycle rechargeable batteries in these units can also be recharged on a plug-in trickle charger.)

Testing the Charge

Essential to making sure the fence will deter predators is knowing how much charge it is carrying, and that requires testing. Obviously, the best way to gauge the net's voltage is *not* by grabbing it for a feel.

There is a country boy version of testing voltage. Pull a long green stem of grass, and lay its far end on the fence. Now slide the stem so the length between your fingers and the wire is progressively shorter. At some point you will feel a tingle of electric charge. Someone expert at reading that tingle will know approximately how strong is the spark in the fence. For greater precision, however, I recommend an electronic tester.

Please do *not* buy one of the dinky little testers with five feeble lights to signal level of charge, probably available at your local farm cooperative—it's impossible to see whether the lights are on or off in full sunlight. Though it will cost you quite a bit more—forty bucks or so—buy a more sensitive tester with an LED readout.

Nature of the Charge

You may be interested to know that the highest voltage I ever measured—on a single roll of net, completely free of weed growth, on the AC charger—was 9,700 volts. If your response is, "*But 9,700 volts would kill you!*" understand that, while the *voltage* from this type of energizer is very high, the *amperage* is correspondingly low—which is to say, the whiplash sting will wake you up but do you no real harm. (The fence does have its hazards, however. See below.)

Another technical matter: Electric net is designed for use with *low impedance energizers*. High impedance chargers, especially of the weed-cutter type, can seriously damage the fence, generating enough heat to melt the plastic strands, burn the wires in two, and—worst case—even start a fire in dry grass.

Robust Grounding

The strength of the jolt from the fence depends on how well grounded the energizer is. Don't rely for ground on a pipe driven haphazardly a foot or so into the soil. The ground for my plug-in unit consists of three 8-foot steel rods driven full length into the earth in the dripline of the poultry house—even the night's dew dripping off the roof keeps the soil moist—joined with heavy-gauge wire, then connected to the ground terminal on the unit. Such a ground will never fail.

Ensuring a good ground for a battery-powered charger, especially when the ground is dry, is more challenging. Some battery units come with a mounting bracket with a metal stake that both holds the charger and serves to ground it. Do *not* rely on such a ground stake, which penetrates the soil only 8 or 10 inches. Use something like a galvanized 3-foot, half-inch grounding rod instead, driving it as deep as required for good ground, deeper in soil that is dry. (Not too deep, however—the ground rod may be hard to pull out when it's time to set up the fence elsewhere.) Pouring water around the ground rod—say, when rinsing a field waterer—will moisten the soil and improve grounding of the system.

Keep It Clean

While solid grounding of the energizer is essential, accidental grounding in the fence will sap its charge. Check the fence line frequently to ensure that limbs have not fallen on the fence, forcing charged horizontals onto the soil. When installing the fence on uneven ground, make sure the bottom charged horizontal doesn't touch the earth.

Inevitably, as grass and weeds grow around the fence, contact with the bottom wire will reduce charge in the fence. When it drops to 3,000 volts, or 2,500 at the lowest—the recommended minimum needed to turn back predators—you should mow the fence line again. Simply stand the fence a couple of yards to the inside of the perimeter, to keep the birds from wandering while you mow. It often happens, however, that the point at which weed growth requires mowing will be about the time you want to rotate the flock to a new paddock anyway.

Challenges to Confining the Flock

Small hatchlings just out of the brooder may slip through the mesh of the fence. If both feet lose contact with the ground as they step onto the bottom wire, they receive no shock. Newer designs of electronet feature bottom horizontals closer together than they used to be, reducing the possibility of small birds slipping through with impunity. But some do—especially chicks—even with the newer netting.

You might prefer to raise your hatchlings inside until they're too big to slip the net. Most of my young birds on pasture have been hatched by a broody hen or broody duck, and tend to stick pretty close to Mama rather than seek adventure outside the fence. I do see chicks wandering outside on occasion, but I don't know that I've had any losses as a result, probably because local predators who have hit the fence avoid it.

As said earlier, chickens can fly as high as 42 or even 48 inches if they want to. Fortunately, once they've hit the fence once or twice with beak or comb, they almost never do. If I do occasionally have a rogue flier, I simply clip her wing to keep her inside the protection of the net. (See the sidebar "Clipping Wings.")

Stopping Predators: Successes and Failures

The fur of many predators insulates against electric shock, and some could certainly jump over it or walk right through it. How does the net deter predators like coyotes, large dogs, and bears? Fortunately, these guys lead with the supremely sensitive nose. Thus, if you set up and manage electronet properly, it gives an extraordinary level of protection.

In contrast with numerous flocksters I've known whose poultry husbandry should more accurately be termed "feeding the foxes," I can report that to my knowledge we have never once lost a chicken to a fox. We often see foxes trotting nonchalantly through the pasture, passing within yards of the flock on the other side of the net, but not even *looking* at the chickens—obviously they've already met the demon in my fence. A large dog once got tangled in my fence while trying to get at my flock. In his panic he broke one of the stout corner posts into *three* pieces before getting free—sans chicken dinner. A friend reports

Clipping Wings

I would never clip the wing of a free-ranging chicken—doing so would make her more vulnerable to a predator. But if I am taking responsibility for protection with electronet, she *must* stay inside the fence. Thus I occasionally clip the wing of a rogue flier with a habit of leaving the enclosure.

Assuming you are right-handed, hold the bird and spread her wing with your left hand as in figure 10.3. Notice that the large flight feathers are in two distinct groups—the primaries in front and the secondaries behind. The secondaries are also called the coverts because they *cover* the rest of the wing when it is folded; thus if we clip the primaries *only*, the clip is hidden and does not mar the appearance of the bird.

Simply shear off the feathers, about an inch from where they emerge from the follicle. If the feather is still new, on a younger bird or one who has just molted, there will still be blood supply into its shaft, and clipping will cause a bit of bleeding, though apparently not any pain. Otherwise, cutting off a feather is like trimming our own hair or fingernails.

Note especially that the clip is made on one side *only*. The idea is not to cripple the bird—a clipped bird will still be able to fly up onto her

Fig. 10.3

roost at night—but to imbalance her enough to discourage taking to the wing.

I make an occasional exception to clipping only primary feathers, and only on one side. Guineas can be much more determined fliers than chickens. If I have a normally clipped guinea who persists in flying out of the net, I first clip off *all* the flight feathers on the one side, both primaries and coverts. If it still flies out—amazingly, it happens—I then do what I call a radical clip, shearing all flight feathers on *both* wings. The guinea is most cooperative about remaining inside the net afterward.

twice seeing her electric net fence turn back a black bear.

The fence can be laid out in any configuration you need, but there is an important caveat: Avoid long, narrow alleys in the layout. I once lost a young goose to a cause unknown. A check of the fence proved that it was plenty hot, there was no limb down on the perimeter—and geese, incredibly clumsy on the wing, should not have flown over it. A few days later I saw the apparent culprits—a couple of dogs using the hunting skills of the pack to attack the geese. They wouldn't touch the fence—obviously they had learned to respect it—but using the pack mentality that lurks below the "good poochie" in all dogs, one of them would rush the geese inside a long narrow corridor between the coop and the main pasture enclosure, intent on forcing the panicked geese, normally reluctant fliers, over the fence—and into the waiting jaws of the other dog. I now configure all my fences so there is always a large interior space for retreat, in lieu of panic flight over the net.

We once lost three hens inside a netted paddock on successive nights, their necks bearing the telltale marks of attack by a blood-feeding weasel. It could only have been a least weasel—incredibly small but admirably equipped to kill—who slipped under the bottom wire over extremely dry soil without getting a shock. That was when I installed the more robust grounding system described previously.

And do remember that keeping the fence line clean means more than just mowing periodically to cut back weed growth. I once had a dozen layers tilling new ground in preparation for putting in a blueberry bed. The ground was on a slight incline—just enough to cause uprooted grass and other debris from the hens' work to sift down in an accumulating little ridge covering the foot of the net. And lo and behold, one morning I arrived to find four splashes of feathers out over the pasture. I grabbed the fence to feel exactly what Mr. Raccoon had felt when he was looking for his dinner—which was *nothing*.

Caring for Electronet

The service life of high-quality electronet is said by the manufacturer to be ten years. My oldest nets are considerably older than that and going strong. The key to long service life is proper care. Never leave nets not in service on the pasture, exposed to weather and ultraviolet radiation—store them properly when not in use. Using a weedeater to clear the line would be lunacy. On a good day I can cut grass up close to the fence with my scythe. If I'm tired, though, not even the sternest reminders to "Be careful of the fence!" will prevent an eventual cutting of a couple of strands.

Netting is surprisingly strong and flexible, and can take a good deal of stretching without damage. There are limits, as we discovered last year after two record-breaking snowstorms that came back to back, causing a number of breaks in the mesh. We've had ice storms that deposited up to a quarter inch of ice on the strands, causing the fence to sag badly, but with no damage. Further accumulation would probably cause some line breakage, of course. Fortunately, breaks in individual strands are easy to repair. You can buy repair kits from your fence supplier, featuring small brass ferrules crimped around spliced-in lengths of fence strand to bridge the gap. It's cheaper, however, to cut snippets from a coil of single-strand—which I keep on hand for other electric fencing applications—and simply knot them onto the ends on either side of a break.

When preparing the nets for storage, I recommend using what I call a lead string to keep the next deployment of the net tangle-free. Start as for moving the fence, with the net folded into a neat bundle of posts on one side, and the panels laid out flat beside them. Tie a piece of discarded baling twine around the bundle—it must be a little longer than the length of the folded panels. Now roll up the net—and note that this is the only time *ever* to roll a net—starting

Fig. 10.4 Hang stored nets out of reach of opportunistic rodents.

with the posts, to make a neat bundle with the posts at the core and the tail of the lead string protruding. Tie the bundle near each end, and it's ready for storage. Next time you deploy the net, grab the lead string as you roll out the bundled net.

As for storage, by far the best advice I can give is: *Do not store nets on the ground, floor, or even on a shelf.* That is, do not store them *anywhere* accessible to your rodent friends—who will happily chew your nets into very expensive nests. The best option is to hang them from the rafters of a storage shed, as shown in figure 10.4.

Hazards of the Net

Potential dangers from electric net fencing may be structural or electrical. Attentive, careful handling of netting prevents snagging and tangling by the user, but animals can become entangled accidentally. For example, I once had to cut a fox out of a net in which it had gotten hopelessly tangled.

With older versions of electronet, young waterfowl would occasionally get tangled after putting an exploratory head through the mesh, as their anatomy is more front-loaded than that of chickens, making it more difficult to reverse out of the net. I had a few fatalities when I didn't arrive on the scene in time to free them. Fortunately, I've never had a duckling or gosling get tangled in the newer nets with closer bottom horizontals.

In the early years there were occasional deaths of wild animals on the fence: One or two box turtles, once even a possum, became jammed under the bottom wire, unable to escape the pulsing current, which burned holes in hide or shell. A couple of blacksnakes crawling over the bottom wire died in the same way. I have not had such fatalities in at least a decade, however. I think the animals in the neighborhood have come to recognize the fence as a threat to be avoided.

Despite what I said above about the low danger of a jolt, it is believed that a shock to head or spinal cord has greater potential for injury. Don't take chances—cut power to the fence when working close to it.

The most tragic example of electrical hazard is that of a crawling infant in wet grass who got tangled in a charged net and died. *Never allow young children near the net unsupervised.* Warn visitors, of any age, who are unfamiliar with electric fencing. Attach highly visible electric hazard signs on your fences, as required by law in some jurisdictions.[2]

11 | MOBILE SHELTERS

If you day-range your flock, or use temporary fencing anchored on the henhouse to rotate the flock over fresh plots, the birds always return to the same shelter at night. If you pasture them farther afield, however, you will need a mobile shelter of some sort to rotate them to new ground, and to shelter them at night or when it rains. I've seen hundreds of mobile coops, and no two are ever the same.[1] The design you come up with will depend on the size of your flock, how you intend to use their services, leftover materials from other projects begging to be used, the nature of your climate and ground—perhaps on how whimsical you happen to be feeling.

The first movable shelter I built was a copy of the

Fig. 11.1 My friend Jon Kinnard combined whimsy, utility, and the urge to recycle into this micro-flock mobile shelter. It is entirely self-contained, with feed storage and nest in the bin under the hinged metal roofing and roosts in the rest of the shelter. PHOTO COURTESY OF DEBORAH MOORE

classic Polyface design—a contemporary example is shown in figure 11.2. If Joel Salatin's mobile pens can produce tens of thousands of market broilers a year to put money in the bank, surely all of us creative amateurs can come up with shelters that allow our birds continual access to fresh grass while protecting them from opportunists on the prowl.

Designing a Pasture Shelter

Below are some issues to ponder as you plan your mobile shelter project. It could help with your planning to have a look as well at appendix C for design and materials considerations and step-by-step construction of my most recent all-purpose pasture shelter.

Pasture Pens and Pasture Shelters

Micro-flocks on lawn or pasture are often confined entirely to the shelter, which is moved frequently to new grass. The larger the flock size, however, the larger the protected foraging space you will want to provide the birds. As discussed in the previous chapter, I use electric net fencing for giving my birds an extensive area to roam outside the shelter. If you do not use electronet, however, you might provide a *pasture pen* using a set of light wooden frame panels with chicken wire, easily locked together using bolts with wing nuts, and just as easily disassembled for moving. Whether you need to attach a frame over the top of the pen will depend on aerial predation where you are.

Cody, a friend of mine, came up with an ingenious pasture shelter-and-pen set for her flock of half a dozen layers, shown in figure 11.4: She mounted a small shelter (2½ by 3½ feet) on a landscaper's wagon, complete with roosts, nests, and a ramp she lowers—using a nylon strap attached to the ramp that runs right through the shelter—to release the flock in the morning. She made a separate 8-by-8-foot pen, 4 feet high and with a cover of wire over the top, and mounted on small wheels for moving. Framed into one side is a narrow opening into which the door of the shelter docks. In the morning Cody moves the pen onto fresh grass; wheels the shelter into docking position; then lowers the shelter's ramp to release the hens into the pen. At dusk they retreat into the shelter on their own, and Cody pulls the ramp into place with its remote-control strap, to guard against unwanted night visitors.

Trade-Offs: Size, Weight, and Stability

The size of the shelter will be determined by the size of the flock it will shelter and its intended use. At the low end of the scale, a shelter could be designed as a

Fig. 11.2 A movable pasture shelter based on the classic Polyface-style broiler pen. Note the 2-by-4 construction–this was a first project for its builders and is overstructured. PHOTO COURTESY OF SAMUEL MATICH

Fig. 11.3 As you can see, it is possible to build the same-sized pen with much lighter framing, making good use of diagonal bracing.

Fig. 11.4 Cody Leeser's ingenious design for a small wagon-mounted shelter and a separate wheeled pasture pen. She moves the pen each morning, then docks the shelter onto the pen and releases her hens for the day.

chicken tractor, holding six to ten tiller chickens and sized to work a single garden bed.

As with the main coop itself, size has everything to do with whether it will be sleeping quarters only for a flock that is ranging outside during the day, or will confine the birds full-time. As said, the first mobile shelter I used was a copy of the classic Polyface model, 10 feet by 12 (see figure 11.2)—I used it to raise fifty comparatively inactive Cornish Cross broilers at a time, about 2½ square feet each. When I later used that same shelter for confined layers, I limited the number of hens to sixteen—7½ square feet each. Remember that you will be more likely to rotate your birds to fresh grass as frequently as you should if it's easy to move their shelter. It might make sense to split the flock into two smaller shelters rather than keeping them all in one large one that is more difficult to move.

The heavier a shelter, the more difficult, and possibly the more dangerous, it is to move. On the other hand the lighter it is, the more likely it is to be tossed into the next county by a rambunctious wind. Of course, it would be possible to anchor even the lightest shelter to the ground; but again, the more difficult we make a move—undoing and redoing a complex anchoring routine—the more inertia will inhibit frequent moves. Shape also plays a part in stability in heavier winds: I have found the boxier-type shelters with a higher profile catch the wind, while hoop or A-frame shapes tend to keep their feet on the ground. (The classic

Chicken Cruiser

Andy Lee introduced the idea of the chicken tractor (or as I call it, a cruiser)—a small, easily moved chicken shelter sized to fit a single garden bed, a key to putting chickens to work in the garden. A few laying hens inside till and fertilize the bed while finding free food in the form of worms and slugs and snails—and laying eggs—but have no access to immediately adjacent beds.

Fig. 11.5 My most recent chicken cruiser, made for maneuvering in tight garden spaces.

Since a tractor gets maneuvered in tight spaces and needs to be moved frequently, it is better to make it small and nimble. Don't forget to provide enough cover on parts of the sides and top for shelter from blowing rain, and for shade in hot weather.

Fig. 11.6 A cruiser keeps the chickens working in a single bed while preventing access to adjacent ones. The lids on this unit–one aluminum roofing, the other wire on wood framing–are separately hinged for access to any part of the interior. Nestboxes are recycled plastic milk crates attached to the framing.

Polyface model, 10 by 12 feet, is indeed rectangular in shape, but it is only 24 inches high and stable even in strong winds.) Materials choices (see below) have the biggest impact on weight of the shelter.

Remember that *diagonal bracing* greatly reduces weight of the frame. I framed my first shelters in 2-by-4s exclusively, all at right angles—clumsy, inelegant, and balky about moving. I discovered that even large shelters could be made with much lighter but well-braced framing, like the Polyface-style pen in figure 11.3. I also found that smaller shelters do not need full 2-by-4 framing even for the bottom rails. For a shelter of this size—at present I have two of about 8 by 4 foot, another 10 by 3—I now rip 2-by-4s to 2¼ inches to use as the bottom rails. The remaining 1¼-inch strips I use for the verticals and diagonal bracing; that is to say, the weight of the entire frame is now not much more than the bottom rails alone before this modification.

Note as well that you can reduce weight by positioning bracing where possible to do double duty as roosts. In a larger A-frame, for example, you will want to include collar ties, the horizontal pieces that tie the rafters together, providing greater rigidity. Position them low enough below the peak to allow use as roosts by two or three hens. Horizontal

Fig. 11.7 Diagonal framing in this A-frame shelter provides rigidity without excess weight. Note the collar ties, set low enough to serve double duty as roosts.

stringers reinforcing the frame can also be positioned for use as roosts.

A final option for reducing weight is to use chicken wire as much as possible in lieu of solid material, consonant with the need for protection from rain, sun, and sharp chilly winds in part of the shelter. In many shelters wire mesh replaces at least part of the roof, and much of the sides. Use of wire has the further advantage of maximizing airflow and sunlight into the interior.

Wheels

I prefer wheels for all my larger shelters. Instead of installing axles across the entire width of the shelter, I permanently install half-inch bolts in the bottom rail at each corner, using nuts, flat washers, and lock washers. It's easy to use a single set of wheels for multiple shelters, popping them onto the bolts and locking them down with wing nuts. If your ground is nice and even, an 8-inch wheel might work for you. I found that, with an 8-inch wheel, the bottom rear rail of the shelter hung up on tussocks of grass. The additional clearance with a 10-inch wheel makes moving much easier on my pasture.

If wheels are to be permanently installed, bicycle wheels—or other large wheels looking to be recycled, like the front wheels from an old tractor—make moving over uneven ground easiest of all.

Does Your Shelter Need a Floor?

The whole idea of using a mobile shelter is to give its occupants access to fresh grass, so it usually makes sense to make the shelter floorless. Some management choices, however, might make a floor advisable. For example, young birds are easier to move with no risk of injury from the rear bottom rail (see below) if on a floor. If you do install a floor in your shelter, I recommend using wire or plastic mesh, as droppings will accumulate on a solid floor, requiring frequent clean-out from the tight confines of the interior.

Predators

If the shelter is inside an electric net perimeter, you will not have to worry about digging predators. However, if there are large owls in your neighborhood, close the shelter at night—nocturnal owls hunt on the wing, but also land and walk around looking for prey.

If the shelter is not inside an electric net, remember that raccoons and dogs may tear a hole in chicken wire—in the case of 2-inch mesh, a raccoon may feed on its victim by tearing it apart right through the wire. If you are designing for such threats, use half-inch hardware cloth instead, well secured to the framing. Foil digging predators with a wire mesh floor (2-by-4 welded wire allows both access to the grass and protection from digging predators)—or by laying 18-inch panels of chicken wire on light wood framing flat on the ground, entirely around the shelter.

The best option of all is to wire for defense, as in figure 11.8: Run some single-strand electric wire around the entire shelter, standing it off from the sides with plastic or porcelain insulators, one at nose level and ideally another about 12 inches up. An inexpensive charger powered by a 9-volt battery is sufficient to charge such a small run of wire. Whether dog or raccoon or digging fox, the exploratory probe of choice is the supremely sensitive nose—once it hits

Fig. 11.8 Wiring for defense keeps predators away.

the wire, the visitor will seek dinner or entertainment elsewhere.

Nests and Other Thoughts

If the shelter will house layers, you should add *nest-boxes*, which can be mounted above ground level on existing framing pieces. A hinged door—to shield the nest from rain but give you access from the outside—is a better option than crawling into the shelter to collect eggs. If hens are inclined to roost and poop in the nest, an additional hinged cover to swing into place at night may be in order.

Install a *door* in the shelter even if you rarely use it (such as when the shelter is inside an electric net). Latching it will help you get ready to move the shelter from one electronetted area to another, do a census or selection, or isolate birds for culling.

In smaller, rectangular shelters, often the only door is a hinged *lid* giving access to the interior. Remember that the lid can be popped open by a wind gust, maybe even ripped off the hinges, and provide a positive catch for locking it shut. (The country-boy version is a heavy rock set on the lid.)

Even a shelter heavy enough to withstand ordinary winds may flip when a gale blows. When weather predictions here are for winds well beyond the ordinary, I temporarily "nail" my shelters down using an

Fig. 11.9 Hinged access from the outside makes it easy for Annecy and Camille to collect eggs from my latest A-frame shelter.

earth anchor—essentially, an abbreviated auger screw on the end of a steel rod with an eye hook on its top end. Once the rod is screwed solidly into the earth, I tie or wire one of the bottom rails to its eye hook. Another way to temporarily secure a shelter is to hang a couple of 5-gallon buckets from the framing inside and fill them with water—that's over 80 pounds—using a garden hose. Just empty the buckets when it's time to move the shelter.

Remember your chickens' need to dust-bathe. Since there is no opportunity for them to do so if constantly on fresh grass, either provide an onboard *dustbox* or set one out for them on the pasture anytime there is no possibility of rain.

Most shelters are designed to be used in the warmer parts of the year only. If you are going to house your birds in the shelter in winter as well, you will need to make at least the part where they sleep a good deal tighter against the winter winds, snow, and rain. As noted in chapter 6, however, the shelter should still allow a lot of airflow.

Materials

Mobile shelters have been made in just about every material other than titanium. Which materials you choose will depend on which you feel comfortable working with, what might be in your recycle pile, and considerations of weight and climate.

Wood

I am more comfortable working with wood, so all my shelters have had wooden frames, with one exception—a hoop structure based on half-inch solid fiberglass rods as purlins and as arches, anchored into a wooden foundation frame. I don't use any pressure-treated wood anywhere on the place remotely connected to producing food. To help prevent rot, I coat all framing pieces in direct contact with the ground with nontoxic sealer, renewed periodically as needed. Using a highly rot-resistant wood—eastern red cedar in my area—would be a better option if you can get it. You might design so that the bottom rails—the parts most subject to rot—can be replaced without taking apart the entire shelter. Or mount the frame on plastic rails. (See below.)

When out of service over the winter, a wood-frame shelter should always be set up on blocks. You might even want to block each corner after each move, to keep the rails out of contact with the ground.

Plastic

Beginners often think of lightweight 1-inch plastic pipe or the like for framing a shelter. I've never seen one that inspired much confidence—such plastic is pretty fragile and breaks down in sunlight.[2] Heavier plastic pipe (Schedule 40 PVC, for example) is another matter—I've corresponded with many flocksters who have used it for shelters that are both sufficiently rugged and easily moved. (See figures 6.8 and 6.9 for examples that could be scaled down for smaller shelters.) I've never used plastic pipe myself.

This year I experimented with recycled plastic decking[3] to make two 6-by-10 pasture shelters for nurturing young birds through the vulnerable (to aerial predation) phase. I framed them entirely in this recycled material, on the assumption that it would last a lot longer than wood. The jury is still out regarding how well plastic decking serves as structural material. It certainly is heavy—the lighter one (24 inches high, to accommodate chickens) will move without wheels, with some persuasion; the heavier (36 inches high, for geese) requires wheels. So far even the heaviest winds haven't fazed them.

I'm especially pleased with the new chicken tractor I made last spring and used the entire growing season. It's mounted on recycled plastic decking boards to prevent rot in its wooden frame and to make it easier to slide the shelter down the garden beds.

Metal

Electrical conduit is light and easily shaped. You may see references to its use for framing mobile shelters, but most reports I've read about it have been negative. Both angle iron and rebar—concrete reinforcing rods made of soft iron—make sturdy frames for those with welding skills and equipment.

I've heard from a lot of flocksters who use cattle panels to frame hoop-style shelters, either secured to a wooden base or welded to metal runners. The standard length of these panels is 16 feet; height varies by species of livestock, but in this application likely 52 inches; steel wire should be heavy enough (likely 4-gauge) to make a semirigid fencing section welded into a 6-by-8-inch mesh. Typically the panels are attached to one side of a wooden frame (or welded onto a metal one); bent in the long dimension into a hoop attached to the frame on the other side; and covered with a tough, flexible, opaque cover. The result has the usual trade-offs among weight, mobility, and stability in the wind but typically has considerably more capacity than shelters framed in other materials.

Covers

If light plastic pipe is a bad idea, such pipe covered with lightweight plastic tarps makes absolutely the worst combination. Not only do such tarps break down in sunlight—and shred, and blow in the wind—but the combination is so light, even a sneeze will move it.

Heavy canvas tarps, like the one in figure 6.8, are tough and weatherproof and make a better choice than plastic tarps. There is one option in plastic covering worth considering, however: 24-mil woven polyethylene—incredibly tough, durable plastic sheeting interwoven with a fiber mesh. It's available in semi-translucent white and a number of colors, including one that is black on one side, silver on the other. The side you face to the outside depends on whether you need to reflect or gain solar heat to the interior—in my climate, putting the reflective side out is the obvious choice.[4] Did I say *tough*? I once had an 8-by-8 A-frame shelter covered in woven poly, which got smacked tumbling by a gust of wind through 30 yards of underbrush. The result was one broken strut only (a testament to diagonal bracing), but not a single tear in the poly.

I have used metal roofing for the solid covering on a number of my shelters. Aluminum roofing is lighter but more expensive; steel, heavier but cheaper. Steel roofing is available either as plain galvanized, or with a baked-on enamel finish guaranteed for twenty-five years. Though the galvanized is cheaper, the paint you would have to apply to extend service life would

Fig. 11.10 An 8-by-8 A-frame mobile shelter covered with 24-mil woven poly. Ten years old at the time of this photo, it is still going strong.

over time cost more than the initial investment in the baked finish.

I used baked-enamel steel roofing as the cover on the shelter in figure 11.9. See as well appendix C for my reasons for choosing metal roofing over 24-mil poly.

Fasteners

I strongly advise against assembling your mobile shelter with nails, which work loose over time as the frame is yanked around; use screws instead. I prefer the self-drilling types such as coarse-threaded decking screws, which don't require pilot holes (as do conventional wood screws) and thus save time. (I do drill a pilot hole for a deck screw going into the last 3 inches of a framing piece, to prevent splitting.) Deck screws with Phillips heads are available galvanized or coated. The best screws of all are stainless-steel decking screws with star-drive heads. Though a lot more expensive than the alternatives, their faster, slip-free drilling and rustproof durability are important considerations for a shelter requiring a lot of screws, and facing prolonged weathering.

Moving the Shelter

Twisted wire or cable, run through a piece of scrap garden hose, makes a convenient pull for moving the shelter. A wire pull can be permanently attached to both ends of the shelter; or a single pull—with twisted loops at either end that slip into open eye hooks screwed into the bottom rail—can be used on multiple shelters.

When moving a floorless shelter with young or careless birds inside, watch the trailing edge of the bottom frame. Usually the chooks come running as fresh grass is exposed, but those who dither at the rear may get a leg caught between the ground and the moving rail. Actual injuries are rare if you pull slowly, and stop and release a hapless bird at the first shriek of distress.

PART THREE
Working Partners

PHOTO BY BONNIE LONG.

12 | PUTTING THE FLOCK TO WORK

The most significant difference between the integrated flock and a flock of pet chickens is enlistment of the birds as working partners in the work of food production in the backyard. And what partners they are! They get to work early and work late, are sharp-eyed and meticulous, and are anything but lazy. And they have a great skill set.

The strategies suggested in this chapter—you'll think of others—are not games (though they are *fun*) but opportunities for getting serious work done on the small farm or homestead. They work because they engage the genetics and galliform lifestyles with which chickens evolved, in ways that are mutually beneficial to them and to us. Really, it all works by making chickens *happy*.

Fig. 12.1 No other factor has been as important in transforming my soil—from the native clay on the left to rich, mellow loam on the right—than the many contributions of my hardworking flock.

The Great Work: Soil Fertility

Soil loss . . . is a problem that embarrasses all of our technological pretensions. If soil were all being lost in a huge slab somewhere, that would appeal to the would-be heroes of "science and technology," who might conceivably engineer a glamorous, large, and speedy solution—however many new problems they might cause in doing so. But soil is not usually lost in slabs or heaps of magnificent tonnage. It is lost a little at a time over millions of acres by the careless acts of millions of people. It cannot be saved by heroic feats of gigantic technology but only by millions of small acts and restraints, conditioned by small fidelities, skills, and desires. Soil loss is ultimately a cultural problem; it will be corrected only by cultural solutions.

—Wendell Berry[1]

There is no higher duty on any farm or homestead than to care for and nurture the soil, not as a *part* of the food independence project—soil care *is* the food independence project. Our soil is our life, and *everything* we do should be in some way an attempt to improve it. A lifetime should suffice.

A troubling but instructive question is this: *Why is it that, in natural ecologies the world over, there is a spontaneous accumulation of soil depth, quality, and fertility over time; while human practice of agriculture has usually led to degradation of the soil—never so much so as in the era of industrial farming?* As we explore the answers to that question, we discover the profound wisdom of imitating natural systems.

Soil Is *Alive*

The soil food web is a complex community of living organisms: bacteria (up to 13 tons in an acre of good prairie soil, which is astounding considering that up to half a million of them could fit on the period at the end of this sentence); actinomycetes; algae; slime molds; protozoans (maybe ten billion in a square yard of meadow topsoil); and much more—and that's just at the microbial level. We think of fungi as mushrooms, but they are also present underground as meandering strands, both invisible (hyphae) and visible (rhizomorphs)—a teaspoon of prairie soil could contain several miles of them. Animal life includes worms (from microscopic up to earthworms almost 10 feet long—well okay, those guys live in Australia), beneficial nematodes, and other invertebrates (several million in a square yard of garden soil); the more familiar ground-dwelling insects and spiders, but as well countless other minuscule arthropods that usually go unnoticed—a full census of this kaleidoscopic community fills whole books.[2]

However little we know about the vast diversity of soil organisms, we know they must have at least one thing in common: They all have to *eat*. And what do they eat? Well, *everything*: Everything that falls on or dies in the soil—dead plants and their roots, leaves shed by perennials, dead animals, poops of living animals—becomes *food* for various players in the soil community. And eating lunch is not a single event when organic matter is consumed. Some classes of organisms (prominent among them bacteria) are first at the table, breaking down the more easily converted compounds, a process that yields energy for their growth and reproduction. But they cannot utilize all the components in these organic materials—some they leave untouched; some that they ingest they excrete; and other soil organisms in turn leap on the unused portions and the excretions as sources of food—of energy. In this way the energy in the parent materials, originally created by photosynthesis in the leaves of plants, passes from one trophic (energy exchange) level to another, and at the end of the process *nothing* leaves the soil. That's right—a natural soil community utilizes the remains of life in and above the soil so thoroughly that *all* residues from the breakdown process are retained. This is the great secret of *spontaneous increase of fertility* in natural soil systems.

As for those final residues, what the soil retains after the decomposition of the original organic matter is *humus*—carbon compounds that resist further breakdown, and thus are no longer available directly as food energy for anybody in the soil food web, in particles too small to see, visible only as a darkening of the soil. Though not used directly as a nutrient source, humus content has everything to do with soil quality and fertility: It increases water retention, assists in the chemical bonding of nutrients with plant roots, and improves soil texture. If we as gardeners want to imitate nature's soil-improvement process, the best way to do so can be summarized quite simply: Feed the soil organic matter in order to increase its humus content. And chickens can help!

The presence of so much teeming *life* under our boots is not merely an incidental fact about the makeup of soil. If you have thought of fertility as something you *add to* your soil, consider instead that the complex interplay of species, their ceaseless exchange of energies available in the various stages of organic matter breakdown—*is* soil fertility. And if you have thought of nature as an arena of relentless competition, here in the dirt is a marvelous vision indeed: *Everybody is feeding everybody else.*

Essentially, the concept of soil fertility underlying industrial agriculture is a good deal more static: Soil is a mix of tiny particles weathered out of the parent rock, mixed with air and water—a medium for anchoring the roots of crops as they grow. There is far too little thought given to *feeding the soil*—the idea instead is that, if we *feed the roots of crop plants* (with purchased, and in most cases man-made) nutrients, they will slurp them up and grow profusely, and all will be well.

Chickens are a good deal smarter than the average Big Ag scientist. Their first task when passing over the soil is to deposit and scratch in a priceless amendment—"holy shit," Gene Logsdon calls it[3]—which adds minerals needed by plants (prominent among them nitrogen, phosphorus, and potassium—the "NPK" with which industrial agriculture is so obsessed) but also, and of far more importance, supercharges the web of living critters in the soil. Just by pooping, they leave the soil more alive, and thus more fertile—more capable of "abundant increase." But chickens can do much, much more to help feed the soil with organic matter, as we will see.

Diversity versus Monoculture

Industrial agriculture grows its crops in huge monocultures—gargantuan fields of nothing but corn, soybeans, and grain—for feeding to gargantuan factory flocks of nothing but layer chickens or nothing but meat chickens. Where in nature do we find such monocultures?

If you've decided to raise chickens, and if you already have a garden (or plan to one day), you're already light-years ahead of monoculture thinking. Almost everywhere we look in nature, we see complex mixes of both plants *and* animals—a diversity that always tends toward enrichment of the soil underlying all. Though plants start the process, animals participate in the great cycles of reproduction and growth and death and decay and renewal. The small farm or homestead most closely imitating nature, therefore, will be based not only on domesticated plants, but on domesticated animals as well. Any strategy for caring for the integrated flock suggested in this book has a tie-in, in some way, with increasing diversity—and hence, fertility—in our soil.

Witches' Brew

Perhaps the starkest difference between natural soil systems and those under industrial agriculture is the latter's witches' brew of man-made chemicals. Some of those chemicals are intended to feed (chemical-salt fertilizers); some, to kill (insecticides, herbicides, fungicides)—and all have serious unintended side effects. Chemical fertilizers are highly soluble, and readily leach from the root zone where they are intended to feed crops into groundwater and natural water systems—where they function more as toxins than as nutrients. The various *-icides* are largely indiscriminate and kill large numbers of nontarget species essential to a healthy ecology.

Man-made chemicals, wherever applied and for whatever purpose, find their way into the soil—*all* have lethal effects on most classes of soil organisms. That shouldn't surprise us—these are compounds living organisms have not had to take in stride in three and a half billion years of life on this planet. Maybe in another billion years soil organisms will adapt to such "gifts" from Monsanto, Cargill, Dow Chemical, et al., but for the present *there is no more sensible start on a soil fertility program than a Great Vow to avoid all man-made chemicals. Whatsoever.*

Fortunately, chickens and other poultry can help—not only with direct contributions to soil fertility in lieu of chemical salts, but also by offering alternatives to "going nuclear" to achieve insect, weed, and disease control.

Soil Disturbance

There are occasions of soil disturbance in nature, sometimes on a massive scale. But the beautiful patterns of interactive soil life that result in spontaneous fertility accumulation over time take place *in the absence of major soil disturbance.* Industrial agriculture, by contrast, is based as much as anything else on soil disturbance. Tillage, a frequent ritual on almost all acreage under cultivation, results in the inversion or mixing of the distinct layers in a natural soil profile—and massive disruption of the living soil communities. If the next round of tillage comes before the soil food web has healed itself, the soil is on a downward spiral with regard to structural integrity and fertility.

Chickens, managed creatively, can accomplish many of the tillage chores required on the homestead or small farm, while at the same time preserving soil structure, with net benefits to soil life.

Tiller Chickens

Like many gardeners I was once a sucker for the blandishments of power tiller manufacturers—yes, I owned and used a power tiller. Until I realized that the tiller actually made garden bed preparation *more* work, not less, when I factored in the additional stress of using a whining, jarring, stinking machine—to say nothing of the irksome job of clearing the tines every fifteen minutes when tilling in a stand of long-stem cover crops. It's been years since I gratefully passed the little demon on to someone with a different perspective on soil care. Since then I have used a supremely elegant alternative for most tillage chores: tiller chickens.

Cover Crops

I do more cover cropping every year, in all parts of the year—perhaps no other strategy does more to feed the soil and improve its structure. There are cover crops for every season, including winter; their possible applications are limited only by your imagination.[4]

Most of the small grains grow well without a lot of added fertility in the cool parts of the year, spring and fall. Crucifers, a large ensemble including mustards and rape, have a cleansing effect on soil (they suppress soil fungal diseases). Crucifers with fat, deep roots such as turnips and forage radishes have a tremendous loosening effect on compacted soil when they decompose. Buckwheat is frost sensitive but grows fast in summer to smother weeds and help keep the soil cool and moist. Sorghum-sudangrass hybrid is a warm-season grass-family cover that gets as tall as corn and produces a tremendous amount of biomass. I discovered this past summer that the ducks and geese *love* it as green forage. Legumes are especially useful, fixing nitrogen in the soil for following crops—up to hundreds of pounds per acre. Many of them are cool-season crops, though some, such as cowpeas and field peas (*Pisum arvense*, not the common garden pea, *P. sativum*), grow just fine in summer. Some legumes—alfalfa and most of the many clovers—are more long-term crops, ideally left in place a full year for maximum nitrogen fixation.

We think of cover crops as providing a big addition to soil organic matter when they are *tilled in*. But don't forget that the root systems of cover crop plants are far more extensive than the plant parts we see aboveground. Roots grow deep into the subsoil or make extensive underground mats of fine rootlets. Alfalfa, for example, pushes its thick taproot 8 to 10 feet into ordinary soils, and rye makes an extremely dense root mass of fine roots growing 4 or 5 feet deep. When the plants die, all those roots decompose in place, not only feeding our buddies in the soil but opening up passages through which air enters (essential for soil organisms as well as critters like us aboveground), and even the heaviest rains can soak in (rather than running away over the soil surface). Note as well that the most intensely bioactive part of the topsoil is the *rhizosphere*, the surfaces of roots and the area immediately around them. This is where plants feed their microbial allies, with exudates synthesized in the leaves and released through the roots, and where those same allies assist with uptake of mineral nutrients in the other direction. Many processes in the rhizosphere produce soil "glues" that help aggregate tiny particles of soil and humus into larger clumps. Soil aggregation is especially important in heavy clay soil like mine, brick when dry and impossibly sticky when wet. Aggregation in effect creates larger particle size, turning clay into a fine loam.

Tiller chickens can be the key to making extensive, continuous cover cropping strategies work. In fact, my soil has improved more since making the marriage between cover cropping and tiller chickens than in all the years before—there is no cover crop they will not till in with ease. Even that notoriously tough, tine-fouling cover crop—4-foot-tall rye—doesn't stand a chance when I sic tiller chickens on it.

Fighting the Jungle

I expect your garden is a showcase of order, but frenzy is more my style—I routinely have fourteen different projects going at once. The result, inevitably, is heavily weed-grown areas I just haven't gotten to. At one time, the more those tough guys dug in their roots, the longer I would put off the assault I call fighting the jungle. Ah, the joys of tiller chickens—these days I just put a tiller flock on such a patch and let them "do chicken."

Actually, if things come to such a pass in your garden as well, remember that weeds are friends too—different weeds specialize in the mineral mix they concentrate and release for use by following crops when tilled in. I'm sure you know they often have bodacious root systems, impossible for you to tear out, but which yield the multiple benefits discussed above when the plants are killed by tiller chickens.

New Garden Ground

Remember why I stressed continual rotation of chickens on pasture? If you leave them in one place *too long*, they will *kill the sod*. Ah, but if you leave them in one place *long enough*, they will *kill the sod*. The promise in that thought should be abundantly clear if you've ever busted your butt behind a bucking power tiller, tearing up a tough established sod over compact clay. (But they said a power tiller makes the work *easy*!) When breaking new ground is the next chore on the list, *chicken power* wins, hands down.

I have three separate times now opened up new ground measuring 1,600 square feet or more (see figures 12.2 and 12.3). To be sure, once the chickens till in the established sod, the ground is still heavy clay—as compact as it comes—but that would be as true using a power tiller. Oh, and a tiller doesn't poop—boosting soil fertility starts right away with tiller chickens.

I like to follow the birds' initial work with some high-calcium lime—which helps microscopic clay particles *flocculate*, or clump together to form larger soil particles—and a cover crop—as diverse a mix as the season will allow. After a second pass by the chickens to till in the cover, soil structure and organic matter content are much improved, and the ground is ready for its first crop.

Managing the Tiller Flock

How many chickens? How long will it take? The answer, as always, is: It all depends. If the ground to be worked is small, and you have only a few birds to put to work, a chicken tractor parked long enough on the new ground will do the trick, discussed in greater detail in the next chapter. For a somewhat larger area, you could use the lock-together wire-on-frame panels suggested in chapter 11. Or you could modify the dueling-gardens strategy in chapter 6: Grow your vegetables in an existing garden plot this year, while the chooks work up new ground on an adjacent one; then next year, switch.

I have occasionally used chicken tractors for this work, but most often I've set my tiller flock, and a mobile shelter, on the new ground and surrounded them with electric net fencing.

A standard electronet is 164 feet long. Set it up as a square, 41 feet each side, and you're ready to till almost 1,700 square feet with chicken power. In one project begun in midsummer, I used three dozen chickens in that space, leaving them for five weeks until they had killed the existing sod cover. Moving the flock elsewhere, I spread high-calcium lime and sowed as diverse a mix of cover crops as I had seeds for: field peas, cowpeas, buckwheat, various crucifers, and small grains. Actually the latter two are not ideal crops for hot, dry weather, but if you keep enough water on them in the germination phase, they'll go.

The cover crops grew well for five weeks—many cover crops do better in compact soil than more delicate garden crops; and don't forget the boost from that great chicken poop—then I returned the chickens to their work. This time tillage required only two weeks—cover crops are not as tough as established sod grasses. As well, the loosening of the soil already

Fig. 12.2 The first step in establishing a stand of pure alfalfa was removing the established sod cover, courtesy of a flock of chickens doing exactly what they love to do.

taking place, together with enhanced soil moisture in the dense shade of a lush cover, made the chooks' scratching more effective. Note that, even though I didn't do so myself in this case, the flockster—like the user of a power tiller—could easily have the birds till in any amount of organic matter available at this point—compost, horse manure, grass clippings.

After moving my tillers back to the main flock in early fall, I sowed a final overwinter mixed cover. By the following spring the new area had a long way to go to become best garden ground, but it produced nice crops of corn, cucumbers, summer squash, and sorghum in its first season.

To be sure, the process I've described took a lot

Fig. 12.3 With the old cover out of the way, I sowed this fine stand of alfalfa.

longer than the single day of backbreaking, nerve-jangling labor that would have been needed to power-till that ground into submission. But didn't someone say patience is a virtue? When opting for tiller chickens, patience is not only good for your soul, but good for your soil—the rapid increase in its quality is an order of magnitude beyond anything you can do with a power tiller.

Shredder-Composter Chickens

If we were satisfied with nothing more than substituting chicken poop for chemical fertilizers, we'd be making the mistake of adopting a modified version of the NPK mentality. When it comes to *feeding the soil with organic matter*, chickens have so much more to offer.

Remember how I advised putting a thick organic duff over the area if you must restrict your flock to a static run? The emphasis there was on responsibly recapturing the fertility in the birds' droppings to prevent runoff pollution. But now let's consider extending that idea to engage hardworking chickens in making compost. *Lots and lots* of compost.

I've already implied above that I'm a lazy old gardener glad to foist off on my chickens any work I can con them into doing for me. So it shouldn't be surprising if I prefer shredder-composter chickens to the unquestionably magical but labor-intensive classic Sir Albert Howard[5] method of making compost.

In the classic method the "greens" (fresh, moist, nitrogenous)—such as spent crop plants, grass clippings, spoiled hay, and manures—are laid down between layers of "browns" (dry, coarse, carbonaceous), such as straw and autumn leaves. Some of the

Fig. 12.4 Shredder-composter chickens on the job. PHOTO BY BONNIE LONG

same microbes involved in breaking down organic matter in the soil start a feeding frenzy, and before long what started as a conglomerate of organic roughage has turned into dark, crumbly, earth-smelling stuff we intuitively know will be well received by our friends in the soil.

There's just one catch. Imagine assembling 2 cubic yards of the compost materials mentioned—that's a heap 3 feet by 4 by 4½, actually a pretty small beginning for serious homestead composting. That heap could easily weigh a ton, assuming a goodly portion of it is fresh manure. Now imagine using a manure fork to *turn* that ton—not once but twice during decomposition. If I put my chickens to work instead, making compost is easier for me, and the chickens are happier as well.

The logistics of setting up the composting project are pretty much the same as for managing tiller chickens—just put the materials to be worked in a heap, provide protective fencing and a shelter if needed (which might be your coop with its fenced run), and send in the chickens (see figure 12.4). Using this approach to composting is extremely informal in comparison with Sir Albert–style composting. I just keep dumping any organic debris I generate here or have ready access to locally—spent crop plants, fall leaves, grass clippings, weeds, deadheaded plants from flower beds, cartloads of comfrey, manure. There is *nothing* compostable that leaves our place—it all goes on the compost heap.

The chickens pretty much ignore the heap for a couple of months, until it comes alive with the usual cast of decomposer organisms. Before long, the first place the chooks head when released from the coop in the morning is the heap, ignoring the offering in the feeder until they've had a run at the good stuff.

Remember what I said about metabolites of microbes—vitamins B_{12} and K and immune system enhancers—generated in deep litter? Chickens foraging a compost heap ingest these compounds as well—along with crickets, pill bugs, earthworms, fat white grubs, and more. They will eat fungal strands or rhizomorphs that meander through the cellulosic parts of the mix, following them with intense concentration—interesting, given that fungi are known to synthesize numerous natural antibiotics.

Fig. 12.5 Composter chickens generate tons of compost for my garden, including this 4-inch application on a bed of asparagus, a heavy feeder.

How much compost can a flock of working chickens make? For my largest production to date, I assembled an impressive heap of all our home-generated organic "wastes," intermixed (as often as we could do the hauling) with six or eight big pickup loads of a neighbor's horse manure—enough to cover more than 1,200 square feet to a depth varying from half a foot to several feet. I released the layer flock onto it for almost half a year. When I moved them to work elsewhere, I sowed a mixed cover crop over the compost-in-making, cowpeas and small grains, assuming that a profusion of roots in the mass would assist the decomposition process (the rhizosphere is intensely bioactive, remember) and that the

Adventures in Slug Heaven: Poultry and Slug Control

by Carol Deppe

When I first moved to Oregon about 30 years back, I lived in the Alsea area in the middle of the coastal mountain belt. I had just barely started gardening when the slug migration started. Banana slugs, that is. Eight inches long. Up to ¾" across. The slugs migrated right through my patch of pea seedlings one night and ate every one. Insects can be pests. But insects are small. Imagine an 8"-long insect. Now imagine a square yard of garden with a couple dozen 8"-long insects in it. Now imagine that they migrate, so that even if you pick that couple dozen this morning, another different couple dozen will migrate in tonight. That will give you an idea of the problem.

I had just moved from the Midwest, where chickens are the main free-range laying bird. Chickens hate cold rain, and there is cold rain six months of the year in the Alsea area. Ducks love cold rain, however, and certain breeds of ducks are the best of all layers. I was still thinking in Midwest terms, though. I wanted eggs. So I got chickens—Brown Leghorns, Barred Rocks, Rhode Island Reds, and a few others. The chickens watched carefully so as to not step on any slugs and get slime on their feet. That was their contribution to the situation.

The banana slugs in my backyard represented a huge amount of protein, enough to provide all the protein my chickens needed if only they would eat it. If I cooked the slugs would the chickens take an interest? If so, maybe I could then ease the chickens by stages—from well-done slugs, to medium-rare slugs, to rare slugs, and finally to Slugs Tartare. So I went out early one morning and picked the banana slugs from a measured square yard of lawn. This gave me a measured quarter pound of slugs. I dropped them into boiling water. They were just as slimy after being boiled. They cooled into a single solidly-stuck-together slug wad.

I took the Slugs and Slime du jour out, scraped it into the chicken food dish, and called the chickens. Most of the chickens just looked at the slug wad suspiciously. One bold hen, however, trusting that if I put something in the food dish, it must be food, marched up confidently and gave the slug-wad a mighty peck. Then she ran off and spent the next ten minutes wiping her beak on the grass and scratching at it with her foot, trying to get the slime off. After that, she was never again quite so trusting.

I also had a flock of geese in that era. Geese are sometimes said to be useful in insect and slug control. They aren't. Geese are strict vegetarians, and eat bugs or slugs only by accident.

So I got ducks. Ducks love banana slugs. They start at one end and gulp and gulp and gulp, with a little more slug disappearing down the gullet at each gulp, until after several seconds of gulping, the entire slug has vanished. It's one of the most satisfying sights I have ever seen.

I now live in the Willamette Valley in Oregon. We don't have banana slugs, but we do have lots of 3"-long garden slugs. Chickens don't eat those either. And garden slugs are one of our most important pests, the most important in many gardens. Some chickens do eat very small slugs, slugs up to about ½" in length. Other chickens avoid even those. In some areas of the country, all the slugs are small; chickens may help you control those. But for serious slug control or bigger slugs, get ducks. Just about

Continued on following page

Continued from previous page
any breed of duck, even the bantams, can eat 3" garden slugs. For banana slugs, you need a bigger duck. However, all the standard breeds of laying ducks from the little Khaki Campbells and Indian Runners on up are big enough to eat even the biggest banana slugs.

I have two basic methods for using ducks for slug control. The first is just to pen the ducks adjacent to the garden and take advantage of the fact that the bigger slugs will travel from the garden to the pen, in order to get at the duck poop, and then get eaten by the ducks. This takes care of the bigger garden slugs. However, there are still lots of tiny slugs that stay in the garden.

My second method for using ducks to control slugs uses the ducks directly in the garden. It takes advantage of two features of duck behavior. First, ducks are the easiest of all creatures to herd. Second, when ducks are allowed into a new foraging area, they eat the meat course first, then the salad and vegetables.

So I arrange for short supervised duck patrols. I open the gate and let the birds into the garden while I am working there (okay, or just lounging in a chair drinking tea and watching). The ducks run all over the garden examining every leaf (top and bottom) and removing every insect as well as grabbing every slug and billing in any loose mulch for more. When the ducks have eaten the meat course and their attention turns to salad, I herd them out of the garden. I just walk in their direction, make scooping gestures with my hands, and say, "That's all, ducks. Let's go. Let's go."

If you just leave the ducks in the garden unsupervised for the day, the ducks will get all the slugs and insects, but they'll get the rest of the garden too. Ducks eat every green we eat, and lots more besides. My ducks are happy, for example, to eat all the leaves as well as the tomatoes off my tomato plants. They eat not just the small squash off the squash plants, but also the growing tips of the vines as well as the young leaves. Ducks of course love lettuce, kale, sorrel,

leguminous cowpeas would fix additional nitrogen in the final compost. This fall we took more than 6 tons of compost off that area for application in one of our two gardens.[6] As I write, about a ton remains in a big heap for use next spring.

In addition, my flock generates compost as an end product of working the deep litter in my 13-by-24-foot poultry house, where there is well over a ton of decomposing litter at any given time. Last summer, the garden flock made an additional couple of tons of compost for the greenhouse in our garden's new Compost Corner, which I describe in the next chapter. So it should be clear that, even aside from the direct fertility value of their poops, chickens can help provide massive amounts of organic matter for feeding your soil.

Insect Control

More than 550,000 *tons* of toxic pesticides are applied in the United States annually.[7] That's about 4 pounds per person, well over 300 pounds per square mile. Per *year*. And while all of us are concerned about the dangers of pollution, we've been convinced that this toxic tide over the landscape is *necessary* to the production of our food. We're made to believe that without use of pesticides, our food crops would be devastated by insects, diseases, and weeds.

But if that were true, how did our great-grandparents grow fruits and vegetables in abundance, long before Monsanto and Cargill and Dow Chemical galloped onto the scene to save the day? The answers go in many directions, but an important one is that back

and every other every salad plant, including garlic and onion greens. About the only edible greens they don't like are the hottest varieties of mustard greens. Ducks will even eat grape leaves and corn leaves up to about 3' high. (Ducks jump.) The key to using ducks for pest control in the growing garden is short supervised duck patrols.

Chickens are often said to be incompatible with gardens because they scratch up seedlings and scatter soil onto low-growing plants. Ducks don't scratch, but they tromp on seedlings with their big feet, which doesn't do the seedlings much more good than being scratched up. Furthermore, ducks have projectile poop that can get on anything within about 8" of the ground. So I don't use duck patrols in an area where I have low-growing salad plants that I want to use raw. I fence off such areas before I let the ducks into the rest of the garden. To keep ducks from tromping down seedlings, I rig a small barrier. A couple of forked twigs poked in the ground with the forks a few inches high can support a bamboo pole running above or just to one side of the row of seedlings. This modest barrier will be enough to guide ducks into walking around instead of across the row.

With a flock of laying ducks, slugs become an asset. During the milder days of winter and spring when the most slugs are available, my ducks can forage essentially all their protein. All they need in addition to the forage is some carbohydrates. During Slug Season, my duck flock lives mostly on forage plus cooked winter squash or potatoes. And for a gardener, what could be better than transforming garden-marauding slugs into delicious eggs?

Carol Deppe is author of The Resilient Gardener: Food Production and Self-Reliance in Uncertain Times *(Chelsea Green, 2010), which has a major chapter on poultry (mostly ducks) that focuses on gardening with ducks, feeding poultry on home-grown produce, free-ranging behavior of various duck breeds, and duck egg cookery.*

then, many backyards and small farms were home to ranging flocks of poultry who helped keep crop-damaging insects in check. Clearly pesticide-free gardens and orchards are more than a fairy tale, and even today, our flocks can help us as well return to such an Eden of natural equilibrium.[8]

In addition to their contribution to runoff pollution, toxic pesticides making their way to ground have a lethal effect on many classes of soil organisms. So using "insecticider" poultry in lieu of toxins is another way to ensure health of the soil food web and boost fertility.

As with tillage, specific strategies for putting chickens to work for insect control depend on context. If you use electronet, for example, you can easily net an entire small orchard and send the chooks on bug patrol. It is a thrill, I can assure you, to see leaping guineas snapping codling moths right out of the air! Do remember about those poops, however. They're a good thing for soil fertility, right? But in the case of fruit trees, which don't need a lot of nitrogen fertility, you could well end with too much of a good thing if you leave the birds on the orchard full-time. I did that one year and had more problems with fire blight in the apples and pears—they're more susceptible to this bacterial blight when growing too rapidly, as can happen with excess added fertility. I began limiting insect patrol in the orchard to spring, when adults emerge from the soil and look for sites to lay eggs—and fall, to break the life cycle of insects that over-winter underground as pupae.

Putting insecticider chickens in the garden is

problematic, since chickens—heavyweight scratchers—would trash it given much time; and other fowl species are likely to enjoy our favorite crops—ripe tomatoes, lettuces, tender brassica transplants—as much as we do. However, guineas are not scratchers, if provided with a patch of soft soil outside their shelter for dust bathing, so they're ideal for insect control in certain crops—especially squash. I don't know how much your garden is troubled by squash bug, but here in Virginia the diseases spread by this insect can devastate squash family plantings. I have found *guinea power* the 100 percent effective, 100 percent organic solution. I move a small mobile shelter to a corner of the squash patch, surround the plot with an electric net perimeter, and plant my winter squash. I monitor the plants through the early stages of rapid growth; and just about the time the vines start to run and I see the first fat yellow blossoms, I see the first of the squash bugs. That's when I put in the guineas—and like magic, no more squash bugs. Note that only a few guineas are needed—maybe three, no more than five.

I've always had excellent results with guineas in winter squash, though some groups have been rough on summer squash. When that was the case, I just surrounded the planting with a close perimeter of 3- or 4-foot chicken wire, staked lightly in place, which kept out the guineas but allowed me to harvest. Tall crops like trellised or caged cucumbers and pole beans, corn, sorghum, and sunflowers also benefit from guinea insecticiders.

Actually it is possible to use chickens to help with insects—or more precisely, with gastropods: slugs and snails—in the *preseason* garden. I sometimes give the chickens the run of the entire garden for two or three weeks before planting the earliest crops. It is several months before the slug population recovers enough to do much damage in my crops.

Many control strategies require giving the chickens enough space for them to chase down their insect prey. But chickens in a chicken tractor can at least assist with slugs and snails. Ducks can help with them as well. Years ago, expert farmer Eliot Coleman graciously showed me through his gardens in Maine. One thing that impressed me was his beautiful Swedish Blue ducks, on slug patrol inside a 4-foot-wide perimeter around his kitchen garden, defined by two low wire fences—only 2 feet high, as domestic ducks usually don't fly at all. Though the ducks were excluded from the garden proper, any slug wandering through their promenade zone didn't stand a chance. You might not think that strategy would help the garden much, with ducks confined to the perimeter only. But plant breeder and duck expert Carol Deppe observes that slugs love duck poop even more than plants in the garden.[9] When they come after it in the ducks' free-fire zone, guess what's for lunch? As somebody said: "There is no such thing as a surfeit of slugs, merely a dearth of ducks."

Those flocksters able to range their flocks completely will get even more insect control from their chickens and turkeys. It is said that a pair of guineas will keep an acre entirely free of ticks.

You will of course discover your own strategies for dealing with your particular problem insects. Grasshoppers, for example, cause no significant crop damage in my garden, but they might in yours. I've corresponded with a flockster in the south of Portugal, where gardens are plagued by swarming grasshoppers. Most gardeners there find that keeping chickens in or around the garden (with movable fences, for example) greatly reduces grasshopper populations.

Other Work for the Flock

The observant, curious flockster finds new strategies for using a working poultry flock all the time. A successful strategy is one in which "What's in it for me?" intersects with "What's in it for them?" Here are three more that come to mind.

Weeder Geese

Before the modern agricultural era, geese were widely used for weeding certain crops. Even today, smart flocksters find that weeder geese earn their keep. I have never used geese as weeders myself—and indeed, a diverse mix of crops in a small space mitigates against their use on many typical homesteads. Someone with a market garden might take advantage of their weeding skills, however. An essential would be separating areas growing crops compatible with weeding by geese (garlic, strawberries, potatoes, cane berries, herbs, tomatoes, onions, carrots, blueberries, asparagus) from incompatible ones (lettuces, cooking greens, and other crops the geese would eat or trample). At the farm scale, geese are extremely effective weeding tobacco, cotton, sugar beets, hops, corn, tree and shrubbery nurseries, vineyards, and plantings of flowers for the floral trade.

Any breed of goose will weed crops, but the smaller Chinese are often preferred for their lighter tread on the beds. Goslings—young geese in their first year—are also preferred for weeding, so weeder geese are usually butchered or sold at the end of the season, in lieu of carrying them through the winter. If a heavier breed is preferred as a better market goose, the more active African would be a better choice than less energetic large breeds like Toulouse and Embden.

Note that success requires more than pointing the geese to the plantings with a cheery *Go get 'em!*—training and knowledgeable management are required to take full advantage of their talents.[10]

Orchard Sanitation

Experienced orchardists know it is essential to clean up spoiled or fallen fruit in the fall, to break the life cycle of diseases and insects that overwinter in dropped fruit. Poultry can perform this chore for us. My experience with fallen fruits is that chickens like them, ducks love them, and geese *inhale* them.

Stacking Species

"Stacking" livestock means keeping two or more species that work well together on the same ground. Usually they are utilizing different parts of the plant and animal life on the area, with the result that you're realizing a fuller yield from the ground as a resource base. The companion species may benefit each other as well.

Geese and sheep pastured on the same ground graze different forage species by preference. Chickens housed under caged rabbits in Polyface Farm's Raken House (stacking *ra*-bbit and chic-*ken*) turn the rabbit pellets and urine into a 12-inch wood chip litter, speeding decomposition and eliminating odor. Gene Logsdon gives his chickens access to the space beneath the floor slats of his raised pigpen, where they harvest fly eggs and larvae out of the accumulating manure pack. Researchers at the Rodale experimental farm built chicken coops over small fish ponds to provide feed for fish—the manure of the poultry, falling through mesh floors—a model that actually came out of traditional aquaculture practice in Asia.[11]

Poultry on pasture with grazing ruminants—cows, sheep, or goats, for example—break life cycles of internal parasites by eating them and their eggs and larvae in the manure. Muscovy ducks and guineas can be especially useful in the control of liver fluke, eating the snails that serve as a vector in the complex life cycle of this potentially fatal parasite. Since avian biology is so different from that of ruminants, the fowl serve as dead-end hosts—that is, they utilize as high-protein food the parasites of other livestock, but are not themselves parasitized by them.

An excellent example of stacking is the Polyface model of following the pastured beef cattle with a big layer flock. Joel actually times the move of the layers in his eggmobile (a henhouse built on a hay wagon chassis) until the fly maggots in the cow pies are "just right"—that is, the point at which they have produced maximum protein for the hens but have not started hatching. The chickens eat the maggots as they scratch apart the cow pats, and in the process expose the manure to air and sunlight—nature's sanitizers. The hens scatter the fertility over the whole sward, rather than leaving the cow pies concentrated in little hot spots.

13 | CHICKENS IN THE GARDEN

Our most recent strategy for putting the flock to work has been such a success, I want to share it with you. I had always assumed that use of working chickens in the garden was necessarily quite limited. That assumption changed after we put up a deer fence around our garden.

Ellen and I had gardened here for twenty-four years with virtually no interference from the deer—but in recent years their pressure has become relentless. I held the deer at bay temporarily with the same electronet I use to confine my flocks, but knew from research that deer are likely to defeat electric fencing eventually.[1] We decided to bite the bullet and invest the money, time, and effort into erecting a permanent wire fence to keep the deer out—12½-gauge galvanized woven wire, 4-inch mesh, 8½ feet high. At the foot of the fence we dug in chicken wire to a depth of 10 inches as an additional barrier to ground-level intruders such as groundhogs and rabbits.

I was not pleased, having to expend so much effort to protect my garden—until the fence started a cascade of ideas about new approaches to gardening within its confines. Prominent among those ideas was bringing in a flock of gardening chickens for the entire growing season.

Compost Corner

When we put up our deer fence, I had been successfully using chickens as shredder-composters for a couple of years. Along the far side of the fence was an area of newly enclosed ground that I planned to use as the garden's compost-making area, and the fence seemed to offer itself as one wall of an enclosure for a flock of full-time composting chickens.

Once I had adopted the fence itself as one wall, it was a simple matter to complete an enclosure with three more sides made of chicken wire on 1-by-3-inch wood framing, ripped on my table saw from rough-cut oak fencing boards. I hinged a wire-on-frame door in an opening on one side, 48 inches wide, allowing plenty of clearance to haul compost materials in, or finished compost out, using either wheelbarrow or garden cart. I put wire over the top of the enclosure as well, to prevent attacks by both aerial predators such as owls, and climbing marauders such as raccoons (see figure 13.1).

Fig. 13.1 The Compost Corner, the wire enclosure in which the garden flock turns garden debris into compost. PHOTO BY BONNIE LONG

As soon as the wire was up, I planted vining beans at the base of the entire Compost Corner fence, except the door. By the time stressfully hot weather came on in midsummer, the beans had overgrown not only the sides, but the top as well, creating a dense, cool shade within, which also helped retain moisture in the compost heap (see figure 13.2).[2] And at the end of the season, the spent vines were a big start on next year's compost heap.

Fig. 13.2 Fast-growing hyacinth bean vines quickly covered the enclosure to provide dense shade for the chickens inside. PHOTO BY BONNIE LONG

Since the chickens would only be in residence in the warmer months, I built a minimalist shelter inside the enclosure, using exterior-grade plywood and metal roofing, to provide sleeping quarters and protection from rain.

As soon as I put the chickens into the Compost Corner, we started hauling in all the organic debris the garden generated—just weeds initially, with soil clinging to their roots to "inoculate" the growing heap with soil microbes, and then spent crop plants as we harvested through the season. We also hauled in horse manure from a neighboring farm, countless cartloads of oak leaves, and many additional loads of fresh-cut comfrey—an excellent composting plant. As the heap became more active with decomposer organisms—pill bugs, crickets, earth-

Fig. 13.3 Hard at work. PHOTO BY BONNIE LONG

worms, together with the usual hosts of microbes and fungi—the chickens spent more time working it. While doing so, they also worked in the ongoing deposits of their own manure and the residues of the generous amounts of cut green forage—from lawn clippings to chard and mangel leaves to cuttings from small grain and cruciferous cover crops—I threw to them daily (see figure 13.3).

By early fall the chickens had turned the heap of garden debris into about 2 tons of compost, applied in the greenhouse for the winter gardening season.[3]

Fig. 13.4 I mounted the new chicken cruiser on two plastic decking runners to keep the frame out of contact with the soil.

Chicken Cruiser

As described earlier, what Andy Lee calls a "chicken tractor" I call a "chicken cruiser," and I have used several of them over the years to employ chickens for tillage chores in the garden. Earlier versions hardly deserved the name *cruise*-er, however—they were a bit too large, clumsy to move within the tight confines of the garden layout, and not sufficiently durable. I made changes to past designs for a more nimble, solid, and durable garden cruiser.

- **Size:** In my new garden layout (see below), I knew that I would be using the cruiser in the garden a great deal, and that I would have to maneuver through some pretty tight turns. The cruiser therefore had to be small. I made it a mere 8 feet long, 40 inches wide—the width of one of our garden beds—and 26 inches high. The total area inside the shelter was almost 27 square feet.
- **Weight:** Another factor favoring maneuverability was shedding as much weight as possible with materials and design choices. I used metal roofing recycled from previous shelters—some aluminum and some baked-enamel galvanized steel—to cover much of the top and half the sides. The rest of the sides and top I covered with 1-inch chicken wire, which helped minimize weight and allowed more sunlight into the interior.
- **Frame:** "Strong but light" was the guiding principle for making the frame. Fortunately, I happened to have on hand a few pieces of high-grade, clear fir, left over from redecking a porch, three-quarter inch thick. Its greater strength, in comparison with common construction-grade lumber, allowed me to use narrower framing pieces, reducing weight. The result, with generous use of diagonal bracing, was a light but rock-solid frame that was easy to maneuver among the garden beds.
- **Durability:** I spent the extra bucks for star-drive, stainless-steel deck screws to fasten the frame together. Given the constant exposure to weather, they should last longer than any other option. In order to keep the bottom rails of the frame out of contact with the highly bioactive garden soil, I mounted them on two runners cut from recycled plastic decking, 5½ inches wide, which should resist rot a lot longer than any wood alternative (see figure 13.4). In addition to increasing the frame's resistance to rot, the plastic runners made the cruiser much easier to slide down the garden beds.

Fig. 13.5 After working up a new asparagus bed in the cruiser, the garden flock will return to compost making in the Compost Corner, to the left. PHOTO BY BONNIE LONG

- **Access:** I allowed for access to the interior, for use when feeding or replenishing the waterer, with a wire-on-frame lid over the section of the top not covered with metal roofing.
- **Nestbox:** I provided the sheltered end of the cruiser with a hinged piece of scrap plywood for easy egg gathering from the straw-lined nestbox mounted on the framing inside.

I put eleven birds into the cruiser, one cock and ten hens—about 2½ square feet for each bird. This turned out to be too crowded, and I began limiting the number of tiller chickens to eight—3⅓ square feet each. Even that spacing is rather tight, but work assignments in the cruiser are periodic—while working in the Compost Corner, on the other hand, the flock has about 225 square feet available.

The first assignment for my birds in the cruiser was to work up new ground for an extension of our asparagus bed. I had made a start the previous fall with a heavy application of chicken-powered compost and a cover crop. Now I tilled in the cover crop by running the cruiser down the new bed, then made an additional pass, dumping in large loads of earthworm-rich compost which the chooks tilled in while picking out the worms. I then moved the cruiser elsewhere and planted my new asparagus plants.

For the rest of the season, the garden flock alternated between the Compost Corner and the cruiser. There was always work to be done on the compost heap, so the assignment at any given time depended on whether there was a bed in the garden needing its cover crop to be tilled in.

Fig. 13.6 At the end of the season, when many gardens are abandoned for the winter, my garden flock (in the chicken cruiser, far left) is still getting beds ready for sowing overwinter cover crops.

A further change in our new approach to gardening was the devotion of fully half our garden beds to cover crops, for substantial and repeated increases to soil organic matter each year. Starting this past season, every other bed in our garden now grows cover crops, and the rest grow harvest crops. Then, in the following year, the beds that grew harvest crops will be planted to covers, and vice versa. Since a cover crop bed may be planted to several covers in succession during the growing season, there is plenty of work for cruiser chickens to do.

On the last pass with the cruiser over beds in the early fall, I dumped into it large amounts of compost made elsewhere and allowed the chickens to till it in, while eating the earthworms in it—then returned the garden flock to the Compost Corner and sowed cover crops to protect the beds over winter. The fast-growing fall covers served as well to capture the soil nutrients in the droppings worked into the beds by the chickens as they tilled, and prevent their leaching to groundwater. The result was that as winter approached—a time when many gardens are abandoned or, worse, bare—my entire garden was lush with cold-hardy cover crops: clovers and alfalfa, winter peas, small grains, various crucifers. Many of the latter—mustards, turnips, rape, tendergreen, kale—we harvest whenever we need cooking greens for the table, in lieu of growing them as separate fall garden crops.

Predators

To my surprise I have had no predator problems with my garden flock whatever. Raccoons are great climbers—but I think the distance established by the fence between the perimeter and the cruiser prevented them from recognizing the chickens in the cruiser as available prey. There was a single attempt on the Compost Corner—most likely by a fox—but the chicken wire we had dug in at the foot of the fence paid off. Digging predators are not stupid—they don't waste energy on intensive digging if there isn't a ready payoff.

A Future Project

We have a minor problem with asparagus beetles each year. In the future, however, I plan to use the chickens in the Compost Corner to help with control of the asparagus beetles as well.

Note in figure 13.6 that the compost enclosure is right beside the asparagus bed. At the end of the asparagus harvest, I will set up a temporary fence separating the asparagus bed from the rest of the garden, using either electronet or chicken wire. I will release the chickens through a small door or pop-hole in the enclosure's wire, allowing the chickens to go after the beetles among the growing fronds, while keeping them out of the rest of the garden with the temporary fencing. In the fall, after I cut the asparagus fronds, I will give them another run, to scratch in the mulch for both the beetles, which hibernate in the winter, and their larvae. The rest of the time the pop-hole will remain closed, and the chooks will concentrate on making compost instead.

14 | A QUESTION OF BALANCE

Up to this point, I've made numerous suggestions for ways our flocks can help increase soil fertility in the homestead or small farm, whether through direct deposition of droppings on the pasture or while tilling in cover crops; through use of deep litter from the poultry house as compost; or with weeds and crop debris that have been chicken-powered into compost. In addition, I operate a large vermicomposting project in order to both feed my birds and to generate worm castings as a soil amendment.

All these strategies are ways to make soil care and keeping poultry a single integrated project. And they really work: In the twenty-seven years I've raised poultry here at Boxwood, the organic matter content of our soils, together with their macro- and micronutrient content—key factors for sustaining soil fertility—have risen to impressive levels. Gardeners in my area start with native organic matter (OM) content in the range of 2 to 2.5 percent. Recent tests of the soil in our two garden areas reveal OM levels of 6.8 and 7.5 percent.

The same tests, however, have been a reminder of what my mother often told me, and probably yours as well: *It's possible to have too much of a good thing.* Indeed, levels of certain mineral nutrients can become seriously excessive if we're not careful.

Tracking Soil Changes

My approach to increasing soil fertility has been marked by a good deal of "irrational exuberance" where use of *manures* is concerned. The manure generated by a flock the size of mine would, over time, yield a tremendous increment to soil fertility. Without tracking carefully enough the rise in soil mineral content, I have applied as well large quantities of horse manure—both to feed my large vermicomposting operation and as a base for chicken-powered compost—free for the hauling from a neighbor with horses (see figures 14.1 and 14.2). The result of years of manure import is that some soil minerals, especially phosphorus (abbreviated "P") and potassium ("K"), have accumulated to levels far above optimum. This chapter will serve, I hope, to encourage my reader to be smarter than I've been, and to track the changes in soil mineral and organic matter content over the years, before running into an issue of excess. If you test your soil regularly—every

Fig. 14.1 It started as multiple pickup loads of manure.

Fig. 14.2 The manure was worked into new ground by tiller chickens, who can help improve soil in a hurry. Be careful about bringing in excessive amounts of other manures, if your flock will be offering much of their own to your soil fertility program.

three years or so, and every two years if soil minerals are approaching excess—you will not be taken by surprise as your soil evolves.

Many of the measures of soil quality are commonsense ones—whether you're getting the results you want with the crops you're growing, and whether your soil retains moisture and resists drought better with each season. Success in these terms is more important than soil tests. Tracking changes to your soil's mineral and organic matter content with periodic testing, however, can help you understand when your soil has reached the point of ideal nutrient storage, at which point there are sufficient mineral nutrients in reserve to match demands from crops.

Probably the easiest approach to soil testing is to use the services offered by the extension service for your state. In my state (Virginia), for example, an inexpensive soil test reports soil pH, phosphorus, potassium, calcium, magnesium, zinc, manganese, copper, iron, and boron, along with giving lime and fertilizer recommendations, plus estimated CEC. (*CEC* is "cation exchange capacity"—the capacity of the soil to adsorb positively charged chemicals. The CEC of a soil is in effect a measure of its ability to store nutrients. Sorbed nutrients are immobilized—less subject to leaching out of the soil—and can later be mobilized by other processes and made available for plant uptake. CEC is controlled by many factors, including the amount and type of clay in soil, as well as the amount and type of OM in soil.) A report of organic matter content can be added for a small fee—and I highly recommend it. Your local extension agent can help you formulate a soil management plan based on your results.

You may find soil testing analysis, interpretation, and advice at greater depth if you work with an independent soil consultant. An organic gardeners and farmers association in your state may be able to give you a reference. I have found working with a consultant to be most helpful in devising a nutrient management plan for my garden and pasture soils.

A soil nutrient management plan cannot be formulated in the abstract—it has everything to do with the specific nature and history of your soil, your climate, and the crops you want to grow. Get to know *your* soil—the way it responds to heavy manure applications and other fertility amendments depends on its makeup. Soluble nutrients leach readily from a sandy soil, for example, but increased humus content in sand reduces the rate of leaching—by increasing the soil moisture-holding capacity and the CEC, and by decreasing the permeability of the soil to water flow. Clay soils are often cursed, but that has more to do with their physical properties—brick when dry, too sticky to work when wet—than with their natural fertility, which tends to be high given the mineral content of clay soils, their ability to adsorb/store nutrients, and the moisture-holding properties of finer-textured soils. Though vastly different from sand in its physical structure, large amounts of organic matter will improve clay soil as well.

Indeed, increasing organic matter when starting a soil-improvement program will benefit almost any soil type. Soil texture, water absorption and retention, numbers and diversity of soil life, cation exchange capacity—all improve as humus content increases. But it is important to be aware of—and plan for—the natural arc that follows rising organic matter over time. Initially, the more the better—and our flocks can help with all phases of nutrient and organic matter accumulation. Once our soil reaches optimal levels of soil nutrients, however, our focus should shift to *maintaining* those levels of stored nutrients while making sure we aren't creating a long-term *increase* to excessive levels; that is, when we reach the plateau of optimal nutrient storage, the goal should become a rough *balance* between inflows (what we add to the soil) and outflows (what crop plants remove).

If manure from your flock is going to be a significant part of your soil fertility program, I would strongly caution you about bringing in large amounts of manure from elsewhere. Small farmers raising large numbers of livestock in relation to the land base should also think about the potential for building excessive levels of nutrients from manures, especially if they are buying in much of their feeds, which increase net gain of soil nutrients over time. This is another case where diversity on the farm is best—in this case, a balance between the numbers of livestock animals, and the amount of row crops or pastureland that can fully take up the nutrients from their manures. Farmers who generate more manures than their farms can utilize could export some of that fertility off their farms, through sales of manure-based composts to other farmers or to local gardeners.[1]

Wretched Excess

I hope that you will indeed track your soil's changes and shape a flexible nutrient management plan that changes with the evolution of your soil's character and fertility over time. To illustrate the problems that result from too much of a good thing, let's look at two soil nutrients in my soil that have accumulated to excessive levels.

Phosphorus

Phosphorus (P) is a soil mineral needed by plants in large amounts and in all phases of the life cycle, for rapid and healthy growth, a strong root system, maturation of fruit, and resistance to disease. When phosphorus is too low, plant growth is poor, and development of yield—bean pods, corn ears, fruits such as tomatoes and squash—is severely limited. Many soils are deficient in phosphorus. Fortunately, the decomposition of organic matter added to the soil releases phosphorus for plant use. Major players in making soil phosphorus—which is relatively insoluble—available to plant roots are *mycorrhizal fungi*, whose presence and vitality are also boosted by increased organic matter. Because of the relative insolubility of most phosphorus compounds in the soil, phosphorus not taken up by plant roots tends to stay in place and accumulate (rather than leach out of the soil)—which is to say, adding too much organic matter from phosphorus-rich manures and manure-based composts can result over time in the accumulation of *too much* phosphorus. Excess soil phosphorus has a number of negative effects, starting with those essential mycorrhizae—it depresses their activity in the soil and leads to the paradoxical effect of phosphorus, in abundance around plant roots, being *less* available without the mycorrhizae to help out. If excess phosphorus kills off mycorrhizal growth, therefore, we can see exactly the effects of phosphorus *deficiency* listed above—weak growth and reduced yield.

Potassium

Potassium (K) is another element used in large amounts by plants to synthesize sugars, to resist diseases and cold temperatures, and to reduce water loss in dry weather. Potassium is less likely than phosphorus to be lacking in a soil—certainly, soils well supplied with organic matter will contain all that crop plants need.

Again, potassium in large, repeated applications of manures and manure-based composts may in time lead to excess amounts in storage. In a sandy soil with little organic matter, potassium leaches readily to groundwater—though it doesn't seem to have the severe water-ecology effects noted below for phosphorus. It is far less leachable in a clay soil with high organic matter—my soil exactly—and the resulting excess can lead to problems such as inhibition of uptake by plant roots of other essential minerals (calcium, iron, magnesium).

Strategies for Dealing with Nutrient Excesses

If you too have been guilty of irrational exuberance in the application of manures and manure-based composts and are now facing a wretched excess of soil minerals, do not despair. With a newly heightened awareness of the potential problems, it is possible to put in place management strategies to ameliorate nutrient excesses.

Preventing Contamination Downstream

Take seriously your duty to help prevent pollution of groundwater and surface-water systems downstream from your chicken run and poultry pasture. The Big Boys of the poultry industry are seeing to the pollution of our waterways quite nicely, and need no assistance from us to make the problem even worse.

Remember that chemical fertilizers, manufactured to be immediately available to plants and therefore highly soluble, contribute to environmental pollution of groundwater, especially by nitrates—and I urge the reader to avoid their use *entirely*. However, anyone using manures in their soil fertility program should also take care to use them responsibly to prevent water pollution. Application of raw manures in amounts too great to be readily taken up by plants—whether harvest crops or cover crops—may cause *leaching* of nitrates, potassium, and other soluble soil compounds to groundwater and streams. Applications in smaller increments, when plants are ready for rapid growth—and proper composting of manures in order to fix their nitrate and other mineral content in more stable forms—should reduce the chance of leaching.

While soil phosphorus is, as said, relatively insoluble (and therefore not subject to much leaching to groundwater), *runoff* at the surface—as in erosion of soil sediments, or washing away of applied manures not yet absorbed into the soil—is a major concern. In a soil that has accumulated excessive phosphorus, any runoff will carry a proportionately heavier load of polluting phosphorus. Leaching and runoff of nutrients from our soils resulting in water pollution are not only bad for local ecologies, but are also indications that our inflows of nutrients are much higher than are necessary for plant use—and therefore a waste of good manure.

Whether from leaching or runoff, unnaturally high levels of soil minerals, *especially nitrates and phosphates*, contaminate drinking water and in lakes, rivers, and estuaries can cause *eutrophication*—algal blooms and other rampant growth of aquatic plants that produce catastrophic ecological effects, including die-offs of fish, crabs, and other water life. Carelessness about pollution doesn't just affect local water systems; agricultural runoff from fields in the Midwest has helped create a dead zone the size of New Jersey in the Gulf of Mexico, a thousand miles away.

In addition to the manure management practices mentioned above, it is good practice to make sure that any garden ground with even the slightest slope is well covered with crop plants, cover crops, or mulch. Make sure that sloped pastures have a tight and vigorous grass cover to prevent the flow of surface water in heavy rains.

Reducing *Net* Gain

Once soil nutrients reach optimum levels, the goal becomes establishing a rough *balance* between nutrient inflows, from whatever sources, and nutrient outflows from the soil—crop yields as food, eggs, and dressed poultry for the table. But in a case like mine, in which I had accumulated excessive phosphorus and potassium reserves, the immediate requirement was to stop the inflows until I could verify that my soil had returned to a balance between nutrient storage and uptake by crop plants.

Much of the increase of P and K stored in the soil I could easily shut off from one day to the next by ceasing to haul in manure from my neighbor's generous horses and spoiled hay in large quantities from a nearby farm. Since my gardens now have more than sufficient P and K, forgoing these inputs has not posed any problems.

The biggest change from ceasing to import manure has been in my vermicomposting operation, which I describe in chapter 19. While I'm still using my extensive worm bins to generate all the feed I can for the flock, I have taken a different approach to the *bedding* for the worm bins (using straw, newspaper, leaves, and cardboard, rather than pickup loads of horse manure) and now ensure that all *feeds* for the worms be produced here on the homestead exclusively. This is a matter for ongoing experimentation. One experiment already begun I will continue especially—the "cooking" of comfrey, cruciferous cover crops, culled roots, and vegetable trimmings in a solar cooker I made last year. Earthworms make much more efficient use of cooked feeds than the same feeds raw—that is, they grow and proliferate faster, yielding more home-produced feeds for the flock.

There are some organic materials I do still bring in from outside. Since deep litter is essential to best manure management in the poultry house, and oak leaves are the best material I've found for litter here, I still bring in oak leaves donated by a neighbor. I still also use purchased straw as a portion of the bedding in the worm bins. But both these materials are low in P and K so will contribute negligible *net* gain to either in our soils.

The biggest net gain to soil P and K here in the future will be in the form of *purchased feeds*. Any phosphorus or potassium in the feeds my birds eat that does not end up in the eggs and dressed poultry on our plates is excreted by the birds as P and K, which must go *somewhere* on the property—that is, must show up as a net *gain* to P and K. I've made a couple of changes in order to reverse that net gain.

It is only *purchased* feeds that result in a net gain of P and K; the more I feed my birds feeds produced or foraged on the homestead, the smaller the net gain. Therefore, my soil imbalances provided a golden opportunity to push even harder to feed my flock more from home resources.

If too much poop has become a problem, the question of too many poopers must be addressed. In other words, *reducing flock size* is a good strategy for reducing net gain of P and K. As with anyone who enjoys keeping poultry, flock creep has been too much a part of my husbandry—keeping more ducks, geese, and chickens than we really need. I have begun a reevaluation of the numbers of birds Ellen and I require at a minimum for eggs and dressed poultry and am in the process of a severe culling down to that flock size. Not only will less poultry manure be generated on our place—whether directly on the pasture or in the deep litter in the henhouse—but there will be a greater share of homegrown feeds per bird, and a further reduction of dependence on feeds purchased in from outside. Calculations of flock size in relation to land base and opportunities for feeding with on-farm resources will be more complex for those with market-sized flocks.

Limiting Use of Chickens in the Garden

I still recommend without reservation using small flocks of chickens as much as possible as partners in the garden—*in the early stages of a soil-building program.* The contributions of the birds will prove invaluable, in terms of both immediate fertility gains from the droppings worked in, and long-term gains in soil humus content. As long as I need to avoid P and K accumulation, however, I will for the time being limit my use of a tiller flock in the garden. I will use tiller chickens only when they are able to subsist *entirely* from the offerings of the bed to be tilled—mature seeds of cover crops, slugs, earthworms, and other available forage—so that their droppings contribute no *net gain* to P and K.

Cropping Out

We know of course that taking crop yields off garden soils removes soil minerals like P and K. Carol Deppe's *The Resilient Gardener*[2] convinced me to greatly expand my planting of crops like potatoes (which she uses to become more largely independent of purchased feeds for her duck flock), sweet potatoes, mangels, and winter squashes to provide more feed for the flock. Growing more of these crops for my flock will both draw down mineral excesses in the garden and reduce their net gain from purchased feeds.

Normally it is desirable to harvest *only* the food parts of plants, in order to limit the amount of mineral fertility removed from the soil—and crop residues themselves should be returned to the soil via composts or mulches. But if the goal is reduction of excess minerals, it is advisable to remove spent crop plants entirely for even greater drawdown of the excesses. So while I continue growing feed corn, sunflowers, amaranth, and sorghum, I now cut the entire plants to offer the flock, after the seeds are ripened but while the leaves are still green. Both chickens and waterfowl eat the seed heads, but the ducks and geese eat the leaves as well, down to the stalk.

Cover cropping, always a key part of the soil-building effort, also offers opportunities to draw down excess mineral reserves. Starting with the past growing season, I have dedicated fully half of both our garden spaces—7,000 square feet in one, about half that in the other—to cover cropping. Cover crops cut and removed from the garden will help reduce excess P and K. Fortunately, the same strategy will produce a great deal of feed for the flock—cruciferous crops, grain grasses, and clovers for cut greens; ripened small grains and buckwheat; sorghum-sudangrass hybrid as green fodder relished by the ducks and geese.

Cover cropping with legumes is an especially valuable option for providing nitrogen (N), especially for heavy feeders such as corn and winter squash. Organic farmers and gardeners sometimes use manures as the nitrogen source of choice for such crops. However, manure applications in amounts to ensure *enough* nitrogen tend to bring in *too much* phosphorus and potassium. Preceding heavy feeders with a nitrogen-fixing cover crop (alfalfa, clovers, cowpeas) is a better strategy for providing the extra nitrogen while limiting accumulation of phosphorus and potassium.

Note that some harvest and cover crops use more phosphorus and potassium than most, among them alfalfa, sunflower, and small grains, so are especially useful for drawing down these mineral excesses. I currently am experimenting with a combination of alfalfa and small grains, sowed together in the fall. In the spring the small grains—wheat, barley, and rye—will nurse the alfalfa planting, suppressing weed competition. Once the grains are ripe and I cut them for feeding to the flock, the alfalfa should grow vigorously in the increased sunlight; and I will take repeated cuttings during the growing season for feeding.

Keeping a High Level of Organic Matter

Did I mention that high soil organic matter is a good thing? But how can we keep organic matter high without adding to P and K—or while reducing an excess of either? Again, cover crops are the key: We can remove their top growth to feed the flock or to use as mulches elsewhere, but their roots will contribute OM as they decompose in place. Therefore, cover crops with a large root mass—alfalfa with its thick, deep taproot; rye with its extensive mat of fine roots reaching 4 or 5 feet deep—serve especially well.

The decomposition of organic mulches also contributes to organic matter. Mulches such as leaves and straw that contribute least to soil P and K are good choices. Mulches grown on the homestead or farm do not contribute to a net gain of soil minerals.

Fertility Transfers

The astute reader has doubtless wondered: *But by cropping out from the garden, aren't you just moving the excess P and K around on the property*? Exactly. The P and K taken up by harvest and cover crops are passed on to wherever else they are used on the property, whether fed to the flock or laid down as mulches. So the challenge is defining the parts of your property that have not been heavily amended with manures and manure-based composts in the past—areas that will actually benefit from P and K and other soil nutrient accumulation, including organic matter. How and where to make such fertility transfers is an important part of dealing with soil nutrient excesses.

Just as soil testing is the key to discovering where excess use of manures has accumulated too much P and K, further testing is also the key to defining where on the property you can make fertility transfers. Areas I have begun testing here include our lawns, pasture on an adjacent property we recently bought, and a part of our small woodlot in the midst of a makeover.[3] As long as continued periodic testing indicates no accumulation of nutrients in storage beyond what is needed for plants grown in these "gimme" areas, I will continue to make fertility transfers into them from areas of excess—in the form of cover crops cut and fed to my flocks, compost litter from the henhouse, and compost made from garden debris.

- **Woodlot:** I have occasionally electronetted this area in the past and ranged the chickens in its weedy cover. This past fall I received about twelve big dump-truck loads of wood chips from a maintenance operation to clear tree limbs near power lines in my area, much of which I spread in a thick layer over the open woodlot ground. By spring it should be breaking down nicely, rife with fungal mycelia and grubs and other critters at or below soil line.[4] I will keep the chickens on that ground for much of the season, offering daily green and grain feeds cut from garden beds—plus weeds, crop residues, and the stalks of crops such as corn and sunflower. What the birds don't eat they will scratch into the chip cover, yielding an increasingly complex and biologically active debris field, increasingly alive with feeding opportunities. Given its size—about a third of an acre—I anticipate, over time, both a significant contribution to home feeding of my flock and a large transfer of excess soil nutrients from the gardens to an area that can use a big boost in fertility and humus content.

- **Pastures:** My existing pasture, over which I have ranged my flocks for years, has also shown a buildup of P and K into the red zone—though not nearly to the same degree as in the gardens. Since I will be keeping the chickens on the woodlot area during much of the season, I will be rotating them over the pasture less than in the past, to slow further accumulation. To balance that addition, I will make daily cuttings of young, tender, rapidly growing grass—highest in protein and other nutrients—to feed the chickens when they are ranging the woodlot enclosure. Grass that grows too long and tough I use as mulches in other parts of the property in need of P and K—nut trees, blueberry and blackberry beds, fruit trees and shrubs, and the forest garden.[5] For parts of the growing season, I will range the chickens on a new pasture plot, on the adjacent property we recently bought.
- **Deep-litter compost:** No organic matter produced on this homestead is going to be wasted. The composted deep litter from the poultry house is an excellent supplement for our young nut trees, hungry for the fertility boost in the early years, and a number of plantings of comfrey, which thrives on just about any amount of fertility I throw its way. Comfrey is a great homegrown poultry feed, especially for the waterfowl.
- **Lawns:** Lawn areas not suitable for grazing by my ducks and geese are also areas that benefit from dressing with litter compost. Three lawn areas around our house are big enough to "pasture" our small waterfowl flock on. I anticipate using these areas as much as I can for this purpose, to reduce further deposition of manure on the existing pasture.

The Silver Lining

Fortunately, excesses of P and K in my garden soils have not yet resulted in any clearly observable growth and yield problems for garden crops. If I continue to avoid additional accumulation, I can continue working with soil supercharged with P and K and can implement other soil-care strategies with no need to build up either of these essential mineral reserves.

The high level of organic matter in my soils is a major silver lining, since it contributes so much to soil quality, while the need to crop out excess P and K from my garden soils is actually an invitation to heavier production of natural, homegrown feeds for my flock.

In addition to the gardener's commonsense measures of changes in soil quality—plant health and vigor, crop yields, resistance to drought—I highly recommend *testing your soil regularly.* Use all these indicators to guide a practical soil nutrient management plan, and to prevent unpleasant surprises in the evolution of your homestead's most precious asset, its soil.

PART FOUR
Feeding the Small-Scale Flock

PHOTO BY BONNIE LONG

15 | THOUGHTS ON FEEDING

It has been hard to decide where in this book to introduce a discussion of feeding. Truly, most of this book is about feeding the flock: Every strategy for making a partnership with the flock—insect control, garden tillage, making compost, raising new chicks using broody hens—is effective mostly because it offers the birds *access to the foods they most want to eat.* But let's think a little more specifically about the principles that guide best feeding of the farm or homestead flock.

The Three Food Groups

For me the starting question about feeding is not "What's on the label?" or "What is the required ratio of nutrients?" but rather, "What would a chicken eat on her own in the wild?" And while I can't travel to Southern Asia to observe the eating habits of *Gallus gallus* (the Red Junglefowl), I do have a more immediate example of natural feeding to contemplate—seeing my grandmother's flock of hens feeding themselves, free-ranging over a 100-acre farm. Granny knew little and cared less about "scientific feeding," yet her largely self-sufficient flock kept eggs and chicken 'n' dumplings on her family's table. I spent many hours of my youth observing Granny's busy hens, and found that they ate a diverse mix of three food groups:

- **Green plants:** Though we do not think of chickens as grazers, a small but important percentage of their diet is green plants if they have access to them. Plants contain fiber that helps with efficient excretion; key nutrients like minerals, vitamins, and proteins; and enzymes—complex proteins that catalyze various chemical reactions in the body. Specifically, plant enzymes assist digestion, enabling more complete utilization of the other nutrients in the chicken's feed.
- **Seeds and fruits:** Chickens avidly seek out seeds and fruits, the parts into which plants concentrate the most food energy as part of their reproductive strategies.
- **Animal foods:** Chickens eat a wide range of animal foods such as insects, worms, slugs, and snails. Such foods are potent sources of both proteins and fats, essential for growth and maintenance, cellular energy, absorption of minerals and fat-soluble vitamins, and functioning of the hormone system. They also are important sources of minerals—for example, the exoskeletons of such insects as beetles and grasshoppers contain a lot of calcium, especially important for strong eggshell formation—and of beneficial enzymes.

What are the two things these diverse natural feeds have in common? They are *alive,* and they are *raw.* Best feeding of our flock means maximizing their access to such living, unprocessed foods. It means taking the feeding of our flock into our own hands as much as we can.

I have found that the idea of making their own feeds introduces a good deal of feeding anxiety for many beginning flocksters. We have been conditioned to believe that manufactured feeds are "scientifically balanced" and fear that in our ignorance we cannot possibly get it right, ensuring the nutrient balance required by our birds. Perhaps the first feeding principle is thus: *Relax!* To be sure, our chickens need a properly balanced diet, but tell me what's wrong with this assumption: The more diverse the live, natural feeds our chickens eat, the more *naturally balanced* their diet will be.

As with my grandmother's flock, free-ranging homestead and small-farm flocks in the past did in fact fill much or all of their nutrient needs on their own, long before the era of "scientifically formulated" manufactured feeds. It helps as well to remember that almost all poultry feeding studies have been done assuming confinement feeding as the norm—of *course* they are going to conclude that ensuring the correct ratios of nutrients is a fine balancing act indeed. Regrettably, few or no large feeding studies have been done of the pastured or free-ranging flock.[1]

The Feeding Spectrum

I find it useful to think of feeding as a *spectrum*, with the feeding of purchased feeds exclusively, to a flock that is totally confined (that is, completely dependent on what we give them to eat), on one end; and on the other end, a completely free-range flock, foraging all their own feeds from biologically diverse ground. Most of what I have to say about feeding in this book is about moving from the former end of the spectrum to the latter. Few of us will have the option, as my grandmother did, of allowing our flocks to supply almost all their own diet, so our feeding program will be somewhere in the middle of the spectrum. *But every flockster, without exception, can find ways to introduce more natural feeds to the flock.*

The more we find ways to give the flock maximum access to live, natural foods, the more we will think of purchased, highly artificial feeds—or even concentrated feeds we make ourselves—as *supplemental*, provided as needed to the extent not enough natural feeds are available.

Feed Costs

Another obvious thing about the feeding spectrum is that it moves from greatest expenditure on purchased inputs on one end, to a flock that requires no cash costs to feed on the other. As the energy/agricultural/financial crises of our time deepen, the question of feeding a home flock affordably will become increasingly urgent.

In 2008, the last time I did a detailed analysis of my feed costs, I found that in just a year's time, the price of whole shell corn had jumped 20 percent; field peas had increased 39 percent; and whole oats, 63 percent. The price of whole wheat rose 40 percent in eight months, then was unavailable from my supplier "until the next harvest." Fortunately I was able to substitute triticale—at 26 percent above the price I had been paying for the more desirable wheat. An acquaintance who produces broilers for small local markets on purchased feeds reported that her feed costs had increased 38 percent in the course of that same year.

As of this writing, weather-related shortfalls in grain harvests in the United States and Brazil—and perhaps most notably Russia, which banned grain exports to world markets in August 2010—are pushing prices of feed grains—and along with them, food prices—ever higher.

Perhaps no other change in our thinking about feeding is more important than getting away from the notion that the feed for our flock must all be *purchased*. Karl Hammer, owner of Vermont Compost Company (VCC), makes high-quality compost for gardeners, farmers, and landscapers. His starting ingredients are food residuals (uneaten food from local food-purveying institutions), dairy manure, hay—and *chickens*.

It is cheaper for restaurants and other institutions in Montpelier, the capital of Vermont, to pay VCC a tipping fee to take their food residuals than to pay the higher fees at the city dump. Vermont Compost makes mountains of compost from the food wastes, the manure, and the hay, and releases onto them a mixed flock whose size varies a good deal, but has numbered as many as twelve hundred chickens, most of them layer hens—Australorps, Buff Orpingtons, Wyandottes, and Rhode Island Reds. The hens do much of the work of turning the heaps, speeding their decomposition into rich compost, made even more potent with the addition of their manure. Karl Hammer sees their work as the major reason to include them in his operation, though their production of eggs—up to fifteen thousand dozen in VCC's peak year, for sales to a co-op, two school systems, farmer's markets, and a couple of CSAs (Community Supported Agriculture)—provides Hammer with an additional source of income. And the cost to feed those twelve hundred chickens? Absolutely *nothing*—they get *all* their feed from the compost heaps, either directly from the edible residuals themselves or from the populations of decomposer organisms as the heaps become increasingly biologically active. (For more on this company, see chapter 30.)

Few of us have access to free feed on VCC's scale. But there are other strategies we can use to cut feed costs.

Challenge the Flock to Do More Self-Feeding

Even those of us who routinely pasture our flocks are a bit nervous, I expect, as to whether our birds will get sufficient nutrition on their own; we tend to feed them more than they really need from the feed bag. Don't be afraid to push your flock to work harder providing their own feed by offering less from the feed bag—some chickens prefer hanging around a feed trough filled with easy handouts. As Karl Hammer observes: "The less we do for them, the more they do for themselves." I've received reports from numerous other correspondents as well that the key to more active foraging is being more stingy with the feed scoop.

Think of purchased feed as supplemental to the flock's real diet of live, natural foods. You can confirm whether your birds are getting enough to eat by occasionally feeling their crops after they have gone to roost: If they are ending the day with full crops, they are getting enough feed however little you're providing.

Practice "challenge feeding": Record current egg production based on purchased feeds. Give the birds maximum access to natural, self-foraged feeds, and reduce the prepared feed by a measured amount each day. The point at which egg production drops indicates the point at which they really do need the concentrate feed to supplement what they are getting on their own.

Cut Feed Costs with the Hatchet

It is common for us flocksters to enjoy our flocks so much that imperceptibly the flock grows bigger each year, beyond the number required to meet our actual home production goals. We can save on feed costs, however, by avoiding flock creep through constant review of the composition of the flock and vigilant culling.

Stacking Species

Among the many benefits of keeping more than one livestock species together is the saving on feed costs, as when sheep and geese share the same pasture, or chickens clean up feed spilled by other animals—or even pick undigested bits of corn, small grains, and acorns out of the manure of cows or pigs. Use sheep to mow your orchard (unlike goats, they won't chew off the bark), then follow with chickens to clean up dropped fruit and any parasites in the sheep manure. Use the Polyface model on a smaller scale by following the family cow with a mobile pasture shelter

Mice with Wings

English, or house, sparrows are an invasive species here in the mid-Atlantic, and likely where you are as well—since being introduced in New York in 1854, they have spread into every part of the United States except Alaska. They out-compete native species, especially cavity nesters such as bluebirds, swallows, and purple martins. For example, I have tried to host eastern bluebirds by setting up nestboxes for them, but the sparrows aggressively move in and claim the space, even when I repeatedly remove their nests.

House sparrows brashly fly in through the open door of the poultry house while chickens are out foraging during the day. If present in any numbers, they gobble a good deal of expensive feed, and are host to many avian diseases and external parasites. Last year they showed up in my henhouse in large numbers, and I decided to start trapping them to thin their ranks. After researching several options for purchased traps, I found one that worked quite well and solved my problem.[2]

Indeed, for the first week or two I caught five or six house sparrows a day. Then for a while it was one to three, then none. Next year I will set the trap up at the beginning of the season. Not only will I protect my expensive feed, but I hope some of the bluebirds nearby will finally get the chance to nest in the nestboxes I've set up for them.

Oh, and the trapped sparrows? I use them to feed my soldier grubs, which I'll discuss in chapter 19.

or electronet paddock with chickens, to glean fly maggots out of the cowpies.

Avoid Free-Choice Feeding

Some flocksters advise that chickens do best if fed free-choice—that is, with feed available in the hoppers at all times. Let me tell you why I haven't found that a good idea.

One winter I started noticing a lot of mice in the poultry house anytime I went in at night—the litter and the wire partitions appeared to be boiling with the little rodents when my flashlight beam swept the interior. I took the infestation as a personal challenge and started setting the common spring-operated snap traps to cut down the population. I lost track of the number of times I caught two mice with one snap—but twice, I caught *three* at a whack! My infestation was your basic industrial-strength rodent rodeo.

Despite trapping a couple dozen mice per night, there was no discernible effect whatever on the population level. Then I noticed mouse droppings in the feed hopper and realized that keeping a concentrated feed source constantly available was an open invitation to the mice to be fruitful and multiply. Creating the solution was as simple as creating the problem: I stopped feeding free-choice. Instead, I fed a calculated amount of feed early—no more than the birds would clean up entirely within the morning, leaving no freebies when the mice came exploring at night. The rodent population plummeted. Since that time, there are always a few mice around, but they have never again approached biblical plague proportions since I started denying them a ready food supply.

English sparrows can also be a problem, when it comes to food theft. While not rodents, I think of them as mice with wings.

How *Not* to Reduce Feed Costs

The one way not to reduce costs is to buy cheap, mediocre feeds. If it is true that "you are what you eat," then you are what your chickens eat. Buy the highest-quality feed you can afford. If you're stuck with commercial

Organic versus Conventional: Feed Grains

In today's world we need to examine the sources of and possible residues in our own food—and so too with the feeds we buy for our chickens. After all, when we eat their eggs or meat, we are eating what they have been eating. Should we seek out certified organic feeds for our birds, even if the cost is far above that of conventional feeds?

A good start might be a study of some of the chemicals routinely used in the production of feed grains. Atrazine, for example, is one of the two herbicides most commonly used in American agriculture. A purchase of, say, conventional marketplace corn has a high probability of having been grown using atrazine. Though the Environmental Protection Agency has resisted calls to ban its use, many studies indicate that atrazine may be carcinogenic and that it can function as an endocrine disruptor. As such it may be implicated in human birth defects and low sperm levels in men. It might also be involved in widely observed malformations among amphibians, especially regarding sexual development. The European Union banned the use of atrazine in 2004.

What is certain is that contamination of streams and groundwater by atrazine (75 and 40 percent, respectively) is pervasive in agricultural areas tested by the US Geological Survey.[3] What is equally certain is that—through the reality of *One dollar, one vote*—if I buy corn grown with atrazine to feed my flock, I become responsible for increasing the dispersal of atrazine into the environment, and for all its effects known and unknown.

At the same time, I simply may not be able to afford to pay the higher prices for certified organic feeds, which would give reasonable assurance I had not contributed to atrazine contamination. Even the purchase of organic feeds exclusively can be problematic, if some of the ingredients are only available from far, far away. A good example is field peas or winter peas, which I use in my mixes in lieu of soy. But those peas come to me, in northern Virginia, from where they're grown 1,000 miles away, in Canada. How sustainable is that?

Even if I spend the extra bucks to buy only certified organic, I may or may not be getting feed grains that truly were grown in strict accordance with organic standards. When I talked with Kevin Fletcher, owner of my feed grains supplier, Countryside Organics, he noted that organic certification is only meaningful if the certifying agency insists on strict adherence to the organic standards. Inevitably his confidence in the integrity of the products he buys from distant growers for sale to his own customers comes down to the reputation of the certifying agency, and the level of trust he has established with individual brokers personally.

Keep in mind that feed grains of all sorts are usually grown as extensive monocultures, and tend to be destructive of soils, whether organically grown or not.

The above thoughts are not intended to make you despair of making feed decisions that are best for your flock, your customers, your budget, and the environment. They are a reminder of the severely compromised agricultural situation in which we live. Our individual decisions are likely to be ambiguous and complex.

As always, the closer we get to the source—in

Continued on following page

Continued from previous page

this case, of our feed grains—and the smaller the degree of anonymity in the transaction, the less the question of organic certification may mean to us. "Eat local" should be at least a goal, for our flocks as for ourselves.

Another question at least as important as certification is acceptance—on the part of both the grower and the purchaser of feed grains—of alternatives to the conventional standards: corn, soybeans, wheat, oats, barley. Kevin at Countryside told me he had recently attended a conference of organic farmers in Georgia. Many of them have found that, while it is possible to grow corn in quantity in Georgia using chemical means, it is extremely difficult to produce reliable corn crops with strictly organic means. So they are interested in finding feed grain alternatives that can be produced with organic methods in Georgia, while filling the needs for livestock feeds. Some of them are researching sorghum and the various millets, far easier to grow organically—and a perfect solution to their problem, so long as livestock farmers to whom they sell are willing to use them as substitutes for corn and conventional small grains. A farmer in my area is experimenting with cowpeas, peanuts, and tartary buckwheat for modest sales to other local farmers.

For me the complexities with regard to sourcing feed grains push me toward feeding my flock as much as I can out of home resources, discussed in chapter 18.

feed of questionable nutritional value, it's all the more important to supplement with vital, fresh, natural—and free—feeds like those suggested in following chapters. The difference that addition of natural feeds makes for the birds' performance is illustrated by a poultry breeder I know. During the confinement period of winter and early spring, his early hatches—from eggs of hens eating commercial feeds exclusively—suffer from low hatchability. After he releases his breeders from the winter house and they start feeding on plenty of new grass, however, hatch rates jump to 90 percent or better. If it's possible, buying feed grains from a local farmer may give you better feed ingredients; it will certainly boost the local economy and help keep that small farmer in operation. It might even be possible to form a buying club of flocksters in your area to contract with a local farmer to grow your feed grains.

The Far End of the Chicken

Dried poultry manure contains 4 or 5 percent nitrogen, 6 percent phosphorus, and 4 percent potassium. The obvious implication is that some of the expensive feed we offer our flock is being lost in transit out the other end of the chicken. That's the bad news, given the high cost of putting the feed into the near end of the chicken in the first place. But the good news is that, 365 days a year, our birds are giving our pastures and gardens a *gift*. If we treasure that gift—making sure that it gets deposited directly onto the pasture, or composted in the deep litter—then we will see that part of every feed dollar is actually an investment in enhanced *fertility* in our gardens and pastures.

When we think of the fertilizer value of manure, we usually think of the mineral values I referred to above. But remember that the remaining 85 percent or so is *organic matter*, which—as any organic gardener knows—is as much the key to soil fertility as any amount of added minerals.

A Paradigm Shift

I've described Vermont Compost Company's approach to feeding, not primarily to encourage you to seek out food residuals as a feeding resource—though that might be an option on a smaller scale for some. Nor do I imagine that more than a few of my readers have the option of ranging their flocks over a 100-acre farm, as my grandmother did. I gave these examples to suggest a paradigm shift in our thinking about feeding: All of us can find ways to increase the flock's access to live, natural foods, and an important way to do that is to find ways to transform what are too often considered "wastes" to feeding resources.

As important as any other aspect of this paradigm shift is a willingness to let the birds guide the learning process. If this implies that our chickens are smarter than the experts, so be it. If the birds resist eating something we offer them, for example, we ought to take that as a hint. I have long noticed that my chickens are not wild about peas, for example. They eat them as part of a coarsely ground mix, to be sure, but if I offer whole peas, side by side with corn and small grains, they eat the latter by preference before coming back to the peas later as second best. If I sprout the peas, acceptance increases, but they're still not a favorite. Taking the hint, I have cut back on the amount of peas I'm using in my rations. As my grub and worm cultivation projects take more of the load of providing protein, I hope to cut out the peas entirely (more on this in chapter 19).

We have already seen how providing for the *happiness* of the flock—pasturing them—also gives them access to good foods. Even use of deep litter for best manure management turns out, amazingly, to have a positive feeding benefit. And *all* the strategies for putting the flock to work are based on pointing them toward *free food.* In the following chapters we will have a closer look at commercial feeds, and the possibility of making our own formulated mixes as a replacement for purchased feed. Things get more interesting as we consider the many ways in which we can increase access to live, natural feeds right in our backyards, gardens, orchards, and woods. Finally, we'll discuss options that really bring out the kid in me: cultivating decomposer organisms in "wastes" to produce high-protein feed for the flock.

16 | PURCHASED FEEDS

Even my readers with no grounding in nutrition whatever probably understand intuitively that the meals at their grandmothers' tables—prepared with care from whole, fresh foods—were not only more delicious and satisfying, they were nutritionally superior to any processed food in the supermarket. Readers more informed about those highly artificial marketplace foods know that they are often made largely from mass-produced commodity ingredients such as corn and soy, tricked out to masquerade as traditional foods; are created targeting cheapest production and maximum profit rather than nutrition; are processed to the nth degree (with heat, pressure, drying, and other brutalities that destroy the more fragile vitamins and enzymes); and contain chemicals never eaten by *Homo sapiens* at any time during the evolution of our species. Most eaters of modern processed foods would admit, with perhaps a bit of guilt, that it's *convenience* that drives their choice, not a conviction that these foods are more nutritious or better support health.

I make this parallel with how we feed ourselves—or should feed ourselves—in order to introduce some skepticism about poultry feeds: *Why should artificial, highly processed feeds be any better for our chickens than artificial, highly processed foods are for us?* Absent such a healthy skepticism, we are apt to be suckers for certain conditioned assumptions:

- That poultry feeding is complex and opaque; and unless we want to take an advanced degree in the subject, it is best to leave its formulation to the experts—and buy what they offer us.
- That the experts know what they are doing.
- That manufactured feeds are scientifically balanced, and we cannot possibly achieve the ideal balance feeding with our own methods and from our own resources.

I question all those assumptions and suggest that you try the following rigorously *scientific* experiment to test them: Put some of that dry, dusty meal from the local farm co-op into a feeder—and throw out beside it a handful of earthworms, crickets, fresh-cut clover, mulberries, and sunflower seeds. If a single chicken even *looks* at the manufactured feed before the flock has cleaned up every morsel of the natural foods, I will eat the remainder of the feed in the bag. Assuming that *Gallus gallus* would never have survived long enough to become *domesticus* if she hadn't figured out what is the good stuff to eat—a scrupulously *scientific* assumption—the proposed experiment would suggest sufficient grounds for skepticism about "scientifically formulated" feeds.

The truth is that manufactured chicken feeds are in many ways no different from manufactured people foods:

- Many of their ingredients are things chickens have never eaten before. Indeed, it's hard to imagine anything farther from the three food groups of a chicken in the wild than feather meal, meal from soybean oil extraction, and stale vegetable oils from fast-food fryers. Such foods are anything but *alive* and *unprocessed.*

- Ingredients are chosen by their market price on the day of purchase, not by their nutritional quality.
- They are highly processed with heat, pressure, and desiccation, with damage to or loss of enzymes and the more fragile vitamins, particularly the fat-soluble ones.
- They contain chemical additives and residues whose full effects—for the chickens, for people who eat their eggs and meat, and for the broader environment into which they eventually diffuse—are not fully known.

If you plan to keep your flock confined, dependent solely on what you feed them, then manufactured feeds might seem a logical choice: As a matter of fact, they are formulated using feeding studies assuming and carried out in precisely those conditions. If, on the other hand, you are convinced about the superiority of natural feeding and management, be aware that no large-scale feeding studies have been done on such feeding practices—for example, the feeding requirements of pastured flocks. Which is another way of saying that the experts don't know much at all about natural feeding—indeed, are likely to know even less than the curious and conscientious beginner willing to abandon the industrial paradigm of totally confined flocks. The feeding practice of such a beginner will be an ongoing experiment on the cutting edge that leaves the experts far behind.

Three Formulations

Remember what I said in the last chapter about feeding as a *spectrum*, with each of our possible feeding strategies coming down somewhere along the spectrum rather than at either end. If you feed your flock commercial feeds exclusively simply for *convenience*, consider that convenience foods for chickens are not likely to be any better than convenience food for people, and explore some of the options for adding in more natural feeds. But of course, many of us—because of limitations of time, space, and foraging resources—will have to base the feeding program primarily on purchased feeds by *necessity.* If that is true for you, it's important to understand a little about commercial feeds.

Manufactured feeds are sold in three formulations, tailored to the varying needs for the major nutrients—particularly protein and minerals—at different stages of a chicken's growth.

Chick or Starter Feed

Chicks need a higher level of protein in their diet, to support their rapid rate of growth. Any commercial starter feed, therefore, will contain 20 percent protein or more, to be fed the first six weeks or so. However, if you are concerned about questionable additives in your flock's feed, chick feed is the place to start. Unless you buy a certified organic starter feed—which may not be easy to find—it is extremely unlikely you will find one that is not "medicated"—that is, one that does not contain antibiotics or coccidiostats. While chicks raised under highly stressful industrial conditions would not survive without being fed such additives, they are most certainly not necessary in the well-managed small-scale flock. Long before I started making my own feeds, I avoided medicated chick feeds entirely and to my knowledge have never had a loss to coccidiosis. Since there is no demonstrated *need* for such additives in the rearing of my chicks, I strongly oppose their routine feeding "just in case." Excess feeding of antimicrobials leads to resistance on the part of bacteria, making them less effective in the long run; may leave residues in the eventual eggs and meat, despite standard reassurances to the contrary; and eventually disperse into the environment, with unknown long-term effects.

What's in the Bag?

Here are examples of the three most typical chicken feed formulations, taken from three labels of feeds manufactured at the feed mill closest to me.

CHICK STARTER: MEDICATED CRUMBLES
Aids in the development of active immunity to coccidiosis under conditions of severe exposure to coccidiosis up to 5 weeks of age.*

Active Drug Ingredient
Amprolium 0.0125%**

Guaranteed Analysis
Crude Protein (min) 20%; Crude Fat (min) 3%; Crude Fiber (max) 6%; Salt (min) 0.2% (max) 0.3%

*Ingredients****
Grain Products, Plant Protein Products, Processed Grain By-Products, Monocalcium Phosphate, Milk Products, Calcium Carbonate, Roughage Products, Choline Chloride, Ferrous Sulfate, Manganous Oxide, Zinc Oxide, Soybean Oil, Niacin Supplement, Sodium Selenite, Vitamin E Supplement, Copper Sulfate, Vitamin A Acetate, d-Calcium Pantothenate, Vitamin D3 Supplement, Riboflavin Supplement, Vitamin B12 Supplement, Menadione Sodium Bisulfite Complex, Folic Acid, Ethylenediamine Dihydriodide, Salt.

Feeding Directions
Start replacement chicks on this feed and offer as the only feed to the chicks until about six weeks of age; at that time, switch to Grower Ration Crumbles.

Warning
Use as sole source of Amprolium

Manufactured by [name and location of feed mill]
163
0251****
1347
1116503

* Who's expecting "severe exposure to coccidiosis"? Certainly anyone raising tens of thousands of chicks in a single enclosed space. But if chicks are being "severely exposed" in your brooder, there is something wrong. *Cocci*—protozoans that cause coccidiosis—are ambient virtually anywhere chickens are found. Chicks that small-scale flocksters raise are routinely exposed to these normal numbers of cocci—and that's a *good* thing: Exposure triggers the chicks' immune systems to develop resistance to the protozoans.

** *Amprolium* is the name of the "medication" used in chick feed as a coccidiostat. If chicks *need* exposure to cocci in order to develop natural immunity, I question the wisdom of routine feeding of coccidiostats like amprolium to kill them off, whenever and wherever they enter

GROWER RATION/PULLET DEVELOPER
Grower Ration (Crumble):

Guaranteed analysis: Crude Protein (min) 16.0%, Crude Fat (min) 3.0%, Crude Fiber (max) 6.0%, Phosphorus (min) 0.55%, Lysine (min) 0.75%, Methionine (min) 0.25%, Calcium (min) 0.7% (max) 1.1%, Salt (min) 0.3% (max) 0.5%

Ingredients: Grain Products, Processed Grain By-Products, Plant Protein Products, Monocalcium Phosphate, Limestone, Roughage Products, Choline Chloride, Ferrous Sulfate, Manganous Oxide, Zinc Oxide, Soybean Oil, Niacin Supplement, Sodium Selenite, Vitamin E Supplement, Copper Sulfate, Vitamin A Acetate, d-Calcium Pantothenate, Vitamin D3 Supplement, Riboflavin Supplement, Vitamin B12 Supplement, Sodium Bisulfite Complex, Folic Acid, Ethylenediamine Dihydriodide.

the chicks' systems. And what about the wider effects? I've seen it asserted that coccidiostats don't count as antibiotics, since the latter can be taken strictly as *killing bacteria*, and cocci are not bacteria. But isn't the distinction something of a dodge? Coccidiostats are fed in order to *kill microbials*—and we'd all be wise to distrust any practice that has the side effect of releasing compounds into the environment that kill microscopic life, with long-term consequences unknown. I doubt those consequences are likely to be benign, given the strong warning farther down the label against administering amprolium to chicks in addition to that in the ration.

*** Exactly what are the ingredients in this bag of feed? The literal answer is: Who knows? The manufacturer is not required to give you that information—and indeed, it would hardly be practical to do so, since the actual ingredients vary constantly, depending on their price on any given day. So long as the manufacturer meets the nutrient percentages listed, the actual stuff making up the feed can be generalized as "Grain Products, Plant Protein Products, Processed Grain By-Products" and the like: whole grains, chief among them corn (wheat in some parts of the country); by-products of the milling of grains for flour and other industrial food products; oilseed meals from extraction of vegetable oils from soybean, sunflower, canola; cottonseed (cotton is among the crops most heavily sprayed with pesticides). Note the presence of "Milk Products"—any by-products of the processing of dairy products such as whey and powdered skim milk—but the absence of "meat scrap meal" or equivalent, which was standard when I first started feeding chicks a couple of decades ago. The scare over bovine spongiform encephalopathy (BSE or mad cow disease) has led to the omission of meat scrap meals as an ingredient in poultry feeds sold to the small-scale flockster. (They are still used to feed industrial poultry, subject to established processing standards.) Note the extensive list of vitamins and minerals. Some are provided by the actual feed ingredients, but, since those are inadequate to filling the dietary requirements, feeds are artificially supplemented.

**** We'd like for our feed to be *fresh*, so where do we see the milling date on this label? We don't—this is another piece of information the

Continued on following page

LAYER FEED

La-Mor 15 Laying Ration (Crumble):

Guaranteed analysis: Crude Protein (min) 15.0%, Crude Fat (min) 3.0%, Crude Fiber (max) 5.0%, Phosphorus (min) 0.60%, Lysine (min) 0.68%, Methionine (min) 0.3%, Calcium (min) 3.0% (max) 4.0%, Salt (min) 0.3% (max) 0.4%

Ingredients: Grain Products, Plant Protein Products, Calcium Carbonate, Processed Grain By-Products, Monocalcium Phosphate, Salt, Roughage Products, Choline Chloride, Ferrous Sulfate, Manganous Oxide, Zinc Oxide, Mineral Oil, Niacin Supplement, Sodium Selenite, Vitamin E Supplement, Copper Sulfate, Vitamin A Acetate, d-Calcium Pantothenate, Vitamin D3 Supplement, Riboflavin Supplement, Vitamin B12 Supplement, Sodium Bisulfite Complex, Folic Acid, Ethylenediamine Dihydriodide, Brewer's Dried Yeast, Dried Fermentation Solubles, Hydrated Sodium Calcium Aluminosilicate, and Methionine.

Continued from previous page

manufacturer prefers that you not have. As a matter of fact, the label is required to carry a date, but regulations helpfully allow the manufacturer to hide it in the lines of code you see at the end of the label, which record details about the batch from which the feed in this bag came, information which the mill could retrieve if required. The manufacturer assigns its own code—and when I called the manufacturer asking for help deciphering it, I was told that the date on this one was indicated by the second line: "251"; that is, this batch of feed was milled on the 251st day of the year. This bag of feed was purchased on October 27, which means it was forty-nine days old on the date of sale. In his excellent compendium on feeding naturally raised poultry, Jeff Mattocks strongly cautions about feeding stale feeds: "Feeds are at the optimum levels for up to 14 days, and are satisfactory up to 45 days after grinding or milling. After 45 days the feed is generally so stale or oxidized that poultry appetite will be severely depressed. Oxidation starts immediately after the grinding or cracking of the grain."[1]

Grower Ration or Pullet Developer

Requirements for protein decrease as the chicks grow, so the second-stage feed—grower ration or pullet developer—contains less protein than starter feed but more than feed for adult layers, about 16 percent. It is recommended for feeding growing chickens from six to eighteen weeks of age.

If you're wondering how to provide for the higher level of protein needed by brooder chicks if you *avoid* medicated chick feed, note that, when I was feeding commercial feeds, I always started my brooder chicks on the stage-two feed from the beginning—and boosted dietary protein with crushed hard-boiled egg, raw beef or deer liver, and fish meal. For years we maintained a productive flock of laying hens and ate many fine broilers and roasters, all started that way.

Layer Feed

Layers need less protein than growing birds, but they need more minerals. A typical layer feed will therefore contain about 15 percent protein, and considerably more phosphorus and especially calcium than in chick feeds. Some layer feeds—higher in protein, say 20 percent, and more densely fortified with minerals—are intended to be fed mixed one-to-one with whole feed grains.

An important point to remember about layer feeds is that *they should never be fed to growing chicks*. The additional mineral content of the layer feed can interfere with proper growth of the chicks, particularly development of the reproductive organs. Begin feeding the layer formulation to pullets when they are nineteen or twenty weeks of age, just before onset of lay.

Organic Feeds

Some readers may have access to certified organic feeds. Though "organic" as a label has come to be a bit more slippery than it was initially, following adoption of the National Organic Standards, an organic feed will not be medicated. Even if it is organic, make sure it is fresh—as said earlier, feed starts to go stale as soon as the feed grains are crushed and should be used by at most a month later. Check with your source about the date of milling.

The Good News

There's plenty of bad news about commercial feeds. But the good news is that your eggs and table chicken will still be an order of magnitude better than supermarket versions. First, feed sold at your co-op for home flocks, with all its flaws, is likely to be superior in quality to the debased feeds used in huge industrial flocks. Second, *every* flockster has the opportunity to feed *some* natural feeds to improve the diet of the flock—with better results if only because the enhanced enzymes in the diet boost the birds' utilization of the bagged feed. So I urge you to explore the many ideas in this book for providing your flock all the natural foods you can within the limits of your own situation; that is, I invite you to a never-ending experiment, which will save you money and improve the diet of your birds, but which is also a kind of play—it's *fun*.

And you probably won't trouble your mind with questions like, "Hmmm, I wonder if they're getting enough hydrated sodium calcium aluminosilicate."

17 | MAKING OUR OWN FEEDS

Making your own feeds may be a possibility for you. To be sure, doing so requires considerably more effort than scooping something out of a bag. So why do I go to the extra effort to make my own feeds? For Ellen and me, there is *no* room for compromise regarding feed quality—if "you are what you eat," then by eating the eggs and chicken on our table, we are what our chickens eat. The feed I make, using certified organic feed grains and supplements exclusively, is fresher and of higher quality than anything I can buy. And as an inveterate tinkerer, I can experiment in response to new ideas, and fine-tune my formulations to target specific groups of birds as needed.

In earlier discussions of feeding, I noted that the major characteristic of natural feeds is that they are *alive*. If we stick mostly with whole seeds, we ensure that our prepared feeds are as close to being alive as possible. As someone whimsically observed: "A seed is a tiny plant, in a box, with its lunch." Any one of my feed grains can be planted to grow into vigorous plants—indeed, if I'm sowing a cover crop of wheat, oats, or peas, I simply draw the seeds from the feed bin and plant.

I could buy organic feed mixes available from my supplier. However, delivery is once a month, so their feeds at the end of that month would not be as fresh as those I make myself every few days. Also, I have found their mixes too finely milled, necessitating having to deal with the "fines" more frequently. (If you remember from chapter 5, these are the feed residues that sift to the bottom of the feeder, where they can become stale and moldy if ignored.)

Should you assume, as most people do, that I make my own feeds in order to save money, let me give you the bad news: Even if you make your own feeds, your chicken and eggs will cost you more than the industry offerings in the supermarket. If you make your own feeds and plan your ingredient sourcing carefully, however, you can doubtless feed your flock more cheaply than with manufactured feeds for the home flock sold at the local co-op. But even if you make your own feeds from *scratch*, if you insist on using certified organic primary ingredients, you will pay more to feed your flock than if you buy already made co-op feeds. Thus *even though we're growing our own eggs and dressed poultry*, they cost Ellen and me *more* than if we bought supermarket poultry products or fed with co-op feeds. Only those who are as uncompromising on food quality as Ellen and I are would be dopey enough to accept that bargain.[1] But remember: "You get what you pay for."

The only ways out of this feed cost conundrum are possibly to source your grain purchases locally, cutting as many middlemen out of the cost equation as possible; to grow your own grains if you have the space, time, and resources; and most important, to maximize the portion of your flock's feed self-foraged on your own place.

Equipment

When I started making my own feeds, I ground the feed grains with a small manual flour mill—a testament more to my stubbornness than to my good sense. Unless you have a micro-flock, you probably won't continue making your own feeds with such a mill for long. When I got serious about making my own feeds—currently a couple of tons a year—I bought a grain mill heavy enough to grind 25 pounds of feed in a few minutes, powered by a 1½-horsepower electric motor purchased locally at an electric equipment supply.[2] This heavy-duty mill requires a solid mount—I bolted mine and its motor to one end of a heavy workbench (see figure 17.1).

You will also need a feed scoop, available from your farm supply (or you could fashion one yourself with a recycled plastic jug with the base cut off), and a scale for weighing feed ingredients—I use an inexpensive, spring-operated, 10-pound hanging kitchen scale, available from a good hardware or kitchen supply. Unfortunately, I have not found such scales durable and have had to replace mine a couple of times. If I find a better version I will buy it, though I have no interest in electric or battery-powered scales.

Fig. 17.1 The heavy-duty feed mill I use for making feeds.

Storing Feedstocks

I made my storage bin from plywood left over from an addition to our house. At 5 by 4 by 2½ feet it will hold three-quarters of a ton of feed. Three partitions divide the interior into four bins—one each for wheat, corn, oats, and peas. The floor of the bin is on a 45-degree slant, creating a gravity flow to the sliding gates mounted on the face of the bin at the bottom.

A grain bin must be rodent-proof. I've seen no evidence of mice trying to chew through my bin's three-quarter-inch plywood, though rats might be able to do so. Rats chewing on a wooden bin could be thwarted with metal flashing or quarter-inch hardware cloth.

It is harder to guard against two insect opportunists in the feed bin—grain weevils and meal moths—whose eggs come in with the purchased whole grains. Don't assume that, because they'll be eaten by the chooks, their presence is no problem—feeding on stored grain by these insects degrades its nutritional quality. There are two things you can do to minimize population levels:

- Sprinkle into the bin a cup or so of diatomaceous earth for every 50-pound bag of grains.[3] *Wear a dust mask!* The diatomaceous earth wears holes in the exoskeletons of adult weevils, dehydrating them, and coats and smothers the larvae, who breathe through their skin. Don't worry—the DE can be eaten by the chickens without harm. Indeed, it's sometimes administered to livestock to treat internal parasites. Since the weevils and moths are inactive in winter, there is no need to add DE then.
- When you see a lot of weevil or moth activity in the bin, use up your stored grains, then clean the bin thoroughly. I remove the interior partitions so I can climb inside my bin with a stiff brush and my shop vacuum (always wearing a dust mask), and remove every last speck

of grain residue, which may contain moth or weevil eggs. If I'm thorough, I only have to do this deep cleaning once, at most twice a year.

Ingredients

I do not use any refined products such as gluten meal, oilseed meals from oil extraction, nor by-products such as feather meal from slaughterhouses. With the exception of a few supplements, I buy whole feed grains exclusively. Note that I am lucky to be in the delivery zone of an excellent supplier, Countryside Organics of Waynesboro, Virginia.[4] Many would-be feed makers I correspond with report terrible frustrations finding some of the feed grains that I buy from Countryside, or convincing indifferent feed dealers to special-order them. Finding high-quality ingredients can become a challenging foraging expedition.

Below is a list of ingredients that I've used or are used by flocksters of my acquaintance who make their own feeds.

- **Corn:** A carbohydrate ingredient that in typical mixes supplies a major part of the energy requirements of the chooks.
- **Small grains:** I add wheat, oats, sometimes barley and rye, always whole. If I could get other small grains I would use them as well—the greater the diversity of feed ingredients, the better. The high fiber of the small grains helps keep the digestive tract efficient. Sometimes, especially during winter, I hold the small grains out of the mix and sprout them prior to feeding.
  - **Wheat:** In some regions wheat might be more readily available for use as the main carbohydrate part of the feed. If that is true for you, note that you should not feed wheat at greater than 30 percent of the diet, in order to avoid digestive problems.
  - **Oats and barley:** Do not feed these grains at greater than 15 percent of the total diet, either individually or in combination—excess consumption causes runny droppings.
  - **Other small grains:** If you have a source for other small grains—such as spelt, an ancient species of wheat, or triticale, a hybrid of wheat and rye—try them in your mixes. If they are good people food or raised for feeding other livestock, you need have no concerns about experimenting with them as poultry feed.
- **Soybeans:** When I first started making my own feeds, I added soybean meal as a by-product of oil extraction to furnish much of the needed protein. When I started getting my ingredients from my current supplier, Countryside Organics, they provided whole roasted soybeans instead, which I cracked when making up batches of feed. In response to concerns from its customers, however, my supplier ceased use of soy altogether in its feeds, substituting field peas instead. It has now been almost a decade since I used soy of any sort in my feeds.
- **Peas:** For many years now I have used field peas or winter peas (*Pisum arvense*—a relative of the garden pea, *P. sativum*) in lieu of soy in my mixes. (Field peas have 22 to 25 percent protein; soybeans, from 30 to 50 percent.) But I have misgivings about that usage as well, given that they come to my feed bin from as much as 1,000 miles away—how sustainable is that?
- **Sunflower seeds:** Some of my correspondents report adding sunflower seeds to their mixes, most often the black oilseed type sold to fill wild bird feeders. They provide proteins, fatty acids, vitamins, and minerals; like small grains, they also furnish dietary fiber to keep things moving through the gastrointestinal system.
- **Alfalfa:** Dried alfalfa can be used as a green addition to feeds in winter. You might start with 100 percent alfalfa pellets (17 percent protein), the kind fed to rabbits and horses;

grind them along with the corn and peas, since chickens resist eating the pellets whole. Note, however, that because the pellets are highly processed in comparison with good alfalfa hay, a better option is simply to feed the latter if you can get it. Suspend the hay in a net, and let the chickens pick out the leaf portions. Feed in small amounts to prevent accumulation of uneaten excess—with its high nitrogen content, it will quickly heat up in the litter and may generate ammonia.

- **Flaxseed:** *Flax* has become something of a buzzword because it boosts omega-3 fatty acids in egg yolks—though so does eating green plants and live animal foods such as earthworms and insects. In our modern diets we tend to get far too much omega-6 in proportion to omega-3; thus any way to get them into better balance is desirable. I no longer feed flaxseeds myself—I try to boost omega-3 production by maximizing access to natural feeds, year-round—but if you do: *Feed them whole*, rather than as purchased ground meal, because flax oil is highly perishable and will go rancid after the seeds are crushed; *do not add too much*, as flax can impart a smell like paint thinner to eggs and meat if fed at greater than 10 percent; and make sure the birds have *plenty of grit*—the seeds are small and hard.
- **Supplements:** In addition to the feed grains that make up the bulk of a feed mix, there are a number of possible supplementary ingredients for boosting protein or mineral content.
  - **Meat scrap:** You may read references to meat scrap meal in discussions of poultry feeds older than thirty years or so. Since the mad cow disease disaster in England in the 1980s, however, there are stringent restrictions on meat scraps in livestock feeds. Though meat scrap meal is permissible in the feeding of industrial poultry, when they meet processing standards, manufacturers of poultry feed for the home and small-farm flock do not want to deal with the headaches involved. As a practical matter, you will see no meat scrap meal as an ingredient in commercial poultry feeds—nor will you find it available for purchase as a feed supplement.
- **Fish meal:** Menhaden is a species taken in quantity not as a food fish but for turning into a high-protein—60 percent—addition to livestock feeds. It is usually recommended not to exceed 5 percent of the diet as fish meal, to avoid fishy off-flavors in eggs or meat. Most people I know who are making their own feeds use fish meal—it's hard to make feeds that are high enough in protein without it, at least if you want to avoid refined or highly processed alternatives like pure lysine from corn—especially for growing birds, whose protein needs are higher than for mature fowl. (Like flax, fish meal also boosts the omega-3 content of egg yolks.) However, it's important to ask: How sustainable is turning countless thousands of tons of fish into feed supplements? Furthermore, though it's a potent source of protein, fish meal is not a fresh, live food, so it will never be as good a food as some of the alternatives discussed in the next two chapters. I am especially interested in the possibility of small-farm conversion of organic "wastes" into high-protein replacements of fish meal. My own cultivation of earthworms and soldier grubs as live protein feeds is driven in large part by their promise as an alternative to fish meal; I try to encourage farmers to give this idea a whirl every chance I get.
- **Crab meal:** Dried, crushed shells from commercial processing of crab and lobster meat are a source of protein (about 25 percent or so) and of minerals like calcium. Though it's a good source of selenium as

well—an essential trace mineral in which many of the nation's soils tend to be deficient—it should for that very reason be fed in small amounts: Selenium is one of those vital minerals needed in trace amounts that actually become toxic at greater concentrations. I limit crab meal to 2 pounds per hundredweight of feed.

- **Cultured dried yeast:** This supplements not only protein (18 percent) but a number of minerals and vitamins, especially the B complex. It is particularly useful in feeds for waterfowl, whose need for B vitamins, and especially niacin, is greater than that of chickens. Contains live cell yeast cultures, which become active in the gut. Together with digestive enzymes in the dried yeast, the live cultures enhance feed digestion.
- **Probiotics or direct fed microbials (DFMs):** It is possible to add minuscule amounts of live microbial cultures as a supplement to boost the microbes in the gut, making it theoretically more efficient. This is another area where the science has been formulated with reference to a seriously flawed paradigm: chickens in high confinement without green forage or live animal foods, eating instead stale feeds based on highly questionable ingredients, birds with compromised genetics to begin with—*of course* the digestive tracts of such birds need as serious a boost as we can provide them. But birds eating more natural foods are likely to have healthier, abundant, and diverse populations of intestinal microbes to begin with and perhaps need little additional boost from us. I stopped using probiotic supplements long ago, as I developed strategies to provide natural foraged feeds year-round.
- **Mineral mix:** Just as makers of manufactured feeds supplement with added minerals, you can buy a mineral mix to supplement your homemade feeds. The best one I know of is Fertrell's[5] Poultry Nutri-Balancer, which I buy in their organic formulation and add to both grower and layer mixes. Not only is Nutri-Balancer a broad-spectrum mix of minerals and vitamins, it contains live *Lactobacillus* spp. cultures that become active in the

Feeding Legumes

At one time farmers grew legumes adapted to their own regions to feed their livestock. Then came the "soybean revolution" of the 1940s, after which the high-protein, widely adapted soybean replaced virtually all other legumes as livestock feed.

More and more people are becoming uneasy about the heavy use of soy in our food supply.[6] Many question the use of soybeans and soy by-products even in animal feeds. Certainly this concern is justified with reference to feeding ruminants—soy feeds have serious deleterious effects in the rumen—though I am less certain about problems of feeding soy to avian species. There is no question that soy is one of the most problematic of all feeding legumes in terms of anti-nutrient factors—components that depress either digestion of feeds or growth. Sprouting or heating soybeans neutralizes a good deal of the anti-nutrients—there is disagreement whether either gets rid of all. In any case, it is universally recognized that livestock should *never* be fed raw soy.

If you want to use soy as a feed ingredient but cannot get whole roasted beans, the most readily available alternative is likely to be soybean meal as a by-product of extraction of oil, used both for industrial applications and food oil. Unless the oil was expeller-expressed—which is not likely—extraction was accomplished using hexane, a potent carcinogen. Hexane is highly volatile, and it's claimed that it is entirely driven off by the end of the extraction process, leaving no residues. Reassured?[7]

Hardly reassuring is the fact that at this point most soybeans in the marketplace are genetically modified.

For years I have used winter pea or field pea (*Pisum arvense*) as the high-protein legume in my feed mixes. However, it is my understanding that the supply of field peas in North America is tight, and I frequently hear from flocksters who want to use field peas that they are unable to find a source. A further dilemma where I live, the mid-Atlantic, is that they cannot be grown in quantity as a seed crop—though I grow them all the time as a cover crop—necessitating their transport from hundreds up to a thousand miles away. I'm always looking for locally grown alternatives.

All legume seeds have some anti-nutrient factors—indeed, some of them are highly toxic (sweetpea and vetch), while others are toxic if fed raw (kidney beans)—with soy at the top of the list. However, some legumes are so low in anti-nutrient factors that they *can* be fed raw—field peas, garden peas, lentils, chickpeas, cowpeas, and others. In the case of avian livestock, anti-nutrient factors in these legumes are likely to be neutralized by the soak in digestive fluids in the crop. More use could be made of these legumes for feeding livestock.

Feeding legumes almost as widely adapted as soybeans are various subspecies of *Vigna unguiculata*, among the most important food legume crops in many parts of the world—Asia, Africa, Southern Europe, and Central and South America. Best known in the United States are cowpea and blackeye pea, with an average protein content of about 24 percent. Flocksters in search of alternatives to soy might seek out local farmers willing to grow cowpeas, blackeyes, or other feed legumes. If home feed makers and small producers of broilers or eggs for market band together, they can offer farmers a significant guaranteed market for such a crop. There may also be opportunities to expand similar arrangements into local production of other feed crops such as corn and small grains, superior in quality to the run-of-the-mill alternatives, perhaps organically certified and non–genetically engineered.

gut. Mineral supplementation is another area where I expect that birds fed a more natural foraged diet have a greatly reduced need for nonfood supplements. However, the amount of forage my birds get varies considerably in different seasons, so at present I do continue to add Nutri-Balancer as just-in-case insurance. At some point in my continuing efforts toward providing mostly natural feeds for my flock, I may cease doing so.

- **Feeding limestone or aragonite:** Even if you are adding a good mixed mineral supplement such as Nutri-Balancer, it's a good idea to increase the calcium available in a layer mix—needed by hens for good eggshell quality. An easy way to do so is simply to add aragonite or feeding limestone to the mix. (Both are different forms of pulverized limestone—calcium carbonate.)

- **Kelp meal:** Dried seaweed meal from the coast of Iceland or Maine is an excellent natural source of minerals. If you're not using a mineral mix, kelp meal—added to your feeds or offered free-choice in a hopper—can supplement a wide range of minerals, as well as several important vitamins and amino acids. I've used two Icelandic sources of kelp meal, Thorvin and Thorverk, both certified as organic and sustainably harvested.
- **Salt:** An essential nutrient for chickens, but usually supplied in sufficient quantity in commercial mixes or a supplement such as Nutri-Balancer. If you're not using either, a high-quality livestock feeding salt includes many trace minerals in addition to sodium chloride.

- **Grit and oyster shell:** Though I don't add these ingredients to my feed mixes, remember that grit is essential for processing feed by fowl of all species and ages, especially those eating whole grains rather than more finely milled feeds; also, crushed oyster shell can provide the extra calcium layers need to make their eggshells. Birds on pasture may get enough grit and mineral on their own. The more confined they are in the winter, however, the more important it is to make sure they have enough by offering granite grit and oyster shell free-choice. Since both grit and shell are cheap, I offer them free-choice year-round.

Putting Together a Homemade Mix

Beginning flocksters tend to be nervous about making their own feeds. Remember that, with a curious mind and a bit of research, you will know more about natural feeding than the experts and can experiment on your own. And the key word is *experiment*—I have been making all my feeds for more than ten years, and I still make frequent changes to my mixes. Indeed, making my own feeds allows a level of *flexibility* I wouldn't have using purchased feeds: I can target a mix to specific species, ages, or types (meat or layer) of fowl, keyed to the season of the year, and to the availability or scarcity of foraged foods.

Formulating a Mix

I recommend basing your feeds on the greatest diversity of feed grains available to you. The more you rely on whole grains as the foundation of your feeds, the less fine-tuning you'll have to do. For example, I started out using complex formulas relating protein, carbohydrate, and fat, to calculate recommended nutrient ratios of the three macronutrients. I soon found, however, that—*as long as I was using mostly whole feed ingredients*—the nutrient ratios came out right if I simply targeted my formulation to the desired protein percentage. If I were using a lot of processed and by-product ingredients, of course, the calculation would become a good deal more complicated. As I discussed earlier, it's best not to feed fish meal at greater than 5 percent of the total diet. But when feeding chicks *on pasture*, I sometimes increase fish meal to 6 or even 7 percent—the additional foraged green and animal foods the chicks eat in effect decrease the *percentage* of fish meal in the *total* diet.

If you are skilled in the use of electronic spreadsheets, you will find it trivial to do as I did: design a spreadsheet for automatically calculating values such as percent protein and cost as I enter varying amounts of ingredients.

Premix

If you add half a dozen different supplements to your feeds, in small amounts per batch, feed making will be more efficient if you combine them all into a *premix*. That is, weigh out enough fish meal, aragonite, kelp, et cetera to supplement, say, 500 pounds of feed. Mix all thoroughly—I just scoop it back and forth between two bins until the mix is even—and store in a covered bin. Now all the supplements are ready to scoop and weigh as one single ingredient as you are making each batch.

Grinding and Mixing

I make feed in 25-pound batches. I grind corn and peas only, setting the grinder to coarse—to crack each corn kernel or pea into several pieces, rather than grinding them into a meal. Then I dump in the premix and small grains and run the next lot of corn and peas through the grinder. When I've layered 50 or 75 pounds of ingredients in this way, I scoop the feed back and forth three times between two bins to mix thoroughly.

Keeping Feed Fresh

You don't have to worry about freshness in your stored grains if you keep them dry, preventing mold—conditions provided in a wooden bin like mine—and *if you keep weevils and meal moths in check*. These insects chew holes in stored grains to lay eggs inside (adults) or to feed (larvae). Once the seed coat is disrupted, the carbohydrates inside oxidize—that is, become stale—and the feed is less palatable and harder to digest.

Make feed in small batches. Once the seed coat is crushed, the seed's more perishable nutrients begin to oxidize and become less available. Feed should remain at an optimal level of freshness for two weeks—and should be fed within thirty days, certainly no more than forty-five. I generally make about a week's worth of feed at a time.

Store prepared feed in a rodent-proof bin with a tight lid. I use large galvanized trash cans. I've read recommendations to avoid metal bins, which may be subject to condensation in certain temperature conditions. While it's true I occasionally see some condensation on the underside of a lid, I use the contents far too quickly for mold to develop. I would not, however, use metal bins for long-term storage of feed grains.

Sample Recipes

At this point I will offer some feed recipes, but only on the understanding that these are *sample* recipes,

Table 17.1: Feed for Traditional Breed Chicks on Pasture (18% Protein)

Ingredient	Amount /100 lb	Amount /25 lb
Premix:		
Aragonite	1	
Nutri-Balancer	2	
Kelp meal	0.5	
Fish meal	7.5	
Crab meal	2	
Cultured yeast	2	
Total Premix:	15	
Premix per 25 lb:		3.75
Grind/Whole portion:		
Corn	27	6.75
Peas	20	5
Wheat	28	7
Oats	10	2.5
Total:	100	25

This is the mix I fed chicks on pasture this past growing season, expressed first in pounds per hundredweight to indicate percentages, then in pounds per 25-pound batch. Since I was feeding the same mix to ducklings and goslings–who need more B vitamins–on the same pasture, I added the cultured yeast. I would have omitted the yeast if I were feeding chicks only. Note as well that the amount of fish meal added exceeds the recommended 5 percent. Since this is a feed for chicks on pasture, eating a lot of green forage, wild seeds, and animal foods, fish meal in the total diet is therefore less than 7.5 percent.

Table 17.2: Feed for Layers on Pasture (15% Protein)		
Ingredient	Amount /100 lb	Amount /25 lb
Premix:		
Aragonite	6.5	
Nutri-Balancer	2	
Kelp meal	0.5	
Fish meal	4	
Crab meal	2	
Total Premix:	15	
Premix per 25 lb:		3.75
Grind/Whole portion:		
Corn	27	6.75
Peas	20	5
Wheat	28	7
Oats	10	2.5
Total:	100	25
This mix contains less fish meal, since adult chickens need less protein, and a lot more aragonite, to increase calcium for eggshell formation.		

based on ingredients available to me, and are merely snapshots of an ever-moving target. I change the way I feed as I learn—from my own experience and that of others, from experimentation, from research, from crazy ideas. I hope you will do the same.

Table 17.1 is the feed I used for young growing chicks—and ducklings and goslings—on pasture this past growing season. Table 17.2 is the mix I fed laying chickens, with less protein and more calcium.

Alternatives to Grinding

Are there readers who are wondering, "Hey, wait a minute, *Gallus gallus* didn't have its feed ground for it, did it?" Others may wonder if they have to shell out big bucks for a heavy-duty feed mill. Isn't reduced dependence on purchased stuff what this book is all about? Such questions beg for a home feeding alternative that doesn't involve grinding.

During the past winter I have been sprouting all the feed grains I am currently feeding my flock: corn, peas, wheat, and oats. I feed the sprouts in shallow containers, sprinkling over them a small amount of mixed fish meal, crab meal, kelp meal, and crushed, dried comfrey and nettle "hay." The dampness of the sprouted grains causes the dry ingredients to stick to them, preventing their sifting to the bottom, and the birds consume most of the supplements before they finish the last of the grain.

The biggest drawback to this approach to feeding is that the chickens resist eating the sprouted peas—they do eat some of them, but they leave some. Unquestionably, they make more complete use of peas when cracked and mixed into a prepared feed.

18 | FEEDING THE FLOCK FROM HOME RESOURCES

Diversity is the foundation of biological systems, and maximizing the diversity of feeds available to our birds should be the key to our feeding program as well. I am convinced that nothing we can buy in a feed bag, with its narrow ingredients base, can match the breadth and depth of nutrition in the many natural foods the birds will forage themselves if given the opportunity.

When I reviewed the list of feed ingredients in the previous chapter, something that stood out was the number of ingredients with one *problem* or other—wheat shouldn't be fed at more than 30 percent; nor barley or oats at more than 15 percent; many legumes can create problems in the diet, some of them serious; fish meal is a high-quality protein supplement with major sustainability issues. These problems inherent in one agricultural product or another make me all the more determined that my chickens get more of the feeds they would provide themselves on their own—they're far more likely to find their way to nutritional balance. Is there a dangerous upper limit to crickets in the diet? Probably not, but even if so, free-foraging chickens are unlikely to hit it.

As always, what goes in determines what comes out. Here at Boxwood, we don't need laboratory analyses to prove that—we *know* that the quality of our eggs and dressed poultry has improved every year, in step with provision of ever-more-natural foods and self-feeding opportunities.

Home Feeding

As said earlier, I think of my grandmother's flock as the model for natural chicken feeding. Her chickens were eating their three food groups, whose defining characteristics were that they were *alive* and *raw*. Now let's think about another aspect: They were eating almost completely *from home resources*. For anyone who takes seriously the scale of the coming changes in the economy, dependence on purchased feeds is a serious dependency indeed. Increased home feeding is a possibility for every flock, wherever located, but the learning curve is steep. Why not start experimenting now? I have heard from so many who have done so and who report: "It *works*—I'm saving real money on feed bills." Which is another way of saying, "My eggs and table poultry are a lot less dependent on purchased feeds."

Fortunately, there are a multitude of the three food groups options on almost any homestead or small farm, and many ways to combine natural feeding strategies.

Green Forages

Every management strategy should provide fresh green plants to the flock *daily* if at all possible. And remember that pasturing the birds gives them access to all three food groups at once, and that even flocksters on small properties successfully "pasture" their birds on their lawns.

Fertility Plants

I grow comfrey and stinging nettle to feed the soil as mulches or composts; not only for their mineral content, but because they are high in nitrogen as well, and thus are quickly converted by soil and compost heap microbes. They can also help feed poultry.

Both these plants have a bad reputation. Many gardeners fear comfrey as *dispersive*, spreading via millions of seeds, or *invasive*, spreading aggressively via underground runners. Actually, it is neither. It sets flowers—thus supports pollinators as well—but does not make viable seeds. And it does not spread—the first patch I planted hasn't moved an inch in more than twenty years. But it certainly is *persistent*—not surprising given the reserves in its thick, fleshy roots, which grow 8 or 10 feet deep—and will not move from where you choose to plant it without a serious fight.[1]

Comfrey is child's play to propagate—I plant more of it every year. I cut and carry it to my flocks—the geese and ducks especially *love* it—or give the birds direct but temporary access to patches of it planted on the pasture (see figure 18.1).

Fig. 18.1 Ducks and geese eating comfrey.

Nettle, however, is both dispersive and invasive and therefore needs vigilant discipline to keep it in place. I cut it as soon as it starts to flower to prevent seed set, and cut it back around the edges of the patch when its roots get adventuresome. Nettle is high in protein and mineral content.

Last summer I dried some comfrey and nettle "hay," which I have been feeding during the winter as a mineral supplement, crushed and sprinkled over the birds' feed, or stirred into it.

Cover Crops

I *love* growing cover crops. If you like jigsaw puzzles or chess, you will too: Since maximizing cover cropping in both space—a garden that is being used to grow food crops—and time—the four seasons—brings so many improvements to soil fertility and texture, no cover cropping opportunity should be missed. Fortunately, cover crops can do double duty as green feeds for the flock (see figure 18.2). We can cut the greenery to carry to the flock—in which case the plants usually quickly regrow—or let the chickens till them in as they dine.

Note that I no longer bother growing fall crucifers—various mustards, winter radishes, raab, kale, rape, turnips—as separate garden crops: I just sow them as fall cover crops, and there's a gracious plenty for everybody—the soil food web, our poultry, and us.

Weeds

Weeds may annoy gardeners and landscapers determined that *nothing* grow on their place they didn't plant, but many wild plants with a mind of their own make valuable contributions to flock nutrition. If you doubt it, remember that common weeds like dandelion, lamb's-quarter, nettle, burdock, and yellow dock are higher in protein than that quintessential high-protein fodder crop, alfalfa. Poultry will eat them all.

Take dandelion, toward whose demise millions of

dollars are dedicated every year, but so nutritious as a cooked or salad green that herbalists debate whether to class it as "superfood" or "medicinal." In addition, it is a "dynamic accumulator": Its taproot grows into the deep subsoil and mines it of minerals, especially calcium, which it makes available to more shallow-rooted plants. Both the fertility plants discussed above—comfrey and nettle—are also dynamic accumulators.

Of course, such friends may not be equally welcome in all parts of the garden, orchard, and landscape, so remember that weeding chores can furnish valuable green fodder for the flock. The most useful weeds where you live will vary. Ascertain which few weeds in your area are poisonous to your flock,[2] then feel free to experiment with any and all harvests from weeding. Weeds vary in mineral content, so the wider the range of weeds available to the birds, the more likely their mineral intake will be in balance.

In most temperate areas, the following palatable weeds should be common. Many of them make fine people food as well.

- Prickly lettuce (*Lactuca serriola*), likely to show up in shady areas.
- Purslane (*Portulaca oleracea*), rich in omega-3 fatty acids, vitamins, and minerals.
- Dandelion (*Taraxacum officinale*), superfood for chooks as well.
- Lamb's-quarter (*Chenopodium album*), one of whose common names, fat hen, hints at its utility as poultry feed.
- Yellow dock (*Rumex crispus*) is rich in vitamin A, protein, iron, and potassium, though

Fig. 18.2 A mixed cover crop is fine feed for a mixed flock. PHOTO BY BONNIE LONG

it should not be overfed because of its oxalic acid content.

- And how do you suppose chickweed (*Stellaria media*) got its name? Common and prolific, both the plant and its seeds provide highly nutritious feed.

Sharing the Garden's Bounty

A useful garden crop to share with the chooks is the *beet*, but note that the one species, *Beta vulgaris*, comes in several variations, increasing its utility. We grow the common table beet as food for us, but we also grow the variants *mangels* and *chard* for the flock. Both the latter are tremendously productive, and you can progressively pull off the large lower leaves—all types of poultry *love* them—which only stimulates the plants to vigorous growth, of more tender chard leaves for us and a mammoth beetroot in the case of the mangel. I grow 12-pound roots, and that's on the small end of the spectrum.

We can share green foods from a number of other harvest crops as well. When preparing my favorite salads, the *chicories*, for the table, I lavishly trim outer leaves and any other less-than-perfect material. All my birds—the geese especially—love these discards. Lettuces and cabbages furnish coarse wrapper leaves the flock is glad to get. Spent harvest plants that are still green—broccoli and bean, pea, and sweet potato vines—and most cull fruits and vegetables are excellent fresh foods as well. Some crops we intentionally start readily reseed and show up in areas patrolled by the flock, or volunteer in the garden as more "weeds." Sylvetta arugula (*Diplotaxis erucoides*) and purslane are examples in my garden.

Squashes offer useful feeds, from the lurking monster zucchini to the protein-rich seeds scooped from the cavities of winter squash. You might even grow an oilseed type such as Lady Godiva, with its nearly naked seeds, easy for the birds to digest and rich in oils and protein. Large winter squash such as Hubbards store well for fresh feeding deep into winter.

Root crops are rarely utilized as much as they could be. In wartime England and other European countries, potatoes substituted in whole or part for scarce grains for feeding poultry. In *The Resilient Gardener*, Carol Deppe makes a compelling case for potatoes as a major contribution to independent feeding, especially of ducks.[3] Inspired by Deppe, I plan to grow a lot more potatoes in the future and increase their use in our feeding program, especially for the ducks. Do note that—though potatoes are fed raw to ruminant species such as cows—they *must be cooked* for feeding to any monogastric species (poultry, pigs, horses, or people).

I mentioned the large roots of mangels, which store well the entire winter in a protected pit in the ground—they're a welcome source of fresh food for the flock in a part of year when it is scarce and provide great entertainment value as the chickens peck them apart.

Every gardener accumulates tail ends of outdated seeds. I like to mix them, perhaps even adding remainders purchased cheaply after the seed-buying season is over, and sow a salad bar for the chickens. It's a bonus if I end up with some prime salad or cooking greens for us as well.

Grass Clippings

If your situation prohibits bringing the flock to the pasture, bring the pasture to them: Lawn clippings—those that have not been treated with toxic chemicals—are excellent fresh forage. Short, rapidly growing grass yields the highest levels of nutrition. Do not feed too large a volume at one time—leftover clippings can accumulate into an anaerobic, slimy mess. If you are using your birds to work compost heaps, however, you needn't worry about feeding too many clippings—the chickens will work whatever they don't eat into the heaps.

Seeds and Fruits

A good deal of the feed we purchase for our flocks is grain, but many wild seeds are equally nutritious and are free for the taking if the birds are given the opportunity to range. As an experiment, I recently offered my laying flock, side by side: cracked corn, cracked peas, wheat, and oats (the four main ingredients of the feed mix I make for them); purchased wild bird seeds (seeds of black sunflower and niger, a type of thistle); buckwheat, clover, and annual rye grass seeds; and seeds stripped off mature pasture grasses. The birds preferred some over others but within the day ate almost every last seed.

Easier Grain Crops

As for conventional feed grains, most are easy crops, and we small-scale flocksters can grow them just as well as big farmers who produce them for feed markets. If we try to do so on any significant scale, however, we are likely to find harvesting, threshing, and storage to be serious obstacles. So it makes sense to look for more simplified strategies for using these crops.

I have grown a *dent corn* called Hickory King for years, and planted others this past season, including Bloody Butcher and Reid's Yellow Dent. I allow the ears to dry on the stalk, husk, and store in rodent-proof bins. I use a simple hand sheller to take the kernels off the cobs for winter feeding.

I also grow *sunflowers*, both the big single-head types (Mammoth Grey Stripe) and the multiheaded ones with smaller flowers (Autumn Beauty). I harvest the whole head just as the seeds ripen and either throw them to the chickens or string from rafters in the chicken house, to protect from rodents, reserved for later feeding. This year I planted a black oilseed sunflower, Peredovik, more useful as a feed seed. Wild birds—and squirrels—like sunflower seeds as well; I monitor the ripening heads daily and harvest as soon as I see evidence of raids on them.

Sorghum (*Sorghum bicolor*) is—like its cousin, corn—a member of the grass family, and as with corn, various strains have been specialized for different purposes: syrup making, broom making, and grain. For several years I have grown a type of sorghum called broomcorn, with big open sprays of shiny seeds, highly ornamental. As with the sunflowers, I either cut the ripened seed heads and throw them to the birds or string them up in the poultry house for winter feeding. My chickens love sorghum, whose nutrient profile is similar to corn's (see figure 18.3).

The broomcorn-type sorghum I've grown for years, with its taller habit and loose heads, is not as productive as the shorter, more dense-headed strains of grain sorghum, commonly called milo; so this past season I planted eight of those varieties. Oddly, none of the poultry—chickens, geese, or ducks—seemed more than casually interested in the seed heads of this supposedly more feed grain type of sorghum. This is a project for ongoing experimentation.

Amaranth is a plant I grow because of its beauty and its nutritious seeds, millions of them per seed head—extremely tiny, but the Aztecs built an empire on their high-protein nutrition. It readily reseeds. For years now I haven't bothered to plant it—I just thin out the abundant volunteers where they're in the way, welcome the ones that fit in, and cut the mature seed heads for feeding to my mixed flock.

There is a sweet spot for the four crops mentioned above—corn, sorghum, sunflowers, and amaranth—when the seeds are ripe but the leaves are still green. If I cut the entire plant at this stage and throw it to my mixed pastured flock, the ducks

Fig. 18.3 Chickens eating mature seeds of sorghum, which is as easy to grow as corn. PHOTO BY BONNIE LONG

and, with somewhat less eagerness, the geese eat the leaves down to the stalks.

Triple-Duty Cover Crops

In the case of other excellent feed grain crops, the work of threshing and winnowing is likely to be an obstacle for the flockster dedicated to home feeding but who lacks big equipment for the jobs. The solution is to turn that work over to the chickens themselves. If you give cover crops such as small grains, buckwheat,[4] cowpeas, even clover—whose seeds are relished by chickens—enough time, they will ripen a crop of fine feed for the flock. Turn the chickens onto the crop and let them till it in while harvesting the seeds (see figure 18.4). While it is true that the proportion of certain feed grains—oats, barley, and buckwheat come to mind—should not be too high as a regular part of the flock's diet, in my experience a brief spike in these feeds as the birds till in a mature cover crop does not create digestive problems.

Your situation may be similar to mine: With just 3 acres, no way can we raise all our own feed grains. But if we get a feeding boost from seeds of cover crops we grow for soil improvement anyway, that's just gravy, right?

Fig. 18.4 The chickens in this cruiser are tilling in a cover crop of oats that has been in place long enough to set its seeds–an additional payoff for their work. PHOTO BY BONNIE LONG

Tree Crops

In more frugal times, farmers released flocks of turkeys (and pigs) into the woods to fatten on the abundance of free feed under oaks, beeches, chinkapins, and persimmons. Planting trees in the backyard with a thought to their contributions to a feeding program is worth considering as well, for the flockster planning long-term.

Acorns

J. Russell Smith points out in his classic book *Tree Crops* that acorns can be used as feed for poultry, quoting a report from England during World War II of acorns being used to replace up to half the feed ration for chickens.[5] Traditionally, farmers fattened turkeys for free on acorns. Acorn production in oaks tends to be cyclical, with relatively few acorns some years but enormous quantities in a mast year. We have just completed such a mast year as I write—the ground was practically paved with acorns. In such a year it would be easy to gather acorns for storage and feeding through the winter. If you do plan to store acorns for later feeding, select only sound acorns with no weevils, indicated by tiny holes in the shells. There is no problem with feeding the acorns with weevils directly to the chickens, who relish the weevil larvae along with the acorn meats.

Since tannin can depress feed intake and digestive efficiency, sweeter (low-tannin) acorns are better as feed. My white oaks produce large, low-tannin acorns, which I used a good deal as feed this year. Because the acorns are so large, my flock will not eat them whole, so I ran the acorns through my feed grinder. Acorns alone didn't feed efficiently into the grinder's burrs, but I found that—if I mixed them with typical scratch grains such as whole corn, wheat, and oats—the auger pushed them into the burrs more efficiently. I set the mill for coarsest grind—my purpose was not to grind the acorns finely but simply to separate the meats from their shells. When offered to the flock broken up, garnished with the usual scratch grains in this way, they were well received by all—chickens, ducks, and geese. A flockster of my acquaintance who doesn't have a grinder smashes acorns inside a stout denim bag she made from an old pair of jeans, using a 5-pound hand sledgehammer.[6]

Don't store large quantities of acorns in a plastic bucket—the trapped moisture will cause them to mold or germinate. Hang them up in a basket or a cloth or mesh bag for maximum airflow while drying.

Hazels

I have planted a number of hazels (filberts) here, both the larger European dessert type (*Corylus avellana*) and the American (*C. americana*). The nuts of the latter are small, and shelling them out for us would be tedious. I anticipate feeding them to the flock instead, putting them through my feed mill as I do with the acorns. Feeding value should be high, based on the nutrient profile of filberts as people food—protein 15 percent, high-quality fat 60 percent.

American hazel, planted as a thicket, would be great forage for turkeys.

Chestnuts

Three years ago I planted three Chinese chestnuts on the flock's pasture, as two-year bare-root saplings; and already this year, two of them produced small crops of plump, pretty chestnuts. Of course, future harvests will furnish provender for our own table, but I expect that three trees will also provide plenty of nuts to share with the birds. It's easy to smash the nuts on a rock with a small sledgehammer to make them accessible—and I expect that a denim bag would be good for smashing them in quantity. I've found that all my poultry love the fresh nutmeats.

Other Tree Crops

Mulberry trees have a wide range; are easily propagated, precocious, and almost completely carefree; and in an extended season produce an abundance of nutritious dropped fruit that chickens relish, in either full sun or the partial shade of a woods edge.

I gather the wild hickories and black walnuts growing on my place and smash them on a rock for my chickens to eat. Cultivated nuts such as pecans and English walnuts culled because of weevils can be crushed and fed to poultry as well.

Three pawpaw trees on our place produce far more fruit than we can use. I have found that my birds don't have much interest in eating them, but I use them as high protein feed in my soldier grub bins, described in chapter 19.

Live Animal Feeds

As discussed in earlier chapters, our worker chickens get a lot of high-octane feeding value foraging insects, earthworms, and slugs as they help out with a variety of projects.

Be on the lookout for other opportunities to offer live animal feeds. If you, like us, are blessed with an abundance of Japanese beetles, shake them off your grapevines or plum trees into a 5-gallon bucket with some water in the bottom—in the cool of the morning or early evening, when they are less likely to fly. Dump them out to the chickens, and step back out of the feeding frenzy. Ducks are keen on beetles as well—they're like little vacuum cleaners. If you take the time to handpick crop-munchers such as squash bugs, plop them in a bit of water as well—one of my correspondents said her chooks think they're bonbons.

Put down some scrap boards and leave them in place until they accumulate a nice assembly of earthworms, slugs, and pill bugs underneath, then flip them over and let the birds feast. Repeat in a couple of days.

When working in the garden, keep a container handy for the fat white grubs that turn up—they're excellent fare for the flock.

There are some animal species we can actively *cultivate* for feeding the flock. These allies offer such a significant boost to the feeding program, I discuss them at length in the following chapter.

Other Protein Feeds

Keeping dairy animals such as cows or goats is like growing zucchini: If you do it, you're going to have a surplus. As well, homemade dairy products such as butter and cheese generate skimmed milk and whey as by-products. These dairy foods make excellent feed—even better if you culture them with natural microbe cultures such as kefir grains before feeding. I expect your chickens would get a lot more benefit from excess milk fermented with various cultures than from purchased microbial powders or probiotics.

Eggs are high-powered feed. Of course, we are not likely to use many of our precious eggs for feeding; but cracked eggs or those that are too dirty to rescue for table use can be hard-boiled, crushed coarsely by hand, and thrown to the birds. I assure you, contrary to what you will hear, feeding eggs in this way will *not* encourage egg eating in the flock. They are especially useful for feeding young growing birds, whose needs for protein are higher.

Offal from slaughtering—of deer, lambs, goats, or other livestock—or from cleaning fish makes excellent feed. Liver, if not favored by your family, would be another protein booster for young growing birds. Blood from slaughtering, if you make the effort to catch it as the animal is bleeding out, is high-protein feed that chickens relish. I have even experimented with feeding poultry blood to my chickens without ill effect. A rogue groundhog persistently raiding the garden—or the eight squirrels I trapped last year to save my shiitake mushroom production, or the odd roadkill—can be opened up a bit with a hatchet and fed. Such windfalls can be cycled instead through soldier grub bins to become high-protein feed for the flock (more on this in the next chapter).

"But What about Gapeworms?"

More and more sensible poultry people these days recognize that *of course* our flocks will be healthier if they're out exercising and foraging natural foods on pasture in the sunlight. Earthworms are one of those high-value, live animal foods that pastured flocks will most certainly be eating, especially on pastures that are improving in quality every year.

Yet we're likely to see stern warnings about the possibility of chickens' getting parasitized by gapeworms from eating earthworms, which—along with slugs and snails—can be intermediate hosts for gapeworms. Infestation is marked by gasping, sneezing, coughing, choking, convulsive shaking of the head, and death. Sounds pretty scary. Maybe I'm stupid not to worry about the fact that my chickens are eating earthworms and slugs and snails—that I even encourage them to do so, every chance they get. To be sure, concern about transmission of parasites is a major reason I rotate my flock over the pasture, rather than leaving them full-time on a single piece of ground. But I think of the eating of earthworms as a positive contribution to flock nutrition and simply don't worry about theoretical threats. My flock has never had a problem with gapeworms as a result. Unless you know there is a gapeworm infestation on ground where you want to range your birds, my advice is to focus on natural feeding rather than worry about what can theoretically go wrong.

Sprouting for Enhanced Winter Feeding

Those of us who have a dormant winter need strategies for producing fresh green feeds even when the snow flies.

Lots of people like to sprout seeds for the table, on the quite reasonable assumption that the sprouts are more *alive*. Sprouting is a possibility for producing more lively food for the chooks as well, especially in winter. Any of the people-food sprouts are fine; the main concern is that you find seeds in bulk—*untreated!*—to keep the expense down. (Feeding sprouts from health food stores would be prohibitively expensive.) Seeds of sunflowers, alfalfa and clovers, broccoli and other crucifers, buckwheat, radishes, beans, and peas—all these salad bowl sprouts would be fine for your birds. Any grain you feed whole to your flock—corn, peas, any of the small grains—can be sprouted in lieu of adding to a feed mix.

Think of sprouting as value-added home feeding, even if you start with purchased seeds. Sprouting causes specific nutrient changes in seeds—an increase in protein, vitamins, and enzymes, though a reduction in carbohydrate content. Of course, you can sprout any time of the year, but I don't make the extra effort in the growing season, when the birds enjoy access to fresh green foods.

While any of the seeds mentioned above can be sprouted, keep in mind in the discussion of sprouting methods below that only those of cold-hardy plants are appropriate for greening up in trays at low temperatures. Even warm-weather species such as sunflower and buckwheat can be sprouted using my bucket system.

Bucket Method

If you don't require a green sprout, this is the simplest method for easily sprouting in any quantity needed. I sprout in my basement, where waste heat from the furnace keeps the ambient temperature high enough to sprout any seeds, even warm-weather ones. The floor is concrete, so a little seepage from the soak buckets is no problem.

Fig. 18.5 Bucket system for sprouting grains.

At its simplest, this method requires five food-grade plastic buckets. One is the *soak bucket*; the other four serve as *draining/sprouting buckets* once I have drilled them with dozens of small holes, in the bottoms and halfway up the sides (see figure 18.5). The size of the holes is important: They must permit the flushing through of dusty debris but not get blocked by a grain of wheat or oats.

The sprouting schedule is as follows:

- **Day 1:** Measure the grains, any mix you like, into the soak bucket and cover with water.
- **Day 2:** Dump the soaked grains into one of the drain buckets and rinse thoroughly; then allow to drain while measuring another batch of grains and covering with water in the soak bucket.
- **Day 3:** Repeat the dumping of the soaked grains into a second drain bucket, rinsing, and starting a new batch of grains in the soak bucket. *Rinse the sprouting grain in the first drain bucket as well.*
- **Day 4 and thereafter:** Now you're rolling. Each day, dump the soak bucket into an empty drain bucket and start a new batch of grains—and each day, *rinse the sprouting grain in all active drain buckets.* The daily rinsing is necessary to prevent growth of molds in the sprouting grain.

Fig. 18.6 A tray of sprouts from the greenhouse finds a customer.

If you grow sorghum, note that the easiest sprouting of all starts with soaking whole seed heads of sorghum. The heads hang together just fine and require less time and effort to sprout. As with the daily rinse of loose sprouted grains, I change the soak water for the seed heads daily. After two or three days the emergence of a white rootlet from a few grains indicates that it's time to throw the entire heads to the chooks.

Tray/Greening Method

If you have the protected space for it, you can grow sprouts on to any green stage you like (see figure 18.6). Indeed, if you time the feeding right, your birds get the benefit of both green feed, the tender sprouts starting to green up—and the sprouted grain itself, with its enhanced nutrition. I use nursery flats that I have saved from past plant purchases. It is better to start the process in this case as well by soaking the grains, at least overnight, maybe longer. Drain and spread in a thin layer over the bottom of the tray. Cover with straw, shredded leaves, even coarser parts of the litter from the henhouse—any organic material that will help prevent drying. Water daily.

If you're growing cold-hardy types such as small grains, you could green the sprouts in a cold frame. If you are willing to make the additional effort, you could even set the trays out during the day, then bring them inside or into the basement at night. If you have an unheated greenhouse like I do, that's an excellent place to grow trays of sprouts. When I put new plastic on my greenhouse this year, I plan to install hooks on the side purlins for hanging trays of sprouts, thus saving growing space at ground level.

A Work in Progress

If you get excited about the open-ended possibilities for feeding your flock naturally, you're going to have lots of fun. And if you are uneasy about feeding your flock exclusively from grain fields far, far away, experimenting based on even your wildest ideas will get you closer to self-reliance in hard times. To conclude this chapter, I will offer some ideas I've read about, am in the planning stage for, or am actively experimenting with.

Duckweed is a prolific pond plant with a wide range. If you have a pond, experiment with harvesting duckweed and feeding to your flock. Making this high-protein plant—40 percent or more, dry weight—an important part of a feeding program is more than a theoretical possibility.[7] My neighbor, another flockster, recently put in a small pond. He has seeded it with a few duckweed plants, and we plan to experiment with feeding when it has become more established.

One of my correspondents reports that she depends on brown-top and proso millets—often planted by hunting aficionados as feeding fields to attract game birds—as key components in the feeding of her flock. Once the plantings have matured their seeds, she rotates her flock over the field in long swaths outlined with electric net fencing. I began experimenting with these two species last season, as triple-duty cover crops.

The flockster can make *hay* from alfalfa and clover, even lawn clippings, and store for winter feeding.

An even more nutritious feed from such plants is *silage*, usually made by farmers with huge tractors for storage in huge silos. The basic process is simple, however, and it can be made in small lots using home methods.[8] Silage starts with a variety of green plants—good pasture grass, kale, cowpeas, or lawn grass, mixed with such weeds as chicory, dandelion, plantain, and even whole plants of small grains or corn that have set seeds but are still green. When I lived on a dairy farm in Norway, I enjoyed helping make silage for the cows from *fôrmargkål*, which I think must have been rape, *Brassica napus*. Fermenting microbes preserve the plant matter by colonizing it and excluding pathogens, while adding nutrients of their own—a parallel with the making of lactofermented foods such as sauerkraut. Well-made silage can contain 20 percent protein or more. Though silage is more typically made for ruminants, I have corresponded with flocksters who made small-lot silage for feeding their chickens, whom they report relish the stuff. Others who feed farm-scale silage to cattle and hogs find that their chickens rush in to grab whatever gets spilled.

As always, we want to pull as many moves as we can into one dance, so look for ways to combine strategies for bonus feeding benefits. We have more space in our 20-by-48-foot greenhouse than we need for winter greens for the table, so we grow cut greenery for the birds—small grains, crucifers, peas. Our multifunctional greenhouse is also the site for our big vermicomposting operation, as you'll see in the next chapter.

I have planted trees on my flock's main pasture—two mulberries and three chestnuts—for shade. The chooks will benefit as well from the mulberries, dropped in abundance in their season, and as they help control chestnut weevils, when the weevils are going to ground for the winter or emerging in the spring. Comfrey planted around the base of these trees increases soil fertility and provides grazing for the chickens, ducks, and geese.

If we need privacy screens or boundary fencing, it might make more sense to grow living hedges, which provide the ecological benefits of edge habitat—shelter and food for birds and other wildlife, enhanced insect diversity including pollinators such as bees, and more—you will never get from a manufactured fence. Figure 18.7 illustrates the possibility of a temporary privacy screen of annuals such as sunflower, amaranth, and sorghum, all of which, as described earlier, mean more free food for the flock. Permanent privacy or boundary fences could be based on goumi, goji, or

Fig. 18.7 Sorghum, amaranth, and sunflower make a privacy screen that is beautiful, supports insect and wild bird diversity, and also feeds the flock.

Siberian pea shrub, all three of which I am experimenting with at present. Goumi (*Elaeagnus multiflora*) is a relative of autumn olive (*E. umbellata*) with equally nutritious fruit, but it's not as invasive as its cousin. Extremely hardy goji or wolfberry (two closely related species, *Lycium barbarum* and *L. chinense*) can be trained either as a shrub or a vine. The tasty berries are highly nutritious—you may have seen them dried in health food stores. I have corresponded with one flockster who tells me that production of up to 100 pounds of fruit is possible from a 30-foot planting.[9] He harvests all the berries he can use, then releases his chickens into the planting to gobble up the rest. Siberian pea shrub (*Caragana arborescens*) is a tough, hardy legume shrub that has been used as windbreaks in some areas. The seeds are about one-third protein and are relished by chickens.

Remember stacking of livestock species, which can yield feeding benefits. One of my correspondents in British Columbia soaks mixed grains and peas for all her livestock, usually for two or three days. When things are busy and the grain has not soaked as long, however, she makes this observation: "Sometimes the less soaked grain goes right through our milking cow or the pigs. The chickens love this manure and pick through it for this grain. If the chickens miss the whole grain, it grows like a weed in the manure pile. The chickens and pigs love this green crop."

Do not doubt that you too can make significant advances in feeding your flock from home resources. One of my correspondents in far Saskatchewan reported that he cut his monthly feed bill from $20 to $5 when he began feeding largely the peas of Siberian pea shrubs, in miles of shelter belts in his area; home-grown alfalfa hay; and chard, fed fresh daily when abundant in summer, and dried, or pureed and frozen in serving-sized portions, for feeding the flock in the winter.

You may have noticed one thing about the strategies discussed above: Many of them come up short with regard to feeds that are rich in concentrated proteins and high-quality fats. Fortunately, we can grow our own, as we'll see in the next chapter.

19 | CULTIVATING RECOMPOSERS FOR POULTRY FEED

More energy is wasted in the perfectly edible food discarded by people in the US each year than is extracted annually from the oil and gas reserves off the nation's coastlines.

—Kerry Trueman[1]

All living things die, and all living things poop. Imagine the state of our world absent the work of specialist organisms to decompose dead plants and animals and their leavings. Happily, key species neatly dispose of such residues, including those that may be vectors of disease, such as corpses and excrement.

"Waste" is not a concept that nature recognizes—*all* organic residues become food for other players in the ecology, who are then fed on by others. The end of the process is *humus*, soil carbon that no longer serves as food for anybody, but which is the heart of spontaneous accumulation of soil fertility over the millennia.

"Waste" always represents a misuse of a potential resource. It is estimated that 12 percent of what we send to our landfills is uneaten food. Add in compostable organic materials such as leaves, lawn clippings, and prunings, and the figure climbs to 25 percent at least. Livestock manures from high-confinement production systems are rarely fully recovered to enhance soil fertility. If we see agriculture as an energy system, failure to convert organic "wastes" represents a sluicing of potential energy resources down a black hole from which they do not return. It is especially tragic that the lost resources end up as toxins in the ecology—increased atmospheric methane and carbon dioxide, and pollution of groundwater, streams, and oceans.

Imitation of nature's use of decomposer organisms—I have taken to calling them "recomposers"—offers us an alternative approach to responsibly managing organic "wastes" and preventing their contamination of the ecology. And this exercise in white magic offers yet another benefit: free, high-quality feed for our flocks. Imagine that we feed 100 pounds of food residuals to a colony of soldier grubs, who reduce it to 5 pounds of high-fertility soil amendment, in the process generating 10 and possibly up to 20 pounds of grubs as live protein feed for poultry or other livestock—in addition to liquid effluent, which can be used like a manure tea to feed crops. Hey, wait a minute—what happened to the "wastes"? *There is absolutely no waste remaining after this conversion—it has* all *been transformed into valuable resource.*

We've already seen examples of getting feeding benefits from recomposers breaking down "wastes." The chicken-powered compost heap comes alive with recomposers busy at their work: shredders such as earthworms, pill bugs, millipedes, crickets, slugs, and snails; predators such as ground beetles and spiders; fungi, whose rhizomorphs chickens love; and fat white grubs of various sizes (pupae of many insect species, awaiting metamorphosis into the winged phase), who serve a number of roles in this teeming ecology but also serve as high-protein fare for the lucky chicken who finds them. Recomposers at the microscopic level include thousands of species of bacteria, protozoans, yeasts, actinomycetes, and more—which provide "free food" nutrients such as vitamins K and B_{12}. And remember the hundreds of laying hens getting *all* their feed eating recomposers in Vermont Compost Company's compost heaps.

For the adventurous, however, there is an option beyond giving the flock access to recomposers that arise naturally in a compost heap—it is possible to *cultivate* some decomposer species as live protein feeds. I work with three of them.

Feed from Worm Bins

Mary Appelhof popularized the idea of "domesticating" earthworms as allies in the recycling of organic "wastes,"[2] educating her readers about the life cycle of composting worms, optimal conditions needed to cultivate them, and instructions for setting up worm bins. Since that time thousands of earnest recyclers have fed their kitchen and table scraps to partner worms, who turn them into one of the most valuable of all soil fertility amendments—worm castings (earthworm poop).

Inspired by Appelhof's book, I made my first vermicomposting bins from 55-gallon steel drums, cut in half with a torch and cradled on their sides on cinder blocks. The basement was the ideal protected location for the project. Working with the worms was great fun, though the bin did generate a good number of fruit flies.

Vermicomposting in the Greenhouse

My next attempt, after we moved to Boxwood, was in a 3-by-4-foot bin, stocked with 10 pounds of redworms ordered through the mail. I used coir, the granular residue of long-fiber extraction from coconut husks, as bedding and kitchen scraps as feed. (More local options for bedding include straw, leaves, and shredded paper.) I dug the bin into the earth floor of our 20-by-48-foot greenhouse. In a climate with winters as cold as ours in Zone 6b, bins in exposed spaces are apt either to freeze in winter or to stay close enough to freezing that the worms' activities slow down to a minimum. I found that, with the bin dug into the earth inside a greenhouse, there was never a dormant period in the composting cycle. However cold the ambient temperatures—or however hot, in summer—the worms in their earth-protected bin continued to feed and reproduce.

I experimented with harvesting worms from the bin to feed my young growing birds, who feathered out earlier and grew more vigorously. That contribution to the flock's nutrition made me greedy for more. A "worms eat my garbage"–sized bin no longer satisfied my ambitions—I wanted to scale up to worm composting sufficient to provide worms as a significant part of my feeding program. In the fall of 2005, replacement of the perimeter foundation of our greenhouse provided the opportunity to take that next step.

We dug out a space for "vermi-bins," 4 feet wide and 40 feet long, right down the center of the greenhouse. Since we needed that central access anyway, we didn't lose much growing space to the new bins (see figure 19.1).

We lined the perimeters of the bin space with

Fig. 19.1 A greenhouse is an excellent site for vermicomposting. Plywood lids over my bins provide easy access down the center of the greenhouse, so I don't lose much growing space.

4-inch hollow concrete block, two courses deep, and placed a cross-wall of block every 8 feet. The result was a series of five bins, 4 by 8 feet each, 16 inches deep (see figure 19.2).

We made lids for the bins from three-quarter-inch plywood on 2-by-4 framing. Since a single 4-by-8-foot lid—one made from a single sheet of plywood—with such hefty framing would be too heavy to move conveniently, we made each lid a more manageable 4 by 4—that is, one sheet of plywood cut in half. The result was two lids over each 4-by-8-foot bin, creating for management purposes two 4-by-4-foot sections per bin, but with no partition between them. Remember that concept—it is the key to some of the management practices we've put into place.

Note that there is no floor in the bin other than the packed red Virginia clay with which we are blessed. This is not a problem, given the species of composting worm we work with and the fact that our clay soil drains well despite its compaction. Note also that our block-walled bins inside the greenhouse have had no incursions of worm-loving moles, which are sometimes a problem for worm bins outside.

Fig. 19.2 The worm bins run 40 feet down the center of the greenhouse. Cross-walls every 8 feet make a total of five 4-by-8 bins like this one, 16 inches deep.

Feed/Bedding

In some discussions of vermicomposting you will see a distinction between "bedding" and "feed" for the worms. In past practice I observed no such distinction: I filled the bins exclusively with pure horse manure from a neighbor who breeds and boards horses. Note that qualifier *pure*: If the manure is mixed with hay, straw, or pine shavings, it will heat up, as in a compost heap—a disaster for the worms. Pure manure will not heat up, or only slightly so. It is an excellent medium for the worms. They live in this medium (using it as bedding) while converting it (using it as feed) into castings.[3]

Manures of ruminants—sheep, goats, llamas—all make excellent worm feed. Cow manure should work fine as well, though its higher moisture content might require some changes in management. I know of one huge, and profitable, vermicomposting operation in North Carolina—where there are a lot of high-confinement pig operations—based on pig manure. Rabbit manure has also been used for worms, sometimes with the bins directly beneath the rabbit hutches to take "incoming."

Another highly efficient feed is kitchen scraps. In this frugal household, that doesn't mean wasted edible food—only peelings and trimmings, tea leaves and coffee grounds, and the like. If you get the chance to receive food residuals from restaurants or schools, however, they make excellent feed for worms.

Other materials can be used for worm bedding and/or feed: shredded newspaper or cardboard, discarded paper towels, and weeds or crop residues from the garden. The worms process this material as well but take longer to do so.

Worm Species

If you've ever dug into the center of an aged heap of manure, you saw *red wrigglers* or *manure worms* at work. It is this type of worm—not the soil-burrowing night crawler types you find when digging a garden bed, or in a bait shop—that is used in the rich, dense feeding medium in vermicomposting bins.

Eisenia fetida, shown in figure 19.3, is the species most often used. This type of earthworm will not burrow down into the compacted soil under the bin, nor will it crawl out the top of the bin. Conditions in either direction are not as compatible, as inviting, as the rich feeding medium in the bin.

Fig. 19.3 *Eisenia fetida.*

Setting Up the Bins

Populating my bins began with hauling in horse manure by the pickup load from my neighbor's place and loading all five bins. In drier parts of the manure, I used a wand on a garden hose to adjust the moisture, aiming for a medium that was neither uncomfortably dry for an animal whose entire body is covered with a wet skin nor sopping wet, a condition that would drown the worms.

After waiting a couple of days to ensure there would be no significant heating, I seeded the bins with worms from the small bin I had maintained several years. If you do not have any other source, starter worms are available for purchase online. They are expensive, but you only have to buy them once—after that, they will be fruitful and multiply, and you should have enough for all future needs. Alternatively, you could visit the aforementioned aged manure heap and dig out a seeding of manure worms for your bins.

In either case it will take some time before your bins are fully populated. Be patient. When you check on progress, you will find more and more red wrigglers—and their small round yellowish egg capsules as well—signs that indeed they are achieving your mutual objective. The small population I used to inoculate five 4-by-8 bins—a huge amount of material—required about a year before I could start harvesting earthworms as surplus.

Of course, if you have access to enough aged manure fully populated with redworms, you could simply fill your bins with it. Your new project would be off and running at full stride from the beginning.

Managing the Bins

It is essential to maintain even moisture in the bin. Before watering, check the deeper levels of the bin, not just the surface. Overwatering at the surface can cause a hidden accumulation of excess water deeper down, especially if drainage from the bin is poor.

At the beginning the horse manure is in the form of the "road apples" so familiar to anyone attending a Fourth of July parade. At the end of processing, the horse manure has been converted entirely to worm castings—a fine-grained, moist, black residue that is one of the best of all natural fertilizers, not only for its mineral content, but because it carries a huge load of beneficial soil microbes added in the gut of the earthworm. Unfortunately, a bin that has gone to pure castings has no living worms—no animal can live in its own wastes.

Thus the trick is to find a way to furnish an ever-renewed source of food for the worms, while separating them from the castings. There are numerous techniques that have been used for doing so, some of them quite tedious and labor-intensive. I don't do tedious and labor-intensive so will tell you the alternative that works well for me.

First of all, as the worms work the manure, they reduce its volume. At some point it is possible to shovel all the bedding from one half of the bin on top of the material in the other half—and still have it fit under the bin lid.

Remember how I said that having two lids allows management of a 4-by-8 bin in effect as two 4-by-4 sections, but with no barrier between them? Now it is apparent what an advantage that is: I fill the emptied half with fresh pony poop. Note that at this point it no longer matters if there is some initial heating in the new material—the worms are safe in the older, established material and can simply wait out the heating cycle before starting to work the edges of the new bedding. As they exhaust the old bedding, they migrate into the fresh material, leaving behind pure castings for the garden, but maintaining a thriving population in the fresh half. I have now established a sort of seesaw for managing a perpetually renewing bioconversion cycle.

Harvesting for Feeding

Harvesting worm poop or castings as top-grade fertility for the garden is great, but for the moment our interest is in harvesting the worms themselves as nutrient-dense feed for the chickens and all domestic fowl except the vegetarian geese, who are appalled. To do so, we intervene at what I call the halfway point. Remember the beginning of the cycle (discrete, clumpy road apples) and its ending (even textured, fine-grained worm castings). Midway along the spectrum is the halfway point: The manure clumps have been pulled apart by the worms into an even mass with plenty of fiber in evidence. At this point, the bedding has been broken down sufficiently to use as a potent fertilizer but still contains plenty of worms and worm eggs. At this point we can scoop up the fibrous bedding and offer it to the flock.

In the winter I release my birds onto the heavily mulched winter feeding yard. Since the mulch is over

Fig. 19.4 Harvesting worms from the bin for feeding.

one of my gardens, I simply dump the bin material with its load of worms onto the mulch, at a different spot each day, in order eventually to benefit the entire area. The chickens scratch the processed manure into the mulch while dining on the worms; and the mulch retains the decomposed bin material for a big boost to soil fertility, come spring. At that time I don't even have to move this chicken-and-worm-powered compost, nor do I have to till—I simply lay out the garden beds, rake them smooth, and spring planting is off to a running start.

Another option in the winter is to dump the bin contents onto the deep litter in my greenhouse poultry pens or in the main poultry house. Again, the birds incorporate it into the deep litter. Periodically, I remove the litter—now something like a mix between a finished compost and a mulch—and lay it down in a heavy layer wherever I need to boost soil fertility, such as for heavy feeders like corn or winter squash—or, as this past spring, to start a new asparagus bed.

Winter typically features a decrease in both production and quality of eggs. During the past winter I fed worms in bedding from the bins in the manner described, in greater quantities and more regularly than ever before. Both egg production and quality were superior to any winter in the past—including a month when four local flockster friends unanimously reported *zero* egg production from their hens. I also found that feeding the worms—together with access to the deep winter mulch, and fresh-cut green forage from the greenhouse and hardy cover crops outside—reduced my use of prepared feed to half what it had been in past winters.

Though most of my feeding from the vermi-bins is in winter, in the summer I sometimes scatter the halfway bedding to feed chickens on the pasture—in a different spot each day, so the chickens can spread it over the sward as a fertility amendment while they scratch for the worms. If I want to reserve the protein boost for younger growing birds, I scatter the wormy bedding inside a creep feeder—a feeder permitting entry by smaller birds while excluding the adults, made from 4-inch mesh wire fencing left over from our deer fence project.

I have encountered concerns that worms raised in animal manures can pose a threat of disease if fed to chickens. Since I have been unable to find an authoritative discussion of this subject—and because I know of so many real-world applications in which chickens have been successfully fed or released onto manure-fed earthworms—I am not deterred by theoretical possibilities. Vermicomposting is proving to be a source of high-quality feed for my flock, and I won't be concerned about potential problems until and if they manifest.[4]

Alternative Approaches to Vermiculture

If you don't have a greenhouse, are there other ways to cultivate earthworms on a more ambitious scale? I've never tried an outdoor bin but would imagine that one or more of our concrete-block bins could be dug into the ground and managed in the ways described above. You should provide a cover to prevent waterlogging from rain, and leaching of nutrients from the bedding, as well as to protect from marauders of earthworms—possums, raccoons, and especially moles—and curiosity seekers such as dogs. At a minimum, freezing must be prevented in the winter—perhaps using hay bales as insulation. Some sort of heat-absorbing cover such as glass (a discarded window sash, for example) would help keep temperature in the bin elevated enough for continued feeding and reproduction of the worm population. Shade in the summer would help maintain more even temperatures.

A more ambitious, farm-scale model might imitate that used on George Sheffield Oliver's grandfather's northern Ohio farm from 1830 to 1890.[5] The farm was 160 acres, and all its added fertility was in the form of large quantities of earthworm castings from a 50-by-100-foot bin, 2 feet deep, at the center of the barnyard. Each day Oliver's grandfather spread the manure from the farm's various livestock over the bin, where it provided ideal habitat and feed for composting earthworms. Periodically the farmer scraped the barnyard itself and added its accumulated manures to the bin. When the bedding attained a depth of 12 to 14 inches, he added a layer of clay soil, high in mineral fertility, and watered as needed to keep moisture suitable for the worms.

At the time of spring plowing, the farmer stripped back the top layers of the bin—those with the greatest concentration of worms—and scooped out the deep layer of worm castings below, which he spread on his fields at the rate of several tons per acre. The material remaining in the bin furnished the teeming population for continuing the cycling of the daily deposits of the farm's manures.

Oliver estimated that the worm population in a cubic yard of the bin's bedding was about fifty thousand, and that they would convert that cubic yard to castings in one month.

He made no mention of protecting the bin from predators—perhaps the farm's dogs and cats kept them at bay—nor whether the chickens enjoyed any dietary assistance from the bin. With some attention to these issues, the small farmer who wanted to experiment with vermicomposting on this scale could find ways to utilize such a bin for live poultry feed. You could release the flock directly onto the bin for brief periods of foraging at a level that would not devastate the population. Or you could spread material from the bin—well along in its conversion to pure castings, but still containing large numbers of worms—into garden or field areas needing a boost in fertility and allow the chickens to feed on the worms while scattering and tilling in the amendment.

Protein from Thin Air

Many years ago I encountered a method for generating high-protein poultry feed "from thin air," based on the use of kitchen scraps to cultivate larvae of the common housefly.[6] I experimented with the technique, using our sparse kitchen castoffs—coffee grounds and tea leaves, peelings and trimmings, but no edible food "wastes." The fly larvae produced were snapped up by my chickens. As in the case of my small vermicomposting bin, the modest level of production made me greedy for a larger harvest.

Historically, European farmers hung slaughtering offal above pens where they were fattening poultry, as well as over ponds with farmed fish. Flies laid their eggs in the offal, the larvae fell out of it to the chickens or fish below, and everybody was happy. The crazy scheme I came up with, then, was not really original.

I asked my friend Sam, the best trapper in our county, to bring me beaver carcasses from his nuisance trapping jobs. Based on concepts I'd encountered in the "protein from thin air" technique, I derived my own system for generating large quantities of fly maggots from the carcasses, using the services of the carrion fly or blowfly.

I drilled dozens of ⅜-inch holes in the bottom, sides, and lid of a 7-gallon, food-grade plastic bucket—large enough for female flies to enter and lay their eggs, small enough to prevent the chickens' pecking at the contents. I then put a large beaver carcass from Sam into the bucket and suspended it above ground level, inside the flock's electronet enclosure.

The larvae of common fly species live, feed, and grow in their feeding medium, then—when they're ready to pupate or undergo metamorphosis into the winged adult phase—instinctively find their way to earth, where they burrow in and await their great transformation. When fly larvae working my beaver carcass were ready to pupate, their location in a *suspended* bucket obliged them to free-fall to reach earth. That's when the sharp-eyed chickens spied them and snapped them up. Not surprising—the feeding value is very high.

That's the heart of this simple system for generating free protein for the flock out of thin air. Let me address questions you may have and explain some alterations I've made in my system.

"But Doesn't It Stink?"

When I loaded that first 7-gallon bucket, I simply dropped in a 35-pound beaver carcass, screwed on the lid, and suspended it in the electronet perimeter. And yes, during the last few days of "processing," it got pretty ripe. So I made four more buckets as described—these were all 5-gallon buckets—and began dividing each beaver carcass into five roughly even chunks.

Then I surrounded the portions inside the buckets with a thick layer of loose organic material such as dry leaves and straw. The female flies had no problem finding their way through this material, but it helped enormously with damping down odor. There was also less odor because the processing of the smaller pieces of carrion was so greatly accelerated, a matter of a few days only. Toward the end of that time, there was a bit of odor when I was passing right next to the bucket, but it didn't carry to other parts of the property. Still, those in suburban settings, or with a neighbor right by the fence line, would do well to seek alternative sources of free protein.

"What Happens to the Remainder of the Carcass?"

When no more maggots were crawling from the interior of the bucket, I dumped the contents out and kicked the wad of padding apart. A few of the larvae had entered pupation in the padding, and the chickens eagerly ate those as well. Nothing remained of the beaver but bones, teeth, and a bit of hair. Those shreds, and the residue of the padding, became part of the organic detritus on the pasture on their way back to soil.

"But You're Just Breeding Flies!"

Correction, this project breeds fly *larvae.* Theoretically, it is likely to *reduce* the ambient population of adult flies. Imagine there are one hundred gravid (ready-to-lay) female flies in the area, and my buckets persuade twenty of them to lay their eggs inside. The larvae that hatch from the eggs become high-quality feed for the flock. Result? The potential new generation of local flies has been reduced by 20 percent.

"Why Not Just Feed the Beaver Carcass Directly to the Chickens?"

Some might wonder if it doesn't make more sense to just feed the beaver carcass directly to the chickens. I tried this, and the birds ate some of the muscle meat—not nearly all of it—but not much of the entrails, other than the heart and liver. Fly larvae convert a carcass almost *entirely* to live protein feed, which seems to have even higher feed value than the carcass itself.[7]

"But Don't the Buckets Draw Invaders?"

True, the buckets will be smelled by eaters of carrion such as possum, raccoon, fox, and dog. But I have always set my buckets up inside an electric net perimeter, which keeps them at bay. In the absence of electronet, defending the system against marauders would be more of a challenge.

As for visitors from the air: A group of buzzards did land to check out that first experimental bucket. But because the bucket had a secure lid, their frustration level visibly mounted as they failed to get at its source. After moping around the bucket a couple of hours, they flew away. I've never had a repeat visit by buzzards since making the modifications to damp down odor described above.

"What about Disease?"

If no adult flies hatch from carrion buckets, they can hardly be a vector for disease. As for the chance that adult breeding flies visiting the buckets will spread disease, my research convinces me that this is unlikely.[8] Recycling of dead animals by fly larvae is a process that goes on in our environs all the time—fortunately—and my bucket system simply capitalizes on that natural process for its feeding potential.

The question of disease among the chickens gets more complicated. On two occasions during the many years I have used this system, chickens who ate larvae from the buckets became ill, and a few deaths resulted. In each case the problem was what the old-timers called "limberneck"—botulism poisoning.

On one occasion my buckets contained offal from a neighbor's recently slaughtered chickens that had not been properly starved the day before slaughter. Their crops were filled with feed, which I believe may have soured and supported the growth of *Clostridium botulinum*, the soilborne bacterium that produces botulin toxin. In another case I had put a couple of groundhogs into the buckets that had been sitting too long before my trapper friend dropped them off—maybe the *C. botulinum* had had longer to get a start than it would have if I were using fresher carcasses.

To prevent botulism I have since made sure that entrails to be cycled through the carrion buckets be from birds who have been thoroughly starved prior to slaughter, to entirely clear the gastrointestinal tracts; and have requested that my friend donate only fresh carcasses to the cause. As well, when I cycle the odd roadkill through my free-protein buckets, I do so only when I know for a fact it is absolutely fresh. Since making these changes, I have had no further cases of limberneck.

My carrion fly buckets may not be a feeding choice for all flocksters, but I offer it as an example of raising a flock in imitation of nature—after all, everything must decompose—and to offer a range of possibilities for converting "wastes" to resources for the flock, and reducing the need for purchased feeds. So if you ever need to tap into "protein from thin air," the strategy will be there, awaiting your experimentation.

For my part, I actually prefer to cultivate the larvae of the black soldier fly rather than of the carrion fly if possible, as discussed in the next section. But I still use my fly maggot buckets, and discuss them here, for several reasons. Readers far to the north of me will not be able to cultivate soldier fly grubs, whose range ends around Zone 6 or so—whereas blowflies are active just about anywhere animals die. If I get a large donation of raw material—say, a couple of large beaver carcasses at the same time from my trapper friend—the carrion buckets can handle them easily; whereas soldier grubs cannot be fed large quantities of such protein-dense feeds all at one time. The maggot buckets yield much quicker production of crawl-off protein, and they are much simpler to set up and manage than bins for soldier fly grubs.

An Alliance with the Soldier

The previous section may have reinforced the repugnance for flies most of us grow up with. Houseflies buzz into the house and onto our food, possibly carrying disease-causing pathogens. Horseflies bite. Blowflies lay their eggs in carrion, and the larvae rid the world of dead carcasses—an essential ecological service for which we are grateful, even if we are repelled by the process as "just *too* gross!"

But none of us within its range are either annoyed or repelled by the black soldier fly, *Hermetia illucens*—indeed, it is unlikely most of us have ever even noticed this innocuous flying insect. Why would we? They look nothing like the flies we find annoying. They do not buzz us or come inside the house. They do not bite. A resting adult looks like a slender black wasp, but without the sting—quite pretty, actually (see figure 19.5).

Fig. 19.5 Adult black soldier fly. PHOTO COURTESY OF BLACKSOLDIERFLYBLOG.COM

Life Cycle

The stages of development of the soldier fly could be a textbook example of the most common insect life cycle: egg, larva, pupa, adult. It starts when the gravid—mated, ready to lay—female finds some succulent vegetable matter or manure and lays her eggs. The eggs hatch in four days to three weeks, depending on environmental conditions, into *larvae*, legless and wingless grubs with a busily feeding mouth on one end, extraction of all nutrients usable for vigorous growth in the digestive tract, and ejection of undigestible feed components out the other end.

Under ideal conditions, the larvae mature in ten days, their tissues developed enough, and with energy reserves enough, to support the next phase—the miracle of metamorphosis into a completely different insect form (see figure 19.6). As with many other fly species, however, *pupation* does not take place within the feeding medium—the prepupal grubs leave it and find a place to burrow into earth. After ten days or so, they emerge as winged adults.

As is the case with many species of butterfly, the adult stage is exclusively sexual—the winged phase is solely about mating, and, for the female, finding the best possible place to lay her eggs. They do not feed at all in this phase, which lasts only five to eight days.

Death quickly ensues for both male and female adults once the eggs have been laid, and the cycle starts anew.

A Useful Ally

Salient points about the soldier's life cycle suggest ways we could make an alliance for mutual benefit: Since the black soldier fly is a specialist in recycling succulent organic residues, we might recruit it in the responsible management of "wastes"—spoiled or unused food, manures, culled fruits and vegetables—reclaiming the residual energy in such materials as additions to soil fertility. Their high level of feeding activity and rapid growth in the larval stage suggest the concentration of considerable nutrients, and the potential for the use of mature grubs as a protein-rich feed for poultry, pigs, or farmed fish. Since the adults do not feed—indeed, they do not even have functioning mouthparts—they do not bite, nor do they come buzzing around us or our houses looking for something to eat.

Of special interest is the grubs' habit of crawling out of the feeding medium when it's time to pupate—suggesting the possibility of directing their crawl-off for *self-harvest* into a collection container.

Other facets of soldier fly biology suggest the potential for a happy working relationship. If there is an inherent problem working with the larval phase of the carrion fly, discussed in the previous section, it is that conditions in the feeding medium tend to be *anaerobic*. Most pathogens, *Clostridium botulinum* among them, thrive in oxygen-starved conditions but do not grow well in oxygenated media. Because the vigorous activity of soldier grubs keeps the feeding medium constantly aerated—and because the adults do not feed at all—this species is not a vector for diseases. Of further benefit is the ability of soldier grubs to inhibit development of larvae of all other fly species—including houseflies, fruit flies, and blowflies. Thus cultivating soldier grubs can actually *reduce* populations of flies with a higher nuisance profile.

Fig. 19.6 Mature grubs ready for pupation. They have just completed crawl-off from the feeding medium in the bin. PHOTO BY BONNIE LONG

Using a Purchased Bin

I began working with this fascinating recomposer in the summer of 2009. I had noticed the flat, segmented, leathery-skinned soldier grubs in my worm bins long before I even learned what they were. As I became more interested in the possibilities for working with recomposers—especially for their potential contribution to a more self-sufficient poultry feeding program—I learned that much creative work has been done toward making a productive alliance with the soldier fly.[9] Since it was obvious there was a wild population

Fig. 19.7 The BioPod is a plastic bin for cultivating soldier grubs. Key design features are the two-part lid, a screw-off jar at the bottom for drainage, the ramp molded into the bin for crawl-off of mature grubs, and a small bucket into which they self-collect. PHOTO BY BONNIE LONG

in my backyard, it was easy simply to set up a tempting feeding medium—castoffs from the kitchen—in a bucket in which ambient soldier females could lay, and soon I had an actively feeding colony of grubs. My chickens, both growing birds and adults, snapped up the mature grubs in a feeding frenzy.

As always, my determination to produce more free, natural foods for the flock was only fueled by this first success, and I bought a BioPod,[10] a well-thought-out manufactured bin that takes maximum advantage of every aspect of soldier fly "lifestyle" and needs and can help the flockster seeking more live protein for the flock to make solid progress toward that goal. See figure 19.7.

A Solar Cooker for More Efficient Feeds

I discovered that food scraps are the most efficient foods for the grubs. But there is no edible food wasted from this household's table, and given my location, I'm not in a position to receive regular donations of food residuals from restaurants and schools in the area. But it occurred to me that I could most closely imitate food scraps by *cooking* the foods I was offering my grubs: culled fruits and vegetables, spent crop plants such as broccoli, and more succulent cover crops such as rape and mangel roots.

Of course, it would be a big waste of energy to cook these items daily in the kitchen. But how about using the same source of energy that grew these plant foods in the first place—*solar power*? The solar cooker you see here is the one I made for cooking feeds for the grubs, increasing feeding efficiency and speeding up production of finished grubs.

The cooker is a double box in three-quarter-inch exterior-grade plywood, with a 2-inch space between the bottoms and sides of the two boxes, stuffed with fiberglass insulation. The lidded cooking vessels—covered Granite Ware roasting pans—sit on a steel plate in the bottom. The interior of the cooker, the steel plate, and the cooking vessels are all coated with a nontoxic flat black paint appropriate to painting a barbecue grill. Sunlight passes through the glass lid—a recycled piece of storm window sash—but is trapped inside by the black, insulated sides, cooking the food as the temperature rises.[11]

In the morning I put foods for the grubs, such as cull pumpkins, into the oven pans along with a couple of inches of water, seal the cooker with its glass lid—and walk away. Which is to say, time expended on "cooking" for the grubs is minimal.

I found that the cooker was softening the foods sufficiently to increase feeding efficiency. I do want to get cooking temperatures higher, however. This coming season I will make a more efficient glazed lid, with doubled window sash.

So my solar cooker is definitely a work in progress, but I anticipate it will be a key to producing better feed for my grubs.

Fig. 19.8 The solar cooker I use for cooking culled garden produce to increase feeding efficiency by the grubs.

Fig. 19.9 Their favorite meal. PHOTO BY BONNIE LONG

The BioPod is a molded plastic composting bin with a metal frame base for stable placement where needed. Since the most efficient feeds for soldier grubs tend to be high in moisture content, a drain with a filter directs liquid effluent into a screw-off collection jar. Molded into the unit is a ramp around the entire inner wall, providing a migration route out of the feeding medium. The ramp ends in a hole, and the grubs in the crawl-off phase have no hesitation launching into space through the hole—assuming that on the other end of their fall is earth, and the opportunity to burrow in for pupation. The positioning of a collection bucket beneath the hole, however, results in the migrating grubs' self-collection. Most active crawl-off is at night, so the flockster simply

A Homemade Bin for Cultivating Soldier Grubs

GW of blacksoldierflyblog.com designed this composter based on years of experience with bioconversion of organic wastes using black soldier fly grubs. The design incorporates all the keys for success: ventilation, access for egg-laying females, drainage, a feeding chamber, and an exit route for the mature grubs to crawl off into a collection container. The unit is small and is not intended for large-scale production of grubs. However, it should handle all the family's coffee grounds, tea leaves, vegetable trimmings, and small amounts of food scraps. Using it will give you experience with a working colony of soldier grubs and get you ready for a more ambitious project when you want to produce even more high-quality protein for your flock.

GW spent about $50 on his project, buying all new parts, including the bucket and lid. If you have scrap PVC pipe and fittings in the recycle pile, you could make one for less.[12]

I have used a number of bucket bins for cultivating grubs. I recommend using food-grade plastic buckets only, which have been easy for me to get at local bakeries and restaurants.

Fig. 19.10 PHOTO COURTESY OF BLACKSOLDIERFLY.COM

grabs the collection bucket in the morning and offers the chickens their favorite meal of the day.

Parameters for Managing a Soldier Grub Colony

Though it is easier to learn how to work with this fascinating species using a well-designed bin like the BioPod (which costs about $200 or so with shipping), you can cultivate soldier grubs in a bin of your own design—sized to suit your production goals—if you pay close attention to the black soldier fly's life cycle and accommodate its needs. The sidebar "A Homemade Bin for Cultivating Soldier Grubs" shows an example of a do-it-yourself bin you can make for $50 or less.

The following are some of the design and management parameters common to all setups.

Range

If you live in climate Zones 10 through 7,[13] there is almost certainly a native black soldier fly population ready to work for you—just set up appropriate feeds in a protected bin that fits their needs, and the gravid females will come. Soldiers can actually survive farther north than Zone 7, though the limits of their range are uncertain. If there is no wild population, you can purchase starter grubs through the mail. How easy it will be to establish a local population if none exists will depend on just how far north of their natural range you live.

Bin

The bin must protect the colony from predation and rain, be readily accessible to gravid females, ensure compatible living conditions, and provide for crawl-off by the mature grubs. It should be in the shade: The high metabolic level of the grubs generates a good deal of heat, so any additional heat supplied by direct sunlight could be disastrous for the colony. Otherwise, placement of the bin depends on your own convenience. It does not have to be stuck off in the back forty, as aerobic or oxygen-rich conditions in a well-managed colony prevent unpleasant odors.

Protect the interior of the bin, the crawl-off route (see below), and the collection bucket from grub eaters such as raccoons and possums, and from dogs, who might make a mess of things just for fun.

Feeds and Feeding

Most efficient conversion to biomass occurs in typical food wastes. But if yours is a frugal household such as ours, there is no edible food wasted, other than castoffs such as coffee grounds—which grubs love—tea leaves, peelings, and trimmings. If you have ready access to food scraps from schools, restaurants, and other purveyors of food, that would be an excellent resource for your project. In the absence of such food residuals, any mix of succulent vegetable and fruit matter works well, such as overmature and cull fruits and vegetables. Grubs love the big outer wrapper leaves of cabbage, and still-succulent spent broccoli plants. Note that feeding efficiency will increase if such garden feeds are cooked.

Some manures make good feed. Low-fiber pig and chicken manures are best—there may be no better way to deal with these manures than using energy-hungry grubs. The larvae will work horse, rabbit, and ruminant manures, but the higher fiber content of these manures reduces feeding efficiency—the larvae cannot digest cellulose in plant stems. Of course, if the manure available to you for processing is a high-fiber type such as horse manure, you could simply increase bin capacity to meet your production goals and plan on cleaning out the bin a little more frequently.

While they make excellent feeds, dairy products and meat and fish scraps—including occasional roadkill—must be fed in limited amounts: Grubs will not thrive with too much high-protein feed. The usual recommendation is to limit such foods to 5 or 10 percent of total feed offered. My experience, however—and it is confirmed by a correspondent with a lot of experience cultivating grubs—is that a temporary spike in protein-dense feeds is not a problem.

Do not use as feed materials that are dry, fibrous, high-cellulose, or tough, such as weeds, grass, leaves, stalks, paper, or cardboard.

Grubs are voracious feeders, and the temptation is to throw it to them. But *overfeeding* creates problems—particularly excess accumulation of undigested (as opposed to undigestible) feed residues in the bottom of the bin, which can unbalance the overall ecology and encourage growth of undesirable organisms. The ideal colony is a pure mass of feeding grubs, their accumulated residues in a thin layer below, and above, an amount of feed they can clean up entirely in twenty-four hours.

Drainage

Look again at the list of appropriate feeds. Note that all have high moisture content, and remember that grubs do not thrive in anaerobic conditions. Therefore: *Efficient drainage out of the bin is essential and will likely be your major management challenge.* If conditions in the colony do become too wet, either cease feeding for a while or add a moisture-absorbing material such as shredded office paper (not newspaper) or coir (the granular residue of coconut husk fiber extraction).

Note that the effluent from the bin can be diluted and used exactly like a crop-feeding compost tea or manure tea. Some sources advise against using it on garden crops, just to play it safe; so if you prefer, use it to encourage cover crops and other nonfood plants.

The Crawl-Off

An essential feature of any bin design is provision of some sort of ramp the mature grubs can use to exit the feeding medium. If the incline is no greater than 45 degrees, they will have no problem wriggling up the ramp. The grubs' migration can be further channeled with a gutter that ends over a collection bucket. In the morning simply pick up the bucket and go give the chickens their favorite meal of the day. The grubs are also relished by guineas, turkeys, and ducks—though not by the vegetarian geese.

A hint at the value of soldier grubs in the feeding program: This past season my buddy Mike Focazio and I split an order for Freedom Rangers to grow our meat birds for the freezer. Though I make my own feeds and he buys his premade, both are based on the same certified organic feed grains from the same supplier. Both Mike and I kept our broilers on pasture after they left the brooder, and we slaughtered together on the same day. My birds were on average approximately a quarter pound heavier than his. The crucial difference, I suspect, is that I had reserved the output of my soldier grub bins for feeding my broilers.

I also scatter a few mature grubs where they can burrow, pupate, and emerge as adults ready to carry on the cycle. If you seed grubs at the end of the season in a place that warms up early—near the foundation of a south-facing wall or, in my case, in the greenhouse—adults will emerge earlier in the spring to start the cultivation cycle.

Productivity

A colony of soldier grubs is like a chicken flock: A well-managed colony—a domesticated one—is a lot more *productive* than a wild one. But what levels of production might we expect? That depends on many factors: bin design, feeds offered, ambient temperatures, and management experience and skills. But as observed earlier, the equation, at whatever level of production, is all positive: The grubs convert the organic "wastes" on which they feed *entirely* to resource. Their work is right at the heart of closing the circle in the self-sufficient farm or homestead.

Please do not imagine that soldier grubs have mere hobby potential—they offer the opportunity for critically important bioconversion on a wide scale. Dr. Paul Olivier has designed systems for grub composting in Vietnam and other developing countries, to reduce the load of food wastes on streets and in landfills while providing needed employment. Academic researchers such as Sophie St-Hilaire and Craig Sheppard have experimented with soldier grubs to manage fish offal from processing of farmed fish, and manures in commercial poultry and swine houses; and to yield high-protein feed supplements for various livestock species, including commercially raised carnivorous fish.

Widespread farm-scale conversion of "wastes" to nutrient-dense grubs—42 percent protein, 35 percent fat, dry weight—could reduce dependence on purchased feeds for poultry, pigs, and fish. Or the grubs could be sold for drying and adding to livestock feeds—as a replacement for the unsustainable harvest of hundreds of thousands of tons of ocean fish annually for fish meal—$1,000 per ton and rising. I am following the development of a couple of large-scale projects. Our productive alliance with *Hermetia illucens* has scarcely begun.

Flexibility

This species is enormously adaptive in response to changing environmental conditions. If their food supply runs out, the grubs go dormant until more food is available. During winter they delay maturation for several months before resuming development. This adaptability gives the operator great flexibility in managing the colony.

Clean-Out

Because of the enormous reduction in volume of the feedstocks offered to the colony, cleaning out the substrate—the undigestible residue at the bottom of the bin, largely the cellulosic portion of plant tissues—need not be frequent. This residue makes a great soil amendment. Or it can be added to vermicomposting bins for even better natural fertilizer production.

As in management of deep litter, it is better not to clean out the bin entirely. As grubs feed and grow in a colony, a whole ecology of ally species—bacteria, actinomycetes, and fungi—emerges. When you clean out, retain some of the existing substrate as an inoculant to speed reestablishment of a total, balanced bin ecology.

Winter

Wild soldier grubs that have gone to earth for pupation in the fall survive the freezing of the ground. However, developing grubs in an active colony will die if they freeze. Since an existing, actively feeding colony will start its work earlier in the spring than a wild one that must start from eggs from newly hatched adults, it is worthwhile ensuring the colony's survival through the winter. Fortunately, it is easier to do so because of the heat generated by their high metabolic rate.

I experimented with my colony's winter requirements during the past winter. Until the heavy frosts came in, I kept my BioPod in the greenhouse. Then I moved it to the basement, where waste heat from the furnace keeps the temperature around 55 degrees. This temperature proved too warm, bringing on a heavy crawl-off—representing a drain from the population that could not be replenished before the following warm season, when mating adults would again be active—and even premature metamorphosis of grubs into bewildered-looking adults. My solution was to remove the colony from the BioPod and return it to the greenhouse, but this time placed into one of the earth-bermed vermicomposting bins, well protected from freezing. For much of the winter, the grubs remained active, eating all the castoffs from the kitchen such as coffee grounds, tea leaves, and trimmings—but without completing development into the pupal stage. For some reason, however, few grubs survived the entire winter; I had to start a new active colony from wild mated females in May. As I write, I am continuing experiments to ensure survival of a robustly active colony through winter.

Partners in Vermicomposting

Those who compost with worms may enjoy expanding their repertoire to include a soldier grub colony. There are important differences between the two: Redworms live and feed in a large, undifferentiated feeding mass; grubs should be fed only what they will consume in a day. The population in a worm bin is self-sustaining, whereas a soldier grub colony requires renewal from an ambient wild population—otherwise there will be no egg-laying females to ensure a new supply of grubs in the bin.

In some ways soldier grubs are superior to redworms for composting; they are more active, yielding greater production of live biomass, and grubs conveniently self-harvest into our collection buckets.

However, we needn't choose between the two on the basis of which is better—actually, the two species make wonderful partners. Grubs digest fresh putrescent matter in a hurry; worms wait until bacteria are consuming them, then feed on the bacteria; redworms convert the cellulosic residues grubs are unable to digest. Studies in Asia demonstrate that redworms grow three to four times faster on the residues from soldier grubs than on food wastes.

If you crave the role of trailblazer, give soldier grub cultivation a try. Though enough work has been done to demonstrate the utility of this species, any of us who ally with the soldier are still making it up as we go. For me it's an exciting exploration, one that brings out the kid in me.[14]

PART FIVE
Other Management Issues

PHOTO BY AUKE-BONNE VAN DER WEIDE

20 | ONE BIG HAPPY FAMILY

The homestead or small-farm flock may start as a simple group of a few layer hens, or a group of hens with a cock. Almost inevitably, though, the adventuresome flockster starts a new generation of birds. Or adds an additional cock or more to the flock. Or brings in new adult flock members from elsewhere. Or decides to add other fowl species such as guineas or waterfowl to the backyard menagerie. This chapter will discuss the management challenges resulting from mixing it up in these ways.

Working with the Cock(s) in the Flock

As the male in the flock, the cock has a key role to play in reproduction. That role is considered in the chapters on breeding—at this point, let us consider some of the management challenges posed by the cock's role in the flock and his distinctive behaviors.

Behavior

We prefer having at least one cock in our flock—we enjoy the way he completes the flock's social patterns, and how he looks out for his hens. The cock has a reputation for being something of a bully, based on his sometimes crass manner with the hens: He may approach one without much apparent by-your-leave and mount her in a way that seems more attack than lovemaking. We sympathize with the hen, shaking herself off as if to say, "Well!—glad *that's* over!"

But the cock is actually quite solicitous of his flock. Alert to the sky, he gives the alarm if a hawk swoops over, setting off a dash for cover. I have heard stories of cocks attacking dogs or foxes in defense of the flock. And if you want to see the true colors of the cock, throw some special tidbit near him—a cricket or grasshopper you've just caught. Does he shove the hens out of his way like a bully and gobble this special treat for himself? Not at all; he calls the hens with a special deep-throated burble used at no other time. It is especially endearing to see him pick up the cricket and beak it to one of his favorite hens. He seems to have the instinctual wisdom to know that the hen needs this nutritional boost to produce the nutrient-dense egg that will carry on their species.

Actually, the more natural style of copulation on the part of the cock is not as crass as too commonly seen. Like the ancestral Red Junglefowl cock, courtship behavior of domesticated cocks involves dancing for the hen that encourages her cooperation for mating that is not brutal or rough. Modern breeding, however, focusing on production traits to the exclusion of all else, has resulted in cocks who have forgotten how to dance—Temple Grandin calls them "rapist roosters." Forcing his attentions on a hen who has not been properly enticed by dancing results in copulation that is violent enough to be injurious.

If the cock in your flock is not a dancer, be on the lookout for *spur injuries* in the hens. In extreme cases, I have seen hens with gaping rents in their sides large enough to show the internal organs. Isolating such hens immediately is imperative—when protected from further injury, their recuperative powers are astounding. Still, I have lost a hen or two to spur wounds, as have others of my acquaintance.

I have not had to do so for years, since I started favoring dancing cocks, but you may find it necessary to trim the spurs of cocks who are injuring hens during mating. Use an ordinary pruning shears to remove the spur's sharp point. Don't prune farther down the shank, or the bird may bleed badly—shear off just enough to blunt the spur. There may be a little bleeding, and you can use a styptic if you like, but bleeding will not be serious if the cut is properly made.[1]

Aggression: Toward People

In most ways our management of the cock is no different from that of the hen—his needs for feed, water, and access to exercise and dust bathing are the same. In one way, however, dealing with the males presents a unique challenge: the cock's natural aggression. True, hens may be aggressive as well, but in most cases only when they are mothering a clutch of chicks. Let us consider the problems of aggression—first toward people, then toward one another.

Make no mistake, an attack from a feisty cock can be a painful and stunning experience. He uses both shank and pinion to "flog" leg or torso with a force surprising for an animal so small, and shocking in the sheer violence of an attack that holds *nothing* back. When a sharp, well-aimed spur strikes home, the result can be a scar that lasts a lifetime, both physical and psychological.

The last time I was hit by a cock in my flock, an Old English Game I had brought in from elsewhere, his spurs drew blood on each side of my calf—through the heavy denim of my jeans.

Over the years I have listened in amazement to horror stories about "mean roosters" who terrorized the family for *years*. I can never believe such behavior was tolerated—at my place, any cock who threatens a person earns a short trip to the stew pot. However, it is my belief that examples of serious aggression toward people are almost always the result of mismanagement on the part of the cock's keepers. It is possible to avoid the development of aggression problems, foremost through respect for the bird, and by understanding the instinctual basis for his behavior.

We call unwelcome attacks by the cock "aggression," but truly the behavior is defensive in nature: The cock is acting out of a deeply felt duty to *defend the flock*. Once he has concluded that you are a threat, he will fight you without hesitation—and believe me, he doesn't care that you are twenty times bigger than he. The key to good relations, then, is to convince him you are not a threat to his flock.

When you are working with the flock, especially in close quarters, keep your movements quiet and gentle, allowing the cocks, and indeed all the birds, plenty of space so they don't feel pushed or crowded by you. Avoid sudden movements. Don't carry large flat objects such as an empty feed bag or a piece of plywood—the birds see it as a flying predator and react in terror. Avoid having to catch one of the hens in the cock's care, especially in close quarters—perform such chores as examining or banding at night if at all possible. If you must get a hand on a hen inside the coop during the day, grab the cock first and settle him somewhere else before proceeding.

I like to offer my cocks special treats by hand from time to time, to encourage trust. If they are with the hens, they will usually approach with them to get the offered treat. If isolated in the breeding pens, they may hang back, uncertain, at first, but eventually will come to get the tidbits from my hand. A crushed hard-boiled egg is a good choice, or a handful of sprouted grain. Soldier fly grubs are like caviar. The more often I take the time to make friends in this fashion, the more mellow the cocks become.

It is especially important that children be taught how to behave around the cock of the flock. We once had visitors whose three-year-old son I allowed into the chicken pen. In his excitement he began running back and forth among the chickens, when suddenly the cock—who had never shown the slightest aggression toward me—jumped up and attacked the boy from behind. I have a neighbor who suspects that his daughter repeatedly teased one of his cocks by rattling a stick on the fence. The resulting tendency to go on the attack whenever a person entered the pen was resolved only when the cock was accorded the place of honor at dinner.

I once had a 10-by-12-foot mobile pasture pen that, though designed for broilers, I began using for a laying flock of New Hampshires with a cock. Since there was no access for collecting eggs from outside, I had to crawl into the confined interior to do so. Naturally the poor cock felt his flock was threatened and reacted accordingly. So intense were his attacks that I had to carry a trash can lid to ward him off as I frantically gathered the eggs. By the time I returned that flock to the poultry house for the winter, the cock's reflexive attacks on me were thoroughly set. I regretfully took him to the chopping block—I *am not* going to be looking over my shoulder around my birds. Still, I felt terrible slaughtering the poor guy, fully aware that his behavior resulted solely from my mismanagement. Having never forgotten that lesson, I now try to anticipate and avoid potential showdowns.

For example, one of my Old English Game cocks attacked my hand while I was filling the feeder in his breeding pen. My first inclination was to jerk him up by the neck and "teach him who's boss!" Instead I closed the door to the pen and thought about what had happened. The cock had been no problem whatever as I serviced the pen for a couple of weeks. But I had just moved a couple of breeding hens in with him the day before. Now he had a different sense of what was at stake when I entered that space. I changed the way I serviced the pen—it was as simple as removing the feeder from the pen before filling it—and had no more conflicts with the cock.

Please be on guard against an intensely emotional reaction when challenged by one of your cocks. True, if it comes to a showdown, you may "win"—I have even heard people claim that they actually did beat a cock into submission. To me that is more admission of defeat than wise management.

Remember, you have more options than he does. Act like it.

Aggression: Toward One Another

Usually, respectful treatment of the cock in your flock will result in amicable relations. If you have multiple cocks in the flock, however, you must also learn to deal with issues of aggression among themselves.

Cocks have a natural urge to establish their dominance over other males in the flock. How do you respect the expression of that urge and at the same time avoid mayhem? The answer is—it all depends: on the total number of cocks, especially in relation to the size of the flock as a whole; the breed(s) of the cocks; how confined they are; and, most especially, the point in the breeding season.

When cocks fight, the intensity of their aggression is strictly a function of how closely matched they are. Understand, however, that by *matched* I refer not only to size but to fighting spirit as well—the will to best the other *at whatever cost*. If the two cocks are *not* closely matched, their fighting will not become serious—after some vigorous sparring, one of them will conclude that he is no match for his opponent and will submit. After that, if the dominant cock flares at him, he will run away shamelessly, and that will be the end of the encounter—the top guy will usually not pursue. If the two are evenly matched, however, they will not back down until one of them is either dead or beaten so near death he *must* accept defeat. We have seen such cocks, utterly humiliated, retreat to a corner, back to the flock and head down, and stay in that position for two days before returning to activity with the flock—the very picture of abject, brokenhearted shame. It is a wrenching sight.

Early on, we tried to intervene—to separate the combatants to help them get over it. Forget it—if interrupted, they simply pick up the contest where it left off, however long they've been separated—they *must* follow the instinctual imperative to either dominate or submit. Our practice after this realization was to monitor fighting closely; when it became obvious we had a deadly serious fight on our hands, we simply picked the cock we wanted to keep and slaughtered the other. It was just too traumatic allowing the fight to go to its unforgiving conclusion.

The easiest way to have two or three cocks in the flock without serious fighting is to add one cock per season. That is, in year two keep a younger cock from the current season's hatch, in addition to the original cock of the flock. The younger cock is almost certain to grow into a position of subordination to the older, without a serious contest to settle the issue. A third can be added the following season, and generally by the next season it's time to cull the oldest cock anyway—most likely before he has to meet a serious challenge as he ages.

The more cocks you have in the flock, of course, the more likelihood that two or more will decide it's a good idea to duke it out. In the past I've had in any given year up to nine cocks in the flock—and, after I implement more ambitious breeding plans, I may have as many as twelve when approaching future breeding seasons. Preventing serious fighting can be a challenge. I have found that the more space cocks have, the lower the level of aggression; and conversely, the more confined, the more likelihood of fighting. In the summer there are usually no serious problems when several cocks are in the flock on large pasture plots, defined by electric net fencing. In the winter the need to provide more space to reduce the potential for aggression is one reason I provide a heavily mulched yard on which the flock can spend most of their waking time.

If the flock must be confined in winter quarters, I monitor daily for serious aggression. One strategy I use is to provide baffles in the corners of the chicken house as refuges for a subordinate cock who wants to retreat from the attack of a more aggressive one. Fair-sized pieces of scrap plywood, for example, make good baffles—just stand them in a corner, so that the corner itself becomes a hidden retreat with narrow access on either side. Once the retreating guy gets out of sight of the dominant one, the aggressor will usually not pursue.

But that is not always the case. I can't stress enough the need to monitor constantly, and be prepared to take action when required. For instance, a few weeks after confining the flock to winter housing one year, I noticed that one of the Old English Game cocks was a little beat up. I set up the baffles, assuming it would solve the problem. It didn't. The next day the beaten cock was in much worse shape. I prepared the Breeders Annex, where I isolate my pairs and other groups for the breeding season; but by the time I had it ready, the beaten cock was in such bad shape that, after a couple of days' isolation, he died. I should have been ready for fighting by having already prepared the Annex. Had I done so, I would not have lost one of my breeding cocks.

In any case, the move of the cocks to the Annex stopped the fighting within the day—with the hens out of the equation, what was there to fight about? For a couple of weeks, it was the Peaceable Kingdom. Then one day I found two cocks with bloodied combs. This time my response was more timely: I immediately isolated the cocks in their separate pens inside the Annex.

There was an interesting thing about that separation, however. Because of limitations on the number of available pens, I had to place two of my Cuckoo Marans cocks in the same space. I was pleasantly surprised to find that the two got along quite amicably, without the slightest trace of aggression toward each other—indeed, they seemed to be great buddies! Clearly a cock's breed has a significant influence on his inclination to fight. My Old English Game and OEG cross cocks are—not surprisingly—a good deal more feisty than a typical barnyard breed such as the Marans. (Of course, when I moved in the breeding hens, I separated the two Marans cocks as well.)

Mixing Young Growing Birds with Adults

Most sources advise against mixing age groups in the flock—that is, raising younger growing birds with adults. That's good advice with regard to fast-growing meat hybrids, especially Cornish Cross—they are *such* wimps in comparison with sturdier breeds. But in the case of any traditional breeds, I have never had a problem introducing young birds—even those just out of the brooder at four weeks or so—directly into an adult flock. Actually, in earlier times it was considered good management to expose young growing birds to disease organisms likely to be present among adult chickens, in order to trigger a healthy immune response in the younger flock members.

If you do introduce a group out of the brooder to your adult flock, the more space you can give everybody, the better. I usually move the new group directly from the brooder to the pasture.[2] Monitor closely the first day or so, but do not be concerned by some hazing of the young ones by the adults—I've never known it to get vicious. The adults are simply putting the young ones in their place—right at the bottom of the flock hierarchy. Since the young birds accept their submissive place in the pecking order—what other choice do they have?—the older ones do not waste energy repeating their instructions. The younger ones, unable to compete with the older, turn instead to establishing their own dominance–submission patterns, the young cockerels especially strutting and sticking out their chests like boys on a grammar school playground.

If chicks are with a mother hen on pasture with the main flock, you certainly do *not* need to worry about the young ones—the hen will kick butt on anyone foolish enough to mess with her chicks.

If you do run younger birds with adult layers, *it is essential that you not feed the young birds an adult layer mash.* A compromise feeding program is in order, but remember that the layers need a higher mineral content (deleterious to the developing reproductive systems of the growing birds), and the young ones need more protein. My solution has always been to offer plenty of oyster shell free-choice, which the layers eat in any amount needed, while the young ones ignore it. To provide a protein boost for the growing birds, I feed higher-protein feeds—whether a fish-meal-boosted formulated mix I've made or earthworms or soldier grubs, cultured milk, or excess hard-boiled eggs—inside a creep feeder, which I'll discuss later.

Introduction of New Flock Members

There may be occasions when you introduce new adult flock members—as replacement stock, perhaps, or to bring new blood into a breeding program. The resulting displays of aggression can make for nervous moments, but it helps to understand that we are seeing a practical and entirely natural behavior, necessary to work new members into the flock's hierarchy. There are equally practical ways we can assist a successful and reasonably smooth transition.

Some flocksters try to short-circuit the barnyard conflict by slipping the newcomers onto the roosts at night, alongside established flock members—the idea being that everybody will wake up in the morning assuming that all the others have always been there. This strategy ignores a fundamental reality: The question for a chicken about any other chicken is, *Who are you, in relation to me, in the flock's social order?* That question is not answered simply by waking up shoulder to shoulder on the roost.

As discussed earlier, young chickens are relatively easy to work into the hierarchy—their first-rung position is a foregone conclusion. In the case of adult newcomers, however, they may well assert their own claims more insistently than birds just out of the brooder ever would. Think of the conflict when the social structure is shaken up and reordered not as *combat* but as *conversation.* The "discussion" sometimes involves a lot of hackle raising and squawking and a little beak stabbing but seldom any real damage before two hens come to an agreement about each's position in the new hierarchy. By the next day they will likely be foraging side by side with no problem.

Chances for an easy reshuffle of the social order will be enhanced if you maximize both the time you can monitor it and the space in which it takes place. Schedule the transition at the beginning of a two-day weekend, perhaps, not in the middle of a hectic week. Ideally, introduce the new flock members in the morning, just after releasing the established flock onto yard or pasture—subordinate members will have more space in which to retreat from a more bossy rival. By the time everybody retires for the night, most of the discussions will have been resolved.

Be sure to furnish extra waterers and feeders during the transition, in order to prevent aggressive competition for these resources.

How serious conflict becomes has a lot to do with how many new birds you're bringing in. If only a few, more of the hazing by the established flock is concentrated on each of the new individuals. If you're bringing in more birds in relation to existing flock size, the hazing per individual is more diluted. In any case, the worst scenario to be alert for is an unusually submissive hen destined to inherit the lowest position in the flock—*if* she isn't identified as the weak sister the others ruthlessly zero in on and eliminate. If necessary, use an isolate-in-view strategy (see below) until the rest of the flock has gotten used to the shy sister.

The default situation in which I'm introducing new members is a pasture area with an electronet perimeter. If your flock is *either* closely confined—for example, restricted to the coop in winter—*or* completely free-ranging, you will have to modify the above strategy. There is much greater potential for serious conflict in a closely confined situation, so it's best to practice what I call isolate-in-view for a few days; that is, give the newbies their own space in the coop, separated from the others by a wire partition. Everyone sees and gets used to one another, even begins relating socially to a degree, but without being able to get physical. After a couple of days, take away the wire partition, and monitor closely.

In the case of a free-ranging flock, remember that the newcomers have no sense of the new coop and grounds as *home*. Until they have developed that sense, they are lost in the world and may wander away in confusion if allowed to roam. Again, keep the newbies inside their own wire-partitioned section until they are well fixed on the new environs as home—a week or so is probably safe—then release to range with the rest of the flock.

If you use electronet, it may be a good idea to clip wings on the newcomers before releasing them with the established flock. If a new member goes over the net in panicked retreat from a bossy rival, she may wander off, disoriented, and not return.

Introducing new cocks is more problematic than bringing in new hens—the chance of a serious fight to assert dominance is fairly high. The cocks of many dual-purpose breeds are more likely to adjust to one another with a minimum of mayhem. A fight to the death may ensue, however, between cocks of more "gamey" breeds such as Old English and Modern Games, Kraienkoppes, Malays, and Shamos. In this case I always start with isolate-in-view, using quarter- or half-inch hardware cloth to isolate cocks, who may fight through a larger mesh such as chicken wire. After a week or so, I release and monitor closely. If necessary, I return the new guy to isolation for a few more days. Providing some hideouts on the pasture, such as wooden apple crates on their sides, gives the new cock a place to retreat from an attack by an established cock.

Introducing new flock members is stressful, even when well managed, so it's not a good idea to make the transition if there are additional sources of stress. If there is a sudden heat spike or a storm, for example, delay bringing in the new birds until conditions are more stable.

What I've said is with reference to chickens only—I've had only one occasion to introduce an adult new member into a flock of another species. I recently added a breeding duck to an established small group of Appleyards, two drakes and two ducks. The meeting was pretty much a non-event in comparison with chickens: The three ducks had an amiable bill-to-bill, get-to-know-ya chat, while both drakes looked on in the background, heads cocked at the same angle, appraising the new lady.

Mixing Species

Aside from cocks as a special case, I've never had any problems keeping different *breeds* of birds together, whether of chickens, ducks, or geese. As for keeping different *species* together, I mix it up all the time and only rarely have problems.

Guineas with Chickens

If you keep guineas with chickens and there is a conflict of interest, the guineas will win, hands down, *always*. Since the chickens have the good sense to defer to the more assertive guineas, however, the two can be housed and pastured together with few issues. The only problem I've had keeping guineas with chickens comes in the late winter as testosterone levels rise with the approach of the breeding season. The guinea cock then becomes merciless toward the chicken cock. The latter quite sensibly retreats, but the former—unlike another chicken cock—relentlessly pursues, exhausting him and pulling out his tail feathers. One spring I actually hobbled the male guineas to prevent such harassment! Keeping the guineas and chickens in separate areas during this period would also be a possibility, though I dislike having to service multiple groups. I've also simply slaughtered the guineas after the winter squash harvest—my main use of guineas is for control of squash bug, as described in chapter 12—hoping to be able to bring in adult replacement guineas by squash-blossom next year.

Turkeys with Chickens

I have too little experience to address this question with any expertise, though I've raised a few turkeys started elsewhere with my chickens. The big question is not a clash of personalities but *blackhead*, a disease caused by a parasitic protozoan. Though they are not usually seriously affected by the parasite, chickens can help pass it on to turkeys, among whom it is most likely to be fatal. In the case of *small* flocks, some flocksters report that keeping the two together works okay; for others, there's simply no way around the blackhead tie-in.

Waterfowl and Gallinaceous Fowl

In my experience, keeping ducks and geese with any of the gallinaceous fowl works fine, so long as you take care to provide a drown-proof waterer that serves the needs of both classes as described in chapter 8. I particularly value the more complete use of feed resources I get, keeping chickens and waterfowl together. For example, chickens will eat the seed heads of sunflower and sorghum, but ducks and geese will also eat the leaves, right down to the stalk—and those as well of corn and the sorghum-sudangrass hybrid I grow as a cover crop. My chickens won't eat the stiffly awned seed heads of mature rye, but the ducks and geese do so with relish—while the chickens till in the rye in preparation for another planting.

21 | PROTECTING THE FLOCK FROM PREDATORS

Sooner or later almost every flockster makes the jolting discovery that somebody out there likes a chicken dinner as much as we do—when the flock suffers its first hit by a predator. Unfortunately, however much we try to anticipate specific threats, most of us go through some on-the-job training, then have to adjust management as we are instructed by predators.

Close Encounters at Boxwood

The following stories are illustrative only, to get you thinking about the challenges and the need for flexible adjustment on the run in a fluid, ever-changing predation environment. We've always seen a success by a predator as a lesson about a weakness in our management. Making the responses dictated in the learning curve has almost completely eliminated losses to predators from our flocks.

Mayhem in Miniature

I hope your first predator experience is not as bad as ours. We started our first batch of chickens—twenty-six New Hampshire chicks received through the mail—in a temporary brooder in the shop. We moved the birds at four weeks to the coop, where they enjoyed having more space, and access to the outdoors. One morning a week later Ellen opened the door to a grisly scene of scattered red-plumed bodies, a few of them with some of the carcass eaten, all of them with grievous neck wounds.

With a heavy heart I gathered the torn little bodies, having no clue as to either what had made the injuries to the necks, or how such a rapacious killer had gotten into the tightly secured coop. Suddenly a head popped up from behind the feeder—a tiny creature no larger than a rat. I had never seen anything like it but understood immediately it was the cause of the mayhem. Grabbing a shovel, I managed to kill the nimble creature, and picked it up to examine it. It had a pelt like silk—and a mouth filled with needlelike teeth. Here was a marvel fashioned to do one thing superbly well: to kill. Using its head as a template, I tested all the small openings up under the rafters—every opening big enough for that head I subsequently closed with blocks of wood permanently nailed in place.

When we had set up the coop, we anticipated predators, of course. But I was thinking raccoon, possum, fox—if *weasel* entered my mind at all, I pictured something larger. What I held in my hand was something I'd never imagined: a *least weasel*, smallest of the weasel tribe, which numbers well over a dozen species—and indeed, at 7 inches or so and a weight of less than 2 ounces, the smallest of all true carnivores.

Such a tiny animal could only eat so much, of course. The reason I was seeing so many bodies is that the weasel's attack reflex is triggered by movement of prey, not by hunger: As long as chicks had scampered about in terror, it had continued to rush in to the kill. It killed a total of nineteen out of our twenty-six young chickens.

The Pack Is Back

One advantage the flockster has is that most predators are nocturnal—thus, if we religiously shut the flock up at night in a secure shelter, our losses should be minimal. A major exception, however, is *dogs*, who could be your major predator threat, not only because they are active during the day but—in contrast to most other predators, who hunt alone—they can cooperate, to devastating effect.

Our second predator experience occurred one season when I was raising fifty broilers in a Salatin-style mobile pen, and the birds were close to butchering size. One morning I went out to find chicken carcasses flung like rag dolls over the pasture. In the early hours, two dogs from a nearby neighbor had both dug under the bottom rail of the pen and torn a hole in the poultry wire, in order to get at the chickens. These dogs were well fed at home, so their assault on my birds was strictly fun and games.

I locked the dogs in a shed and called the county's animal-control officer. Though their owners paid me for my losses—as required to get the dogs out of the pound—no stack of dollars could compensate for the two-month effort I had put into providing a cache of dressed poultry for the freezer.

Be aware that many jurisdictions, both state and local, strongly support the landowner whose livestock are being threatened by dogs—laws in Virginia, for example, even give the landowner the right to kill harassing dogs "running at large." Study applicable laws carefully, especially if you have dogs of your own.

Masked Marauder

I guess I'm a slow learner, because I continued to use the unmodified broiler pen to raise batches of meat birds. My second lesson in its vulnerability came one night when a raccoon tore a hole in the poultry wire enclosing part of the pen and made off with eleven young broilers. This time I took action by wiring for defense: I mounted single-strand electric wires around the outside of the pen, one at nose level and another higher up. Charging such a short run of wire required only a light-duty fence energizer powered by a 9-volt battery. My electric defense was thus entirely self-contained, with only the energizer's ground rod requiring pulling up and resetting when it was time to move the pen. Until I quit using that mobile pen—when I switched to electric net fencing—it never suffered another loss to a predator.

A Sly Fellow

I'm not someone who is spooked by snakes, but I have to admit it gives me a start when I reach a hand in to gather eggs from a nest—only to find a blacksnake coiled up inside! Fortunately that has been a rare occurrence here, though you should be aware that snakes will eat eggs, as well as young chicks. Or ducklings. Three of mine once went missing when I was brooding a batch. Under a piece of scrap plywood I had foolishly left on the litter, I found a large and very satisfied blacksnake, with three bulges down its length leaving no doubt what had happened to my ducklings.

Since I leave my poultry house open during the day, there is no way to prevent a snake from wandering in. Thus the best way to discourage its doing so is to decrease the appeal of the henhouse—that is, to prevent a *rodent* population inside: If there are rodents in the coop, a snake may come on the prowl for them and then, always the opportunist, grab a snack of egg or hatchling if available. It helps as well to be more diligent than I was about leaving anything about as a place for a snake to hide. You might check the rafter plates as well: If a snake is using that as a place to enter or laze about after a meal, there may be streaks of chalk-white snake poop down the wall to give away its hiding place.

We value the role of blacksnakes at our place, so my practice has always been to carry them out of the henhouse and release them elsewhere on the property. But it has been several years since I've had any incursions whatever, perhaps because of the success of my efforts to keep the mouse population to a minimum.

Leprechaun

We always love seeing the local leprechaun, the red fox, on our walks through the surrounding pastureland—and in our backyard. I mention the fox here because, though known as the quintessential chicken thief, to my knowledge we have never had a single loss to one. A combination of digging the poultry wire 8 inches into the soil at the foot of the fence, when we were still using a static run; shutting the flock up at night; and now, routine use of electronet, has allowed us to enjoy the beauty of the foxes that come through our property even as we serve ourselves a second helping of roast chicken.

Aerial Predators

Most predators on the ground can be deterred with electric net fencing. But it does nothing to deter aerial predators—hawks, owls, eagles, and other raptors.

A surefire way to deter aerial predation is to net the pastured flock entirely, with some sort of netting above the flock as well as at ground level. For example, I usually keep young birds without a mother in an enclosed shelter with wire on top until they grow large enough to be less vulnerable. But the larger the flock and the larger the plot they are foraging, the less practical the netting option. An interesting alternative is not to net the area entirely, but to string monofilament fishing line here and there in the airspace above the flock—40- to 50-pound test. Make the lines as tight as you can manage, in random runs among trees, fence posts, corners of buildings, or stakes. Some flocksters using this strategy believe the lines more effectively spook raptors flying into the space if they are invisible; others increase visibility by tying on streamers such as lightweight strips of cloth. The effectiveness of this spiderweb strategy was demonstrated by the experience of a member of the APPPA, who reported: "I was having problems losing juvenile turkeys to an owl, and tried tying fishing line from the top of the pen out to the posts of the electronet. (I ran about 10 or so lines.) It worked great. I'd untie the lines from the net once a week when I moved the turkeys, which was a little annoying, but not that bad. I continued using the fishing lines about a month till the turkeys got big enough that the owl gave up." However, another APPPA-er reported that while monofilament line, strung lavishly above her flock, was completely effective at first, eventually hawks learned how to get past it. This is clearly a "your mileage may vary" option for you to experiment with.

Happily, in the case of hawks, adults learn that full-sized chickens are really not easy prey and tend to focus on small mammals like chipmunks and voles instead. Thus I frequently see hawks flying over on the hunt, but attacks on the flock are rare. It is the juvenile raptor—something like a teenage boy with more bravado than smarts—who is apt to make the most frequent and determined attempts on the flock. Almost all my airborne losses have been to juvenile Cooper's hawks.

Several kills by the same individual usually reveal its modus operandi. For example, I had several losses one summer, all in the early hours before I got out to feed. The defense was as simple as it was effective: I started leaving the chickens shut in their house until the morning was well begun. By the time I released them, the disappointed raptor had flown elsewhere looking for lunch.

There was a time when I left my mobile pasture shelter open at night, since it was protected inside an electric net perimeter. But then I had a couple of losses to a barn owl—who will land and walk about on the ground seeking prey—and started shutting up the pasture shelter until I got out in the morning to feed.

Unlike chickens, who come in to roost with the approach of dusk, ducks and geese after they are well feathered may prefer to stay outdoors rather than returning to the shelter. As a result, waterfowl may be especially vulnerable to raptors who hit in the night or in the dawn hours. I therefore *train* them to return to the shelter at night. It takes herding them into the shelter every night for a week to ten days—a bit of a pain, but not *nearly* as bad as trying to herd chickens—to get them accustomed to going inside when night falls on their own.

Be aware that some flock members may be quite capable of seeing to their own defense. A mother hen with chicks has no fear whatever of the taloned monster from the sky. Once when I was doing the morning feeding, a Cooper's hawk, impudently indifferent to me, made a dive on a chick right by my boot. Its mother, an Old English Game hen, responded in outrage before he could even make contact—and kicked his butt right over the electronet fence.

We were once looking out over the pasture at lunch when a hawk dived after a chick on a pasture plot with not only several mother hens, but half a dozen geese as well. Did they all retreat squawking and honking in terror? They did not—both geese and mama hens *converged* to face the intruder with an emphatic *Bring it on!* The hawk made an equally emphatic U-turn in mid-dive, having decided it had urgent business elsewhere.

Other strategies: Be sure your birds have cover to which they can retreat—the coop or pasture shelter, shrubs, fruit or shade trees—to escape attack. Flocks on pasture with larger livestock such as sheep seem to be less subject to aerial predation. Visual distractions that move in the wind—aluminum pie plates, discarded CDs, old clothing, toy pinwheels—may deter aerial predators, who are typically cautious. Switch the locations of the whirligigs from time to time. Most flocksters who use them report that guardian dogs with poultry are extremely effective at deterring aerial predators.

Though attacks out of the blue are unnerving, as a matter of fact even the simple expedients I've mentioned can be quite effective. I've received numerous reports from proactive flocksters that they manage to avoid losses to raptors for years at a stretch.

The Ball Is in Your Court

I have related some of our predator stories not on the assumption they can prevent all your own potential losses to predators—perhaps there is no more "your mileage may vary" aspect of husbandry than predation—but to get you thinking about the nature of predation in your area and some commonsense ways of thwarting attacks. There are a number of predators I didn't mention because we haven't had a problem with them—but you might. *Rats* can decimate chicks in a brooder. *Coyotes*, *bobcats*, *minks*, and *wolves* can cut great swaths through the flock in a successful incursion. *Opossums* and *skunks* are mostly egg eaters but will kill chickens as well, especially when they are asleep on the roost at night. *Cats*, both domestic and feral, kill young growing birds and attack and injure older ones—and like dogs, they may be active either day or night. *Crows*, *ravens*, and *black vultures* are not primarily predators, but they are opportunists. We once had crows descend on a weakened gosling who had gotten tangled in the electronet. After they discovered what a tasty treat they were, they began swarming healthy goslings and ducklings. I had to return the young waterfowl to the safety of the coop until they grew large enough to be immune to the crows.

A few further thoughts about preventing attacks: I have emphasized electric net fencing to deter predators. If you are using a static fence, be sure to dig the bottom of the wire about 8 inches or so into the ground at its base. Digging predators such as foxes are not stupid: If they find their strenuous effort to dig their way into your flock thwarted, they will not continue wasting energy on the project. I do *not* recommend you put your trust in purchased, high-tech gizmos advertised to have an almost magical deterrent effect on predators: Electronic devices that emit ultrasonic sound or vibrations, flashing lights, or chemical odors have little proven effectiveness—save your money. Since local predation is so variable, your most important first step could be consulting with experienced flocksters in your area about predators to look out for and their methods for coping. Some flocksters find that guardian dogs and other guardian animals provide excellent protection for their flocks.

Your first loss to a predator will be upsetting, but

Pasture Pen to Protect from Raptors

A fellow member of American Pastured Poultry Producers Association designed a 20-by-55-foot movable pasture pen to protect his pastured broilers from predators. Steve Blake of Squirrel Hill Farm raises Freedom Rangers for local markets near Ellicott City, Maryland. His pen is quite mobile for its size and impervious to aerial attack while protecting from predators on the ground as well. Steve explains his design:

> My movable 20x55-ft pen protects my pastured Freedom Rangers. It is enclosed up top with netting, and has electronet fencing for the guys on the ground. So far both have stopped all losses either by land or by air. The plentiful hawks still hover and occasionally swoop in only to be deterred and frustrated.
>
> After a lot of research and a little trial and error, I was able to bring together many different pasture pen designs and develop the system you see here. My many years in the motion picture industry helped tremendously as we were well schooled in building strong yet lightweight scenery and props that could be quickly moved with minimal effort.
>
> The design is quite simple. The pen consists of these independent mobile components: The standard 164-ft length of electronet fencing energized with a solar-powered charger; an arched hoophouse based on the standard 16-ft by 54-inch arched cattle panel (mine is a "Sap Bush" 10x12-ft); and modular frames about 15x20-ft made up of a combination of standard Schedule 40 1-inch PVC pipe and fittings—tees, side-out elbows, 4-way crosses, and 5-way fittings—glued together with heavy duty cement. Fittings are spaced such that all frames are modular—that is, any frame can be used in any position, up or down.
>
> Set-up is simple: Just sleeve 5-ft pieces of pipe into the fittings to join one frame as top to another as bottom, and secure the frames to one another with bungee cords or baling twine. I use the hoophouse itself as support for the upper frames in the center.
>
> When it's time to move the pen to new ground, I don't have to disassemble entirely—I leave the two end cubes assembled as I easily pull them to the new site, then reassemble the center section using the hoophouse as support. At the end of the season all the verticals, none of which are glued, are removed, and the modular frames store flat.
>
> Any type of non-conductive mesh covering can be used over the top. The netting I used is a black synthetic 2x2-inch mesh called "Flight Top Netting" I ordered from Murray McMurray Hatchery. I stretch it over the frame tops and secure with detachable 1x4-inch clips made for attaching poly tarps or sunscreen cloth onto a greenhouse or high tunnel. At ground level I erect the electronet and clip onto the frame sides and bottoms for extra support and security. I tie the netting in with the electronet using plastic shower curtain rings and bread ties. Again, any non-conductive material can be

substituted, preferably one that attaches/detaches quickly.

It takes about one hour to break down, move, and re-erect this particular system, which is more labor intensive than most, but it is also very lightweight (except for the hoophouse) and can be easily performed by a single person. Two people can cut that time in about half.

The pen shown in figure 21.1 cost about $230, a significant investment. However, saving just ten or so market-ready birds from predators covers my initial costs. These savings, and the fact I feel a duty to protect my chickens to the maximum extent I can, make this design a cost effective, logical choice for my purposes, in an area with a lot of aerial predators.

Though it comfortably holds about 110+ broilers, this pen offers little wind resistance, so there is never a problem with moving in high winds—even severe storms that took down four very large trees nearby this past summer.

Next year I plan one modification: I will cut the electronet fencing into pieces I can secure for the whole season to each of the two end "cubes" which I move as assembled units. Attached alligator clips will allow passing of charge from one section of net to the next. Not having to deal with the netting as a separate component should save 20 minutes when I move the pen.

In the coming market season we anticipate increasing our broiler production to 100 birds a week, using eight pens like this one. The best thing about this modular design: It allows me to mix and match all of the parts for each frame, to change the size and configuration as needed.

—Steve Blake,
squirrelhillfarm.com

Fig. 21.1 Steve Blake's pasture pen is based on modular frames, which are easy to set up and to disassemble. In an entire season of use, he had zero losses to predators either on the ground or from above. PHOTO COURTESY OF STEVE BLAKE

Fig. 21.2 A closer view of the modular frames, secured with baling twine or bungee cord, then covered with game bird netting above and electronet at ground level. PHOTO COURTESY OF STEVE BLAKE

Reading a Kill

An effective response to an attack on the flock requires determining the predator involved. Each has its own modus operandi, and in some cases we can tell who made the hit by telltale characteristics of the attack—reading the kill. If you know someone in your area with a lot more experience with local predators, pick their brains for tips on preventing attacks. If you do have an attack, perhaps your more knowledgeable friend can read the scene for clues about the perpetrator. For me that friend is Sam Poles, the best trapper in my county, with a vast store of animal lore that has helped me solve predator problems over the years. Of course, the predators you have to deal with may be different from the ones challenging the flock at Sam's place or mine. But Sam's observations may get you thinking about the kinds of signs to look for following an attack.

A hit by a fox may be the easiest to read. A fox will make a single hit—not multiple kills—and will not eat the kill on-site. It will carry it away to eat elsewhere. An odd thing I've seen with fox kills over the years: The fox will stop every 30 or 40 yards. I guess it shakes the carcass roughly at those points, making sure it's dead—anyway, the resulting trail will be a single feather here and there, then a cluster of feathers every 40 yards or so, all the way to where it had its meal. In the spring, when the vixen is denning up to raise her pups, the trail of feathers will lead to the underground den. Other times, the trail will lead to a thicket or a crevice among big boulders or similar hidden place where the fox felt safe to let down its guard and eat.

A fox is a digger, but it may have to return to where it is digging under a fence several days in a row to complete a hole big enough to get inside the fence. If it's a fox who's been digging, the dirt will be in a neat, concentrated pile—because the fox has been kicking the dirt back between its hind legs. If the dirt is scattered all around, the digging was by a groundhog. If a fox hits an underground barrier, like dug-in chicken wire, it won't keep trying to dig an entrance hole.

A mink or weasel will eat its kill on-site, starting with the head or neck. They will return to a kill the next day and then eat other parts of the carcass. Either mink or weasel will kill by reflex, as long as it sees prey on the move—so they may leave behind a lot of dead chickens, many with no parts eaten at all. Both these predators are long and skinny—even the larger mink can get through an opening as small as an inch and a half. Hardware cloth might be a better choice than chicken wire to keep these guys out of the chicken coop, especially in the case of the least weasel—really tiny but quite a killer.

A possum will often stay close by after a kill—I've found them on a rafter or even in a

don't waste time blaming yourself. Instead, understand that the ball is in your court, and that *you must learn and adjust*. I get so frustrated with flocksters who report despairingly, "I can't keep chickens—they keep getting eaten by foxes!" when they have made no changes in management to thwart demonstrated avenues of attack. Such slackness is a serious failure in one's duty to the flock and reminds me of

corner of the barn, sleeping it off after a nice meal. They don't carry the carcass away, and will return to the kill next day. A possum is not a fast, aggressive predator like a dog or fox—if he can get into the coop at night, he takes advantage of the fact that the chickens are "zombied out" on their roosts. A possum will start eating on the head, or will eat in through the anus to get the internal organs, which it prefers to muscle meat.

Raccoons eat their kill right away, on-site. Most of the year they will just make a single kill, and they like the meatier parts—they will start eating the breast meat first. They will return to the kill next day to continue eating. In the late spring or early summer their attacks can be harder to distinguish from a dog's—I think the mother is teaching the young ones how to hunt at that time, so they may kill multiple chickens for practice that they do not eat.

Dogs kill just for fun. A dog who is well fed at home will probably not eat any of his kills—but he can sure leave a lot of dead bodies behind. Dogs may attack day or night.

If you find a young bird who is dead and looks slimed—covered all over with the kind of shiny dried slime that slugs and snails leave behind—that means a snake tried to swallow it but found it too big, and gave up.

A bear can be a headache for the chicken keeper, but usually because it's trying to get at the feed rather than the chickens. I once had a bear pay my flock a visit four or five times in two weeks. It ignored the chickens, but each time emptied the range feeder. I use electric net fence but I guess the bear didn't hit it with his nose—he just walked right over it. If you're in bear country, it's best not to leave a feeder out with a big reserve in it—it can draw a bear like a magnet.

A hawk will leave a big splash of feathers where it hit a chicken. It prefers to fly away with its prey if it's small enough and eat it up on the branch of a tree, where it feels secure. But a lot of times it can't lift a chicken carcass, so it will eat on-site. It will start with the neck and crop, and return to the kill next day to continue eating. How serious the predation threat is from hawks depends more on species than size. In my area, for example, the Cooper's hawk is more likely to be a problem for the chicken flock than the red-tail, a much bigger hawk.

The worst predator of all is the rogue predator—the one who figures out he has an easy meal ticket at your place and keeps coming back for more. So as soon as you have a hit, you need to figure out who the predator is and take steps to keep him away from your flock. Occasionally that might mean you have to take him out to keep him from coming back.

—Sam Poles

Einstein's definition of insanity: "Continuing to do the same thing while expecting different results." I wonder why they don't take up growing petunias instead.

The good news is that, if we learn from each lesson from predators, losses will likely be quite rare. The pattern here has been a hit from an unexpected direction, an adjustment in my defenses, and never a

Guardian Dogs

I noted that some members of the flock can give protection against predators—turkeys, geese, mama hens. There are other species as well whose natural protective and territorial instincts can be used to defend the flock. Llamas and donkeys have been used for this service. Both are relatively low maintenance, since they can graze the same green sward as pastured poultry. Both are especially inimical to canids—foxes, dogs, coyotes.

The most commonly used guardian animals, however, are dogs. Among the breeds used on farms are Great Pyrenees, Anatolian, Maremma, Komondor, Akbash, and Kuvasz. Their use as livestock guardians is relatively new in the United States, though they have been used for centuries in Asia and Europe. Dogs known more for shepherding, such as Border collies, can also form a close bond with poultry and be trained to look out for them.

Success with guardian dogs depends as much on training as genetics. It should be started early—the puppy should be raised in close association with the birds to be protected from eight weeks on, with not much contact with humans other than the trainer. Temperament is important—fierce toward animals approaching the flock, but not threatening to children or visitors. Most owners with experience with guardian breeds warn that they should be thought of as working partners and *not* as family pets. However, I have received a number of testimonials to the Great Pyrenees as excelling in both roles.

Since most predators are nocturnal, guardian dogs who spend the night with their charges are especially useful for protecting the flock.

Given the considerable investment of time and money in purchasing and training guardian dogs, their use might be of more interest to owners with larger flocks, such as those producing for local egg and broiler markets. For example, Joel Salatin has been using guardian dogs to protect his chickens at Polyface Farm for four years. They now use three, and "they've been worth their weight in gold," says Joel.

However, I have heard from a number of owners of small flocks who started using guardian dogs simply because predation had made keeping poultry impossible for them otherwise—and they were just determined to eat chicken and eggs. Juanita Boutwell of Northern California, who keeps only a couple of dozen chickens, says, "I wouldn't trade my dogs for anything—I live in an area where it was almost impossible to have chickens not totally confined before I got my Great Pyrenees. The hens were under constant attacks by coyotes, bobcat, and hawks.' In the last four-plus years, I have only lost one to a coyote, and it was because the dogs were inside at the time."

repeat loss to that particular modus operandi. Note that the heavy hits I recounted above were all in the early years of keeping poultry here. With the exception of a rare loss to a raptor, we have not lost any fowl for many years. And our experience is not unusual—most of my correspondents in the American Pastured Poultry Producers Association, who manage larger flocks than most of us homesteaders, report that they rarely have economically significant losses to predators.

The Best Offense

Note that in my advice above, I've said nothing about *killing predators* as a strategy for protecting the flock. Though there can be exceptions, going on the warpath to meet the challenge of predation doesn't make sense. My reluctance to kill them is based not on squeamishness, but on two practical considerations.

Killing predators is not *effective* as a strategy. We might terminate the threat from an individual predator by killing it—but only for a while. It's a law of nature that, when the established occupant of a territory dies, another claimant quickly moves in to fill the space. To prevent predation in the long term, therefore, *the best offense is a good defense*—changes in management as required to keep the flock safe in the face of ambient predator populations.

Good Neighbors

The second reason to avoid killing off predators is a fundamental ecological fact: *We need those guys*—they are an essential part of natural balances on which good farming depends. Seeing to the health—which is to say the balance—of our local ecology is *everybody's* job; failure to do our part is both myopic and stupid.

Joel Salatin has for years been my mentor in this approach to dealing with predators. Counterintuitively—since he pastures hundreds of layers and thousands of broilers—Joel goes to considerable lengths to make sure there are large swaths of viable habitat for predators scattered over Polyface's 1,000 acres: brushy and wooded areas, numerous ponds, tall trees for perching raptors. Joel knows that Polyface *needs* its predators, lest it become overrun with rodents and rabbits.

Joel knows as well that some predators blunt the threat of others—with a logic similar to "The enemy of my enemy is my friend." For example, if Polyface is generous with habitat for hawks and owls, they help ensure that minks and weasels seeking dinner out in the pastured flock become dinner themselves.

While working with predator populations for a more balanced ecology is simply good sense, there is no need to be a silly sentimentalist. Occasionally we, like Joel, may encounter what he calls a "rogue predator"—one who makes repeated culinary trips to his flock. In response to such an individual—as opposed to a population—Joel has no qualms about going out on the pasture at night with a strong light and a scoped rifle to protect his flock.[1]

Since I've told you that, in the heat of the moment, I did indeed kill the weasel in my first encounter with a predator, I'll end by telling you about another encounter that illustrates the delicate balance between protecting the flock and respecting the predator. One summer it was clear that an aerial predator was making repeated attempts on my flock. There were no successful strikes, but at least once and sometimes several times a day the flock would make a squawking dash to the henhouse. Following one of those panicked retreats, I heard a chicken, obviously in distress, near the edge of our patch of woods. Grabbing a garden stake—"Okay, I'm gonna *whack* that guy!"—I ran toward the hen's shrieks in some underbrush. Suddenly the red-shouldered hawk who had made the attack bounded up out of the brush and landed on a fallen branch, and we were eye to eye at a distance of about three feet. I stared transfixed—I could no more have struck out with the stake than I could have struck my grandmother. It spread its wings into flight, leaving me entranced in a moment of transcendent beauty—and wishing it well.

22 | HELPING THE FLOCK STAY HEALTHY

The birthright of all living things is health.
—Sir Albert Howard

This is a good-news, bad-news chapter. Might as well get the bad news out of the way first: Regrettably, I am *not* the guy to help you out much with curing a sick chicken. The abundant good news is Sir Albert's observation above. We tend to get far too concerned—with regard to both our chickens' health and our own—with the myriad things that can *go wrong* and pay too little attention to the reality that *the natural state of a species that has been honed by natural selection is near-perfect health.* That's why this chapter is titled "Helping the Flock Stay Healthy" rather than "Dealing with Health Problems": The best way we can ensure flock health is to make sure, with proper housing and feeding and provision of opportunities to work, that our chickens' robust good health is supported rather than compromised.

Health Crises at Boxwood

Since the reader may find my thoughts on dealing with health issues a bit cavalier at best, and callous at worst, let me review the actual health issues we've had here in almost three decades of homestead poultry husbandry—while raising seven fowl species, dozens of breeds, and thousands of individuals. The following list is, as best I can remember, *all* the health problems we've had.[1] When I advise that you can trust in the natural good health of domestic fowl, and should do all you can to support it—instead of buying magic-bullet cures for sickness—that advice is borne out in our actual experience here.

- **Spur wounds:** For a while we found that hens were being badly injured by the cock's spurs as he vigorously clasped their sides during mating. "Badly injured"? In the worst case, there was a gaping hole in the hen's side through which I could see her intestines. I isolated the hen and she quickly recovered, without infection or other complications—and without antibiotics. (We haven't seen spur wounds here in many years, which just may have to do with favoring dancing behavior when selecting cocks.)
- **Botulism:** We've had a couple of outbreaks of botulism poisoning—what the old-timers called limberneck—over the years, unquestionably associated with my infamous maggot buckets, described in chapter 19. It's possible to have a problem with limberneck as well if you leave spilled feed on the ground where it can get wet and serve as a medium for growth of *Clostridium botulinum.*
- **Eye infections:** One year when the deep litter was unusually damp in the henhouse, we had a number of eye infections, a few of which resulted in blindness or death. We have always since taken great care not to let the litter get

damp, even in the wettest seasons, and have not had such infections in many years.

- **External parasites:** During that same damp-litter episode, the birds were unable to dust-bathe properly, and some of them became seriously infested with mites and lice. Since that time we've provided a dustbox as an always-available backup to ensure access to dust bathing. Our chickens have taken care of this essential hygiene themselves, and we have had no further problems with external parasites.
- **Cannibalism:** Injurious pecking of individuals by other flock members can be a problem in highly stressed flocks—debeaking in large-scale industrial flocks is routine for exactly that reason. Our only experience of cannibalism arose solely out of stress during an unseasonable heat spike. Our solution was *not* to debeak future flocks but to ensure that they would not be unduly stressed even in unexpected weather extremes.
- **Broken leg:** I once accidentally ran over a six-week-old broiler when moving a mobile pen. When I examined it I found that one thigh bone was broken—imagine holding a pencil in your hands and snapping it in two. It was an unusually busy time, and I told the poor guy, "Sorry, you're on your own with this one"—and walked away. Now this was a Cornish Cross, mind you—one of the most compromised examples of chickenkind on the planet. Despite that, I found that he was getting around on the pasture better every day. Long before slaughter six weeks later, I could no longer pick him out of the crowd.
- **JCOS:** It's been years since we've had a loss to any of the health issues discussed above. The only losses we typically have—one or two a year or so—are to a condition I call JCOS (Just Crapped Out Syndrome): I walk into the poultry house or onto the pasture and find a bird lying dead. No drawn-out crisis, no anxious interventions, just: Bye-bye. Or I might find that a bird, though still living, is seriously ill. I take such a bird to the chopping block immediately—if it has an infection, I do not want contagion in the flock.[2] (And no, I do not eat any bird who may have any sort of illness.)

When I reflect on this short list of health problems we've had over the years, a few basic thoughts about health come to mind that may be worth sharing.

"It's the Management, Stupid"

That's the message to myself I keep tacked to the wall in my henhouse. Almost all health issues we've had to deal with are amenable to prevention by management changes. Every practice recommended in this book is intended to give the birds the most natural and thus most healthful life possible.

Remember what was said in the chapter on best manure management with deep litter, the alchemy that transmutes lead (the potential for the poops to be a vector for disease) to gold (good health that's supported by what the chickens find to eat in a mature litter). Make sure the litter stays dry, to prevent growth of molds. High levels of ammonia lead to conjunctivitis—also known as ammonia blindness—and can increase the chance of infectious bronchitis as well. Prevent ammonia generation by adding plenty of fresh high-carbon material *before* it becomes apparent—if you can smell it, there's already too much for the chickens' respiratory health.

Maximize the flow of fresh air through the henhouse. Assuming you protect them from direct cold winds, your winter flock is far more likely to become ill from damp, stuffy conditions than from the cold.

Chickens are subject to a number of respiratory infections, such as coryza, Newcastle disease, and infectious bronchitis—as well as adverse reactions to breathing *Aspergillus* molds—but it is highly unlikely that small, uncrowded flocks with abundant air exchange over well-maintained deep litter will ever suffer from any of them.

We don't have problems with cannibalism, and neither will you if your birds are well fed and have plenty of space. It's a good idea as well, especially if they're confined to the coop in the winter, to hang up cabbages, mangels, nets of leafy alfalfa hay or barley, or bunches of nettles—which provide both good food and entertainment as the birds peck them apart.

I've discussed the threat of rodents to the feed supply. Rodents can also carry diseases that affect poultry—among them fowl cholera, chronic respiratory disease, and infectious bronchitis. Few poultry coops are likely to be entirely free of rodents, but take all measures you can to minimize their presence.

What about the role of wild birds in the spread of poultry diseases? Though we hear much of this possibility every time there is a big avian influenza scare, the chances of infection of your farm or homestead flock from wild avian species are remote. Though wild birds inside the henhouse should not be ignored, as carriers of disease they are mostly a threat only in the huge industrial poultry houses. But those places are *incubators* of disease because of the crowding and filth.

An excellent illustration in my region occurred a number of years ago, when an outbreak of avian influenza led to a panic in the broiler industry in Virginia's Shenandoah Valley—millions of broilers and turkeys were bulldozed, a more straightforward word for "depopulating" flocks of thousands in a last-ditch effort to prevent the spread of devastating infectious disease. I made no changes whatever in the care of my flocks as a result of the AI scare; neither did two sustainable farmers of my acquaintance in the Shenandoah, Joel Salatin and Timothy Shell, with flocks far larger than mine. They were *already* managing in ways most likely to keep their flocks disease-free, and sailed through the crisis with no problems.

Do note, however, that I am careful to make sure wild birds are not able to *nest* inside the flock's housing, going so far as killing especially persistent individuals. The starlings I've had to deal with seem to have gotten the point and have not made the attempt in a number of years. Last year English sparrows didn't try to nest inside the henhouse but were unusually insistent on getting into the feeders. I used a sparrow trap, which I found quite effective, in order to defend my flock.

External parasites—lice and mites—weaken chickens by feeding on their blood. Standard advice for getting rid of them is treating with an "approved, medicated" (read "toxic") powder. As noted from our own experience above, though, chickens with adequate opportunities to dust-bathe almost always take care of the problem entirely on their own. Note, however, that an external parasite that dust bathing does not prevent is scaly leg mite (*Knemidocoptes mutans*), a microscopic mite that burrows into the skin between the scales on the shank, causing the scales to thicken and crust over. You can prevent them on your birds by adding red or yellow cedar chips to the nestboxes. (To treat, see below.)

As for internal parasites—various roundworms (nematodes), flatworms, and protozoans, particularly the cocci that cause coccidiosis—rotating the flock over good pasture is the best prevention. A completely confined flock, or one kept continually on the same ground, is more likely to suffer from internal parasites.

For maximum health of pastured flocks, be sure as well that they always have access to shade in summer, and plentiful access to fresh water. And keep an eye out for foreign objects the chickens might mistake for grit and pick up—screws, small nails, pieces of glass. Such items if ingested can pierce the crop or wear a hole in the gizzard, with lethal results.

Illness versus Injury

The recuperative powers of a chicken following even quite serious *injury* are astounding. They are extraordinarily capable not only of healing themselves, as did the young broiler with the broken thigh described above, but also of resisting septic infection while doing so, as did the hen who recovered from the gaping spur wound.

On the other hand, where *illness* is concerned chickens pretty much have two settings: "On" and "Off." Once a chicken has become ill, the chances of recovery—while not impossible—are so low it makes more sense to cull immediately. Which is to say: "Sometimes the best medicine is the hatchet."

The Best Medicine

If you want to treat a sick chicken and try to nurse it back to health, I wish you success. Certainly, if you have a flock of half a dozen hens, the loss of one will be a proportionately greater loss than in my flock of dozens. But the theme of this book is husbandry of the productive small-scale flock, and therefore using poultry as a serious partner for food independence; that is, management decisions look to the future as well as the present.

Seen in that perspective, my decision to cull a diseased bird from the flock immediately is hardly as callous as the reader may at first assume. My concern is not only to provide as healthful and enjoyable a life as I can for today's flock but to help ensure the well-being of all its future members as well. Especially for someone who breeds some of his own stock as I do, is it kindness or cruelty to rescue a sick bird and thus pass her genes on to future progeny?

What Is Disease, Anyway?

If we understand disease as *caused* by germs, then buying some miracle drug to kill the germs makes pretty good sense. And that's all pretty plausible, until we reflect on the fact that the infectious agents associated with coccidiosis or Marek's disease are universally ambient—that is, they're present anywhere chickens are—sort of like the virus for the common cold among *Homo sapiens*. So if they caused the infections they're associated with, seems to me all the chickens would be dead.

It makes more sense to me to think of disease as an *imbalance* in the organism—an imbalance that changes the default state—near-perfect health—and gives the disease agent its opportunity to proliferate. So the problem is not the germ but a failure of the conditions that support the *default* abundant good health of *Gallus*—problems with the quality of feed, undue stress, or chronic unsanitary housing conditions: damp, dark, with accumulations of poop that are unable to enter the natural cycles of return.

One way of expressing this idea is the proverbial three keys to success, in this case: *prevention*, *prevention*, and *prevention*. But I'm not entirely comfortable with that summary: I don't want to be obsessed with *prevention*—of *disease*. My primary goal is really *promotion*: promotion of the conditions that support the natural abounding good health of our chickens. The distinction is between focusing on *keeping bad things from happening* and *helping good things happen*. The three keys to success then become: Keen observation of the birds ("The best medicine is the eye of the flockster"); the robust good health and natural immunity of the birds themselves; and husbandry practices that support rather than compromise that natural good health.

It's a truism that we should "treat the disease, not the symptom." But in this context disease *is* the symptom; our challenge is to treat the true *cause* of the disease—that is, whatever in the environment or our management is getting in the way of the natural good health of our flock.

Assisting Natural Immunity

Far from trying to put our flocks in a bubble to isolate them from disease organisms as "the enemy," we should recognize the benefits of exposure to disease organisms in the environment. For example, you could prevent coccidiosis in the flock with vaccination; or by feeding coccidiostats through the entire brooding period; or by ridding the environment of cocci entirely by cleaning out brooders between batches, sterilizing, and fumigating. A natural means of dealing with the threat, though, is simply to *expose chicks to cocci in the environment.* While it's true that massive exposure can overwhelm the chicks' immune systems (especially if they are weakened by stress), exposure to levels of the protozoans (cocci) normally present where chickens have been raised for any length of time actually stimulates resistance in the growing chicks. Just as children's immune systems become stronger by being *challenged* by germs in the environment—as opposed to an attempt to keep them in a bubble—so chicks' immune systems develop resistance to cocci by exposure to them. Chicks "learn" resistance in the same way by exposure to other pathogens as well—those of Newcastle disease, chronic respiratory disease, infectious bronchitis, infectious coryza, infectious bursal disease.

It's ironic: We are almost universally advised to avoid mixing age groups in our flocks to prevent certain infectious diseases. Yet we look back on the practices of earlier eras—mixing age groups precisely to *expose* growing chicks to pathogens and thus trigger immune response—and call it "natural vaccination"![3]

Treating Illness or Injuries

If you want to try curing a chicken with an infectious disease, it is imperative to isolate it from the main flock—you certainly do not want to run the risk that contagion will spread in the flock. Further, the bird itself may have little chance of recovery if left with the others: Chickens have a natural instinct to eliminate sick or weakened members from the flock—in the wild, such weaklings could make the entire flock more vulnerable to attack by a predator—and they can be ruthless and utterly unsentimental about doing so. Isolation gives the bird's natural recuperative powers the best chance to come into play. If it does heal, return it to the flock under close supervision to make sure it is not unduly harassed while reestablishing its position in the flock's hierarchy.[4]

If a bird is badly injured, it should also be removed to its own quiet spot to recover. Yes, I know I said earlier I left a young broiler with a broken thigh to manage on his own. I was lucky in that case. Such luck would *not* be the norm. Other flock members will zero in on an open wound especially. If a wound is not too serious, however, I sometimes smear it with a thick salve strongly scented with balsam of fir, and leave the bird with the flock. The dark salve hides the color of blood in the wound, and the scent repels curious fellow members of the flock.

A wound can be cleaned with hydrogen peroxide. I have never applied any sort of bandage to wounds and have never had an injured but otherwise healthy bird succumb to septic infection.

Dust bathing will prevent infestations of external parasites. If on rare occasions you do have to take temporary action yourself to rid chickens of lice or mites, diatomaceous earth—applied liberally by hand—has always worked well for me. I have also used finely powdered garden lime. *Wear a good dust mask.* As said above, dust bathing will not prevent scaly leg mites. Still, on the rare occasion when my chickens have been infested with them, I've found them easy to treat: I make a mix of mineral oil, Vaseline, tea tree oil, and oil of oregano, and scrub it vigorously on the entire shank and foot, using an old toothbrush and wearing disposable latex gloves.[5] Repeat after a few days. Not long after the second application, the unsightly old scales slough off, and the hens' shanks shine with pristine new skin.

The Future of *Gallus gallus domesticus*—in My Backyard

I say again, unapologetically, that I never doctor a chicken who seems sick with an infectious illness. The fact that an individual succumbed to disease, while her sisters receiving the same care and facing the same environmental challenges did not, is the best possible reason for culling her from the flock. It's easy to say we plan to manage the flock "naturally," but natural management means adopting nature's sternest measures where infectious disease is concerned. Disease is nature's hoe for weeding out weaker individuals—meaning that over time more individuals in the species will be robust and naturally resistant to disease.

A good illustration of the "bio-logic" involved here is Marek's disease, universally present where chickens are raised. Apparently *Gallus* has already made a good deal of progress developing complete immunity to this viral disease—some individuals have an inherited immunity, though others are susceptible and will succumb if under stress or heavy exposure. We can help the advance of natural immunity to Marek's in *Gallus* as a whole by culling all susceptible individuals who develop the infection. Looked at in this way, vaccination of whole flocks might be the *worst* choice: It confers immunity to *all* potential breeders and thus masks the carriers of immunity and of susceptibility alike, making it less likely that the flock as an ongoing entity will increase natural immunity over time.

To be sure, there may be exceptions to the above bio-logic: You may need a vaccination program, for example, if you exhibit your birds; if serious diseases have occurred on your place, either under your management or that of a previous owner; or if there is a concentrated potential disease incubator close by, such as an industrial chicken operation. *In no case should you ever vaccinate against diseases you don't have good reason to think are serious threats to your flock.* With that caveat it seems to me that the more we use vaccines, antibiotics, or other artificial means to prop up whole flocks who do not have, or are too stressed to express, natural immunity—the more we ensure future flocks with weaker immune systems who *require* such supports.

23 | MANAGING THE WINTER FLOCK

Readers who don't live in Florida or Southern California will find that winter has its challenges. Despite the loss of summer's greenery and insect provender, determined flocksters will find strategies for natural management and feeding in winter as well.

Culling for Winter

Without the availability of as much free food provided by the growing season, feeding nonessential members of the flock doesn't make sense—costs for purchased feeds will be higher, and amounts of natural feeds available on the small farm or homestead will be reduced per flock member. The approach of winter is thus the time to cull the flock down to its mission-capable minimum. Core size in our flock has to do with my plans for the coming breeding season; the number of hens I estimate will keep us and a few friends supplied with eggs, factoring in the usual winter decline in laying; the aging of layers who will not be as productive next year; and the number of reliable broodies I will need to hatch out new stock next spring. Thus in my final fall culling, I butcher all males I will not be using as breeders in the coming breeding season; older females of declining productivity; and any individuals with undesirable flaws. As I write, I am preparing for my final culling in preparation for winter, and I anticipate retaining about three dozen chickens (two dozen layers, about six each breeding cocks and reliable broodies), five breeding Appleyards (three ducks and two drakes), and three Pilgrims (two geese and one gander).

Unless you simply practice a culling program such as "two years and out" with your laying flock, you will need to discover "who's laying and who's lying" among your hens. One option is *nest trapping*—the use of nestboxes with doors that fall or swing shut when hens enter to lay. The hen is thus trapped inside until you release her and record her performance. (See appendix A for instructions.)

Effective nest trapping requires frequent checking of the nests—if urgent hens are blocked too long by closed nest doors, they will lay on the litter instead—and there goes your careful record keeping.

If you do not use trap nests to determine layer productivity, there are a number of physical clues that help to determine productivity of individual hens—all a result of increased hormone levels associated with egg laying—and it is worthwhile to learn to read them.

- The vent will be large, oval, soft, and moist (as opposed to dry, puckered, and round).
- The greatly increased size of the ovaries will round and extend the abdomen between the tip of the keel bone (breastbone) and the tips of the two pubic bones (rising to sharp points on either side of the vent). When palpated, the abdomen will be large and soft. The distance between the pubic and keel bones should

be about four fingers, at least three, pressed down flat and side by side.

- The pubic bones of a productive hen will be wider apart and flexible to accommodate passage of the egg. You should be able to insert at least two and ideally three fingers between the tips of the pubic bones.

The above are the surest signs that a hen is laying well. Other indicators that are not as reliable but are worth noting: Comb and wattles in a high-producing hen will be larger, softer, brighter, more pliable (as opposed to small and pale)—in a single-comb breed, the comb may even lop over to one side. The nonlaying hen will actually look nicer than the productive one: Since the latter is putting more resource into making eggs, her plumage will look more worn and disheveled and her shank more bleached—the nonlayer's plumage will be more sleek and pristine, and her shank more intensely colored.

If you are serving a market for eggs, the calculation about culling gets more complex. You might delay your heavy culling of older hens until the spring to keep winter egg production up, though in that case you'll be investing more feed per dozen eggs produced. On the other hand, if you sell through farmer's markets, most of which are closed in the winter, the off-season might be the best time to cull unproductive hens. If you supply eggs through a CSA, your customers should understand that the weekly "basket" you furnish depends on what is currently being produced.

Managing Winter Laying

Any sudden or extreme weather fluctuation can cause a temporary dip in egg production—a heat wave or a storm, for example. But there is one fluctuation that brings a decrease in laying for an entire season—winter.

All breeds reduce egg production in winter, some more than others. Actually, it isn't natural for the hen to lay *at all* in winter—domestic hens lay in winter only because selective breeding has unmoored their ovulation from a seasonal reproductive cycle—so we should be grateful for *any* eggs our hens give us in their off season.

Reduction in rate of lay in winter has two causes: The *molt* involves the replacement of every feather on the hen's body by shedding all the old feathers and growing new ones, as discussed in previous chapters. Since feathers are almost pure protein, replacement of her entire set requires a *lot* of resource for the hen—no wonder she takes a break from laying to accomplish the task.

The other reason rate of lay declines in winter is the decrease in day length. Obviously there are lingering effects of the fact that laying is about reproducing—the decreased photoperiod in winter is still to some degree nature's signal to the hen that it is not the sensible time to hatch chicks. Selective breeding may have partially defeated that inhibition, but it is still operative in the reduction in laying.

Not surprisingly, techniques have evolved to further manipulate the reproductive biology of hens to induce them to lay more in winter. The practice of controlled or forced molting is practiced on most industrial laying flocks; the birds are starved for a period ranging from five to twenty-one days, either by complete withholding of food or sometimes the feeding of nutritionally deficient feeds—water may be withheld for briefer periods as well—together with light manipulation and feeding of drugs, hormones, and metals such as dietary aluminum and zinc. This program of "shock and awe" in the layer house has the intended effect of reducing body weight of the hens by 25 to 35 percent, which induces a more sudden and short-term molt and an earlier return to laying. Not only is the procedure highly stressful—indeed, it is utterly cruel and inhumane—but it also induces a greater susceptibility to the spread of salmonella in force-molted flocks and an increased risk of contamination in the supermarket egg supply.

Though there are home versions of forcing molt that are not as extreme as in commercial flocks, to my mind there is no place in small-scale poultry husbandry for the inducement of such high levels of stress in our birds. On the other hand, increasing the amount of protein in the diet will increase the speed at which the hen can regrow her feathers and return to normal laying. Undoubtedly, the improved rate of lay in our flock last winter had much to do with supplementing protein by feeding from the vermi-composting bins.

The other approach to increasing rate of lay over winter is the manipulation of apparent day length through use of artificial lights in the henhouse. That is, if the hen's circadian rhythms respond to the decreasing day length of winter by decreasing egg production, we can fool her system into thinking that longer days have returned using supplemental lighting.

I do not myself use artificial lighting in winter—it seems ungenerous to push the hens toward greater production when they have perfectly good reasons to be cutting back. However, I think that if your layers are well fed and in good condition, a program of supplemental lighting to increase production will not harm your flock. Note that you don't need much light to do the trick—a 25- or 40-watt bulb should be sufficient in most small-scale henhouses. One of my correspondents uses a 10-foot string of the tiny Christmas icicle lights for softer and more uniform lighting in his poultry coop.

Note as well that the supplemental lighting should be set to a regular schedule, not turned on and off randomly when you think about it. The usual recommendation is to provide artificial light to bring the total apparent day length, natural and artificial light combined, to fourteen hours. The most convenient way to do so is to put the artificial light on a timer. And note that the best strategy is *not* to provide the additional lighting in the evening hours—when the timer abruptly shuts it off well after dark, the hens will be caught unawares by the sudden darkness, and find it difficult or impossible to get onto their roosts. Instead, set the timer so that the light comes on in the wee hours of the morning.

Though chickens themselves are hardy in temperatures far below freezing, do remember a couple of problems low temperatures can create. If eggs remain in the nest too long in low temperatures, they will freeze, then crack because of the expansion of their contents. It takes temperatures below the freezing point to freeze eggs—28 degrees or lower. In my experience even at temperatures in the low 20s, eggs don't readily freeze in the daytime. In any case, collect eggs more frequently in winter, especially if temperatures dip into the teens, and make certain none remain in the nests overnight. If an egg does freeze, you can thaw it and use it right away—or pass it on as a special treat to your pets.

Your flock needs water in the winter as much as in the summer. If your hens suffer from water deprivation created by freezing of their waterer, egg production will decline drastically.

Though I offer supplemental oyster shell and granite grit free-choice in the summer, doing so is mainly insurance that the flock will have plenty of calcium for eggshell formation and grit for the gizzard. I don't remember my grandmother offering her flock either, and they took care of these needs with what they picked up on their own, roaming over a 100-acre farm.

For the flock confined to winter housing, however, it is essential to offer grit and shell (or another calcium boost such as aragonite or feeding limestone). A hopper mounted on the wall high enough to keep it free of litter scratched up by the chickens is the most convenient way to offer these supplements for self-serve intake.

Give some thought as well to the mental health of the flock, if confined to winter quarters. Chickens get bored when confined to a relatively small space. Entertainment is to be had first of all in a fluffy deep litter in which they can scratch to their hearts' content. The birds will also enjoy pecking at special treats hung up in the henhouse—mangels, cabbages, alfalfa hay in a mesh net, ripened seed heads of small grains, sunflowers, or sorghum.

Winter Feeding Strategies

For most of us winter is the biggest challenge to keeping an array of live, natural feeds available to our flocks. Some simple strategies can help. *Sprouting* represents a kind of value-added feeding we can practice to increase the enzyme, vitamin, and protein content of any seeds we grow or buy to feed whole. In my area dandelion and yellow dock stay green deeper into the chill than any other forage weeds—as long as I can get a spading fork into the ground, I dig these plants by the roots and throw them by the bucketful to the winter flock. They make short work of the green tops, and the chickens eat as well part of the dandelion roots, though not the tough roots of the dock. The roots of both species get turned under the litter by the chickens' scratching, where they sprout new growth—similar to the forcing of Belgian endive from roots in a cellar. When they surface again, the chooks have second helpings. Since our greenhouse—at 20 by 48 feet—has more space than needed for winter greens for Ellen and me, I grow cut-and-come-again greens for the flock as well. Finally, for me winter feeding means taking advantage of the thriving populations of earthworms in my vermicomposting bins. Last winter we fed worms in greater quantity, and more regularly, than we ever had before. Egg production remained higher than in any previous winter—and this in a winter when four local flockster friends got zero egg production for a full month—and there was no decline in egg quality, as is typical in winter, with its scarcity of fresh foods.

Remember that fat hens do not lay well, so ensure your hens do not get too fat. At this time the hens are likely not getting nearly as much natural feeds—a great antidote to putting on too much fat—nor as much exercise. Be careful especially not to feed too much corn, which is high in carbohydrate. Feeding some oats—higher in fiber and lower in carbohydrate—is a good alternative, especially if the grains are scattered over the litter to encourage more active scratching.

It is natural and desirable that hens put on some fat, especially in preparation for winter. But it's important to learn how much is just right. Do a hand examination of your hens, keying on the pubic bones, beneath which are the major fat deposits. Pinch one of the protruding pubic bones: If the fat surrounding the bone is half an inch thick or more, the hen is far too fat. If it is extremely thin—if you're feeling essentially just the bone itself—she is far too thin. If the fat layer around the bone is an eighth to a quarter of an inch thick, the hen has enough fat for good condition but not so much as to interfere with laying.

The best antidote of all to excess fat is continued access to live, natural feeds, especially fresh green forage plants, and plenty of exercise. Thus the best winter strategies are those that keep the chickens working through the winter—don't forgo possibilities for doing so just because it's cold and many species are dormant. Nothing keeps chickens more happy and content than having good work to do.

The Garden's Overwinter Cover Crops

Any and all cold-hardy cover crops I grow to improve soil fertility over winter—crucifers, small grains, and peas, for example—make good cut 'n' carry green fodder for the winter flock. I end the gardening season with almost every garden bed under a lush green cover and am able to cut fresh greenery for the flock every day, starting with the crucifers, most of which succumb when the ground finally freezes solid. In the bleak days of midwinter, I am still harvesting generous daily portions of small grain grasses, wheat and especially rye (not oats, however, as they winterkill as well when the ground freezes solid). We are now a month past the solstice, and I expect many beds I have will soon resume growth with the lengthening days. If you live in a zone that sees a lot of snow

Fig. 23.1 Cold-hardy cover crops over the entire garden have furnished cut green forage every day of a mid-Atlantic winter.

in the winter, of course, you will have more problems getting to such cold-hardy forages as rye and kale beneath the snow, even if they survive the low temperatures.

The Winter Feeding/Exercise Yard

There is no better provision for the winter flock than a heavily mulched winter yard.

Use any and all organic residues you have on your place or can get elsewhere—anything a compost heap would eat. Capture the fertility in the birds' manure while preventing its runoff as pollutants to groundwater. If the mulch is heavy enough to prevent freezing, the continued activity of animal life such as earthworms and slugs at the soil line remains a foraging opportunity for the flock.

As discussed in chapter 2, I provide a feeding yard for my mixed flock living in one end of my greenhouse over winter. In my case, the birds are released onto a garden space deeply covered with organic litter. At the end of the winter, I cover the garden with the made-in-place mulch/compost, and I am ready to start the gardening season at full stride.

Flocksters with large enough mobile shelters for their flocks sometimes simply park them over winter on a garden space, lay in a deep litter, and let the birds do the advance work on soil preparation for the coming year. A sizable hoophouse can also be assembled for the overwinter flock on a garden or new-ground space and disassembled and stored during the growing season.

A Winter Salad Bar

Always leave room for serendipity in your flock management. Last winter I exhausted all the sources of fresh greenery discussed above to feed my ducks and geese and was getting desperate for more. Suddenly I noticed how green and lush was a large patch of winter rye near the main poultry house. The area had gotten worn bare as a result of poor management on my part, and I had sowed it to rye simply because I hate leaving ground bare over the winter. The rye had established well, then gone dormant—though not actively growing, it was thick and green. It was the work of an hour or so to set an electronet around the patch and move my few geese and ducks onto it—a breeding trio of each. Because the stocking density was so low, I was able to keep the waterfowl on the rye full-time until it began growing again in the late winter. I moved the birds elsewhere until the rye matured, then returned the ducks and geese, along with the chickens, to the area to till in its cover. The chickens showed no interest in the mature rye—it was gobbled down greedily enough by the waterfowl—but busily tilled in the plants themselves and prepared the ground for a planting of buckwheat.

One success cries out for another, so this past fall I sowed that same ground again to serve as a winter salad bar. This time, however, I sowed barley and wheat in addition to rye. I also sowed a mix of crucifers—mustards, turnips, rape (which all have a cleansing effect on soil with regard to fungal diseases, and offer excellent grazing)—and alfalfa. If I exhaust the fresh greenery from my garden cover crop beds, I will release my waterfowl breeders onto that patch to graze. I won't put the chickens on it full-time, but will allow them a taste late in the day, when the coming of dusk will summon them back inside the henhouse before their scratching damages the winter cover.

Fig. 23.2 A deeply mulched yard gives the winter flock access to live animal foods at the unfrozen soil line.

I anticipate that in the spring the alfalfa will grow slowly, but without serious competition from weeds, in the shade of the grains as a nurse crop. Once the flock harvests the mature grains, however, the alfalfa should come on strong in the additional sunlight. I will then have a stand of pure alfalfa to experiment with in making alfalfa hay for feeding the flock next winter. The alfalfa will also set nitrogen in the soil, getting it ready for whatever comes next.

A Preseason Sweep of the Garden

The actively growing garden is not compatible with free-ranging chickens. However, chickens can do useful prep work in the garden during the preseason.

If parts of the garden have been under a protective mulch for the winter, the birds get a big protein boost while knocking back the slug and snail population for months to come, and shred the mulch, softened by winter rains, to speed its decomposition to feed the soil.

Other parts of the garden will have been under cover crops for the winter—such as small grains, crucifers, or peas. Less cold-hardy species will have winterkilled and collapsed into the same protective mulch—but one the flockster-gardener didn't have to collect and haul into place. Cover crops that survived the winter will serve as much-needed green forage the birds eat as they till the remains into the soil in preparation for planting.

24 | OTHER DOMESTIC FOWL

I enjoy keeping a mixed poultry flock; species other than chickens can give us broader coverage of resource use and of other skills for work on the homestead or small farm. For example, a couple of years ago I made the serendipitous discovery that my geese and ducks, with their powerful serrated bills, were able to strip the kernels from whole dried ears of corn still in the husks. I was thus able to make good use of the Hickory King feed corn I grow—without having to harvest, husk, store, and shell it by hand to feed. From that point on, I simply chopped down a few stalks of dried corn per day and left it for the waterfowl to clean up. At the same time my hens were unable to get fat on the Hickory King—a condition inimical to good egg production—as they were unable to peck the kernels off the cobs.

On pasture different species prefer different forage plants. The geese *love* wild chicory, which they chew down to nubs, but I've never seen the chickens showing it any interest. Last year I grazed the flock on a stand of rye—the chickens would have nothing to do with the mature rye seeds in the head, but the ducks and geese devoured them.

Many flocksters have noted that geese kept with their pastured chickens help prevent attacks from aerial predators, which has been our experience here as well.

Throughout the book, I've referred to other fowl species where their needs or contributions differ from that of their chicken cousins. Now let's look at them in more detail.

Waterfowl

If you have water on your property, keeping waterfowl is a natural choice. But if you don't, there are also some good options for keeping them happy. When I refer to *waterfowl,* I could be referring to as many as four different species, however, so I will start with observations of this group as a class, then consider them individually.

Waterfowl need plentiful access to *water.* If you don't have a pond, at least give them an opportunity to bathe, so they can rid themselves of external parasites—in the same way chickens do by dust bathing—and for their mental health. You'll know what I mean when you see them taking their first exuberant splash. I find it easy to provide bathing to my waterfowl, using a stock watering tank, even in the middle of winter. As a far-distant third best, you *must* at least provide waterfowl water deep enough to dip their heads and rinse eyes and nostrils. Note as well that waterfowl may choke on feeds, especially fine-textured commercial feeds, if they do not have simultaneous access to water.

If you keep waterfowl only, their housing can be quite simple. Indeed, if winter temperatures do not drop below 0 degrees, a windbreak made of straw or hay bales may be adequate, so long as protection against predators is assured. An open-front, roofed shed can also be used. The same basic principles of housing apply as for chickens: a lot of airflow, deep litter to absorb the droppings, and protection from predators. Allow adequate space in the housing, depending on the size of the birds—always remembering that "adequate" depends greatly on whether the birds just sleep inside the shelter or are confined to it full-time, as in the winter. For geese, for example, I would advise a *minimum* of 10 square feet per bird in the former case, 20 square feet in the latter.

Usually I keep my waterfowl flock in my main poultry house with the chickens. The only problem with that arrangement is if I try to water them inside—but these days I do all watering outside, regardless of the season. In winter I release all my birds, chickens and waterfowl alike, onto their winter feeding/exercise yard.

In contrast with wild Canada geese and mallard ducks, who amaze us with their migratory feats, domestic geese and true ducks—bred for greater size and weight—are no longer capable fliers and are thus easy to confine, even with a fence only a couple of feet high. A possible exception is Muscovies—some flocksters report that their Muscovies are forever flying over their fences, even roosting in trees at night. I have personally never had that problem with my Muscovies. True, they proved a little more capable of flight than mallard-type ducks, but I almost never had them fly over my electric nets, which are 42 inches high. Just as with rogue chicken fliers, you can clip wings of Muscovies who need a bit of discipline to stay where you want them.

When handling chickens, it's common to catch or carry them by the legs. You must *never* do either with waterfowl of any type, as *their legs are easily injured.* To catch waterfowl I either use a poultry net on a long handle—like a larger version of a butterfly net—or, if in close quarters, simply grab the bird briefly by the neck until I can gather it with hands around the wing shoulders, and tuck it under an arm. Watch those claws—especially in the case of Muscovies, they are *sharp.*

Culinarily, all the waterfowl have the same gifts to offer. Their flavor is rich and delicious. Their fat is easily extracted and as easily rendered for a superior cooking fat. Some waterfowl make significant egg contributions as well.

Fig. 24.1 I made this outdoor nest, containing separate nests for two broody geese. It's inside an area protected by electronet, with overwintered rye as a cover crop. PHOTO BY BONNIE LONG

If you are interested in breeding waterfowl, sexing (distinguishing the genders) is more or less difficult depending on species and breed. Vent sexing, if properly done, will give certain gender identification in all cases. Remember that the heavier breeds of both geese and ducks may require sufficient water to swim in for successful mating—the male may be incapable of mounting the female on the ground.

Broodiness among true ducks and geese varies a great deal. Broodiness is likely to be retained in virtually all Muscovy ducks (females), however, so they could be used as foster mothers where incubation or nurturing skills are lacking in other waterfowl females. The trait is also likely to be strong in all the bantam mallard-type ducks.

So long as they shield the broody females from wind and rain, there is no problem with placing nests outside—after all, wild waterfowl nest in the late winter or early spring when it's still cold. Set the nests up well before the laying season begins, to accustom the females to laying in the nests in which they are going to go broody. An outdoor nest unit I made of plywood for broody geese is shown in figure 24.1. Unlike chickens, none of these broodies can be moved once broodiness commences without breaking them up. Onset of broodiness is signaled by lining of the nest with feathers pulled out of her breast by the broody female, and by staying on the nest full-time. Access to bathing during incubation is a plus—the broody female who bathes during a nest break will take some water back to the nest on her feathers, helping to ensure proper incubation humidity.

In my experience all the waterfowl species are really easy to raise, and fun. They also make full use of good pasture; I even graze my waterfowl flock on four lawn areas around our house. Why mow?[1]

Fig. 24.2 Two African ganders.

Fig. 24.3 Domestic geese descended from the European Graylag are represented by the Buff Pomeranian on the right; from the Asian Swan Goose, by the African on the left.

Geese

Domestic goose breeds represent two different species, descended from separate wild ancestors. Embdens, Pilgrims, Romans, Toulouse, Pomeranians, and other common breeds descended from the European Graylag (*Anser anser*); while Chinese and African geese descended from the Asian Swan Goose (*A. cygnoides*). An identifying characteristic of the latter two breeds is the large, forward-inclining knob that develops where the upper bill meets the skull, shown closer up in figure 24.2. (Two feral species—the Canada and the Egyptian—have also been domesticated and are kept by some hobbyists.)

Geese are not kept for their eggs—their laying season is quite short—but the eggs are perfectly usable. Like duck eggs they are especially prized for baking. The best fit for geese in a small-scale flock is as meat birds and sources of cooking fat, who can grow without large feed inputs if they have plenty of access to good grass—geese are excellent grazers. Indeed, once past the brooder phase, they can thrive quite well on good pasture alone—though they will fatten better in the fall if given some grain feeds as well.

Sexing geese will be essential if you want to try your hand at breeding them. In most breeds gander (male) and goose (female) are colored the same; and the secondary characteristics of size, carriage, posture of the neck, and other distinctions often recommended for sexing are *not* reliable. A partial exception is sexing by reference to knob size in Chinese and Africans—the gander's knob grows earlier and larger than the goose's—but that difference is no help in sexing goslings. Even in adults there is so much variation among strains and individuals that knob size will not yield certain gender identification in all cases. Complete certainty is possible only with vent

Fig. 24.4 Pilgrim goose family. Pilgrim ganders are almost all white, and the geese a soft gray. Even the goslings are autosexing–note the darker female gosling between the two lighter males. Ganders participate fully in parenting. PHOTO BY BONNIE LONG

Fig. 24.5 Pilgrim goose with just-hatched gosling.

sexing—the complete eversion of the cloaca to expose the genitalia. Do note that qualification *complete*. If you fully evert the cloaca of a male, what pops out—an odd corkscrew-shaped organ with little rubbery knobs—makes identification certain: It's a boy! But there is a little "gotcha": If you do not fully evert—and it takes some skill and expertise to do so—the penis will lurk hidden inside the cloaca and give much the same appearance as the genitals of a female.

Note that vent sexing any waterfowl is stressful at best and can even injure the bird if handled clumsily. Learn this skill from someone experienced if possible; otherwise, study directions in a good source on this procedure.[2]

A few breeds of geese are autosexing or dimorphic—the goose and gander are distinctly colored, even at hatch. The most available of the autosexing breeds is the Pilgrim. Other dimorphic breeds exist—West of England, Shetland, Normandy, and Cotton Patch—but are quite rare. Do note that the autosexing trait remains useful *only* when breeders take pains to maintain breed purity. I once found, to my dismay, that some cull Pilgrims I was slaughtering—white and blue-eyed, supposedly dead ringers for ganders—in fact revealed tiny egg clusters in place of testicles on evisceration. Crossing breeds is common in commercial goose breeding, and an Embden x Pilgrim cross would be especially likely—the most probable explanation for Pilgrims who were not reliably autosexing—so be sure your source is breeding conscientiously to maintain breed purity if you want to raise autosexing geese.

I have found geese the most challenging of all the domestic fowl I've tried breeding. Last spring, though, I finally had a bit of luck.[3] As in wild geese, the gander helps with parenting. Indeed I found in my Pilgrim family—a gander and two geese—that they all shared equally the parenting of the goslings one of the geese had hatched (see figure 24.4).

Geese are the most long-lived of all domestic fowl. A neighbor of ours had an African gander, Peepie, known to everyone in our village, who lived to be thirty-five. I have read of a goose in England who, at her death at 104 years—in a farm accident, mind you, *not* by natural causes—was still laying and rearing a clutch of goslings each spring. Average life span is twenty years or more.

Geese can be pretty talkative, so they may not be a good choice for flocksters who need to minimize noise from their flock. On the other hand, that very trait may be welcome if you'd like alert watchfowl on your place.

We have raised Pilgrim, Chinese, African, Gray Saddleback and Buff Pomeranian, Roman, and

Embden geese. All have been fast growing and easy to raise. Geese are also chock-full of personality and great fun. Be warned if you adopt them into your productive farm or homestead flock, though—those very qualities make slaughter day more wrenching than is the case with any other domestic fowl.

Ducks

All true ducks share the same ancestor—the wild mallard (*Anas platyrhynchos*)—though as noted above they have been bred for greater size and weight and hence have largely lost the ability to fly.

You may prefer to raise ducks primarily for eggs. As with chickens, selection for laying has resulted in ducks that lay much of the year, not just in a narrow breeding season. You may be surprised to learn that the most productive laying ducks—especially Campbells and Runners—lay more eggs than the best layer breeds of chickens, from 250 to 325 eggs per year or more.

Heavier breeds such as Aylesbury, Pekin (the fastest growing of all duck breeds), and Rouen lay many fewer eggs and are raised primarily for meat. Just as with chickens, there are also dual-purpose breeds that perform well as both layer and meat birds—Saxony, Swedish, Orpington, and Magpie. My current favorite of all dual-purpose mallard breeds is the Silver Appleyard—or simply Appleyard—offering the best combination of beauty, egg laying, and fast growth to good slaughter size.[4] While not in the same league with Muscovy or bantam-breed ducks as broodies, Appleyard females are more likely than most heavier mallard-type ducks to retain the broody trait and make good mothers.

Sexing adult mallard-type ducks is usually easy. Among many breeds bill color is different in ducks

Fig. 24.6 Appleyards are among the most beautiful of all ducks. The female's coloration is more subdued, while the two drakes exhibit the green head and more showy plumage typical of many breeds of mallard or "true duck" males. Look closely to see the curled tail feathers of the drake on the left, another key to his gender. PHOTO BY BONNIE LONG

Fig. 24.7 Appleyard ducks tend to retain the broody trait and make good mothers. PHOTO BY BONNIE LONG

from that in drakes. The drake is noticeably brighter among particolored breeds than the more subdued duck. Except during molt, the drake has distinctively curled feathers on the tail, which are lacking in the female. Vocalization of the sexes is distinctive as well—the male's is a lower-pitched, more throaty *queeg-queeg*, while the female's is the brassier *quack-quack* we commonly associate with ducks.

Sexing ducklings can be more challenging. In some breeds differences in bill color between duckling and drakeling are definitive. Vent sexing as for geese is recommended if you need absolutely certain gender identification.

Ducks will make good use of all the green forage you can give them, but as grazers they are not in the same class with geese. They are, however, voracious eaters of worms and insects and can be particularly useful for controlling slugs and snails.

Muscovies

Though referred to as "ducks," Muscovies have an entirely different wild ancestor, *Cairina moschata*—considered by some more closely related to geese than to true ducks. They can mate with true ducks, though the offspring are likely to be sterile, like mules. In southwest France Muscovies and Pekin ducks are crossed to produce prized Moulard ducks. Though they are comfortable in water for brief periods, their plumage is less water repellent than that of the mallard-type ducks, so their needs for shelter in extreme weather are greater. Though like geese, they are great grazers, like mallard-type ducks they also eat an abundance of animal foods, such as insects, worms, and slugs. One of my correspondents reported flocks of between fifty and a hundred Muscovies on her parents' farm, who earned their living entirely through their own efforts in the growing season and were only fed in winter. Flocksters concerned about their noise profile should note that, in contrast with both geese and ducks, Muscovies are almost entirely mute.

Muscovies are a prized meat breed, and the females—as said, among the best natural mothers of all waterfowl types—may even brood more than one clutch of ducklings in a season. If I had to choose a single rugged, self-sufficient, easily reproducing waterfowl type, my choice would probably be the Muscovy.

Sexing of adult Muscovies is no problem, not only because of differences in size—the adult drake is almost twice the weight of the duck—but also the greater size of the bare skin patches or "caruncles" on the head and face of the drake. Ducklings must be vent sexed for complete accuracy.

Fig. 24.8 Muscovies are not true ducks. Like geese, they are excellent grazers who make good use of pasture. The drake is considerably larger than the duck, with larger face patches.

Guineas

Guineas are actually a kind of pheasant that evolved in the central and western plains of Africa and was domesticated four thousand years ago by the Egyptians. Both the Greeks and Romans prized them as table fowl, and the Romans carried them into their expanding empire.

Guineas go feral more easily than perhaps any other domestic fowl, and may need a good deal of encouragement to fix on the poultry coop as home—as opposed to setting up housekeeping in the woods, roosting in trees. A flockster friend of mine set an open platform on a pole for his few guineas to roost on, thereby accommodating their desire to roost well off the ground while keeping them close by. A further refinement would be to put a roof and wire mesh sides over the platform, complete with a door you could close at night if you wanted to get a hand on the guineas.

Most flocksters of my acquaintance agree that guinea hens are terrible mothers—allowing their babies (keets) to get wet in the early-morning dew, chill, and die. Others report success using guinea hens to hatch and raise keets and that they often nest and rear the young communally. But I have always played it safe and had broody chicken hens hatch out my guinea eggs when I needed more guineas. It is said that guineas will grow up tamer if hatched and brooded by a chicken mother.

Fig. 24.9 Guinea keets. PHOTO BY BONNIE LONG

Guineas ranging outside will hide a nest so well it is extremely difficult to find. One possible giveaway is that the guinea cock sometimes stands guard while the hen is laying her egg, giving a clue to where the nest is hidden. When you do find it, leave some fake eggs when you gather the eggs—otherwise, the guineas will make an even more securely hidden nest elsewhere. Laying is seasonal, from around mid-May to mid-September here.

We prize guinea eggs for the table. As with any smaller egg, the yolk of the guinea egg is larger in proportion to the white, making for richer omelets and scrambled eggs. The eggs are wedge-shaped in profile, pointed on one end and bluntly rounded on the other, and are softly banded in earthen colors, or finely speckled—a basket of them makes a striking centerpiece for the table. They are extremely hard, and fun for practical jokes: An unsuspecting guest asked to help break the eggs for the omelet will conclude before managing to crack one that you have substituted ceramic eggs for the real thing!

As a table fowl "guinea hen" is considered a gourmet item in some quarters. The meat is all dark with a hint of game in the flavor. Younger guineas are usually preferred for the table. I do not find guinea any better for the table than our pasture-raised, traditional-breed chickens—but both are of course orders of magnitude better than supermarket chicken.

Cocks and hens are colored the same, making gender selection challenging. You may read advice about sexing adult guineas with regard to body size, size of the "helmet," and the like, but such physical characteristics vary widely enough to make them unreliable for determining gender. Still, there are two foolproof ways to sex guineas:

- **Vocalization:** It is said that the distinctive two-note call of the guinea is *buck-wheat! buck-wheat!* (or *come-back! come-back!* or, according to my father, *pot-rack! pot-rack!*);

Fig. 24.10 Sharp-eyed guinea on the prowl.

but actually, that two-note call is the call of the guinea *hen*. The cock's cry is a harsh staccato shriek, *chi-chi-chi-chi-chi!* Be aware, though, that guinea hens will make the same cry as the male if alarmed. So the trick is to listen for distinguishing vocalization when the flock is relaxed and on routine patrol.

- **Tidbitting:** If you throw a special treat such as a cricket to your guineas, guinea cocks will exhibit the same gentlemanly behavior as chicken cocks; that is, rather than gobbling up the special treat himself, the guinea cock will call one of the hens to come and eat it instead, maybe even passing it to her with his beak.

The caveat to both the above methods for sexing your guinea is that now you have to catch it!

I have had some groups of guineas who only had something to say when on alarm, while others kept up a shrieking chatter *all* the time. You'll have to decide for yourself whether guineas would fit the noise profile for your situation. If on the other hand you're looking for watch fowl, guineas are hard to beat.

Guineas are among the best of all fowl for insect control. I have read that a pair will keep an acre entirely free of ticks. I use guineas for control of squash bug, nemesis to squash family crops in my area.

Guineas can become major players on the landscape when grasshoppers swarm to biblical-plague proportions. A correspondent in the Midwest reported seeing a flock of a hundred guineas in a line, taking out great swaths of grasshoppers down a field, then reversing to take out a swath in the other direction. An aficionada from Texas got her start with guineas when she faced a similar infestation. Since that time she has more than once had the experience of seeing the landscape around her as 360 degrees of brown, stripped clean to dirt by grasshoppers, with her own property a guinea-protected green emerald in the center.

Turkeys

Turkeys are a fun addition to the small farm or homestead. A friend's big tom turkey follows the family around like a dog, gobbling and displaying importantly, endlessly entertaining. Turkeys also have a lot to offer in terms of homestead self-sufficiency—I've seen how my same friend's turkeys glean insects and other free feed on his place.

Unfortunately, I have had dreadful luck trying to raise day-old poults ordered through the mail. On my first attempt I lost every last one I ordered. On the second I did only slightly better, with a surviving

Fig. 24.11 Narragansett male turkey displaying.

Fig. 24.12 Standard Bronze hen turkey.

fifteen out of an attempted nineteen. So I can only be honest with the reader that my sad experience bears out what a lot of flocksters find—that turkeys are fragile and die readily in the brooder phase, though they are tough as nails after that. (The same is true when starting guineas as purchased keets.) I would bet that poults would do better with a real mother—perhaps even a good broody chicken hen as a foster mother—than with any poor effort I can make in a brooder.

Once into the tough-as-nails phase, turkeys can roost outside in even the harshest weather, with just enough cover to keep the rain and sharpest winds off them. Do remember how vulnerable to predators they are when asleep, and give them adequate protection at night.

A major question for the small-scale flockster considering turkeys, particularly one who wants to keep chickens as well, is *blackhead*, a disease that is usually fatal for turkeys. Blackhead is caused by a parasitic protozoan, *Histomonas meleagridis*, likely to be present almost anywhere chickens are raised. Symptoms among turkeys include lethargy, retracted neck, drooping wings, suppressed appetite, and, especially, sulfur-colored runny droppings (sometimes dry, solid black droppings with waxy yellowish streaks). The head may darken—though usually it doesn't, so the common name is something of a misnomer.

Fig. 24.13 My friend Mike Focazio houses his small flock of Midget White turkeys in this open shelter, protected by an electric net fence, right through the harshest of winter weather. PHOTO BY MIKE FOCAZIO

Chickens themselves are highly resistant to the parasite and rarely suffer ill effects from it—though they may succumb to it if under great stress. However, chickens are carriers—that is, they help perpetuate the complicated life cycle of the protozoan and pass it on to turkeys, who are highly susceptible. For most flocksters, therefore, turkeys and chickens don't mix. If you raise both, keep them strictly separated, and run the turkeys only on ground chickens have not ranged. I have heard from a few flocksters with *small* flocks that they keep turkeys and chickens together without problems of blackhead. If you start with a clean slate where blackhead is concerned, you may be able to keep it that way. The key is whether or not your chickens have ingested the cecal tonsil worm—chickens that are not infected on clean land cannot infect turkeys.

PART SIX
Breeding the Small-Scale Flock

25 | BREEDING FOR CONSERVATION AND BREED IMPROVEMENT

The ambitious small-scale flock is a *productive* flock. But who can ensure that the birds we raise in the future will be equally productive? And what do we mean by *productive* anyway? Questions about the continued productivity of our flocks over time are ultimately questions about *breeding for improved production.*

If we define *production* narrowly—for example, how much meat and eggs our birds can put on our table in exchange for the time, effort, and expense required for their care—then we could hardly find better models of breeding for production than those in the large-scale poultry industry. Imagine a fast-growing meat hybrid ready to slaughter in as little as forty-four days to fill supermarket coolers. That's the Cornish Cross, a wonder of the modern world produced by the billions annually (a child growing at a comparable rate would weigh 349 pounds at age two). Industrial layer hens are equally impressive, coming into lay at sixteen weeks old and laying more than three hundred eggs a year. But of course, these miracles of the breeder's art are bred to a narrowly defined production model that is unsanitary, wasteful, polluting, and destructive of agricultural soils; they also require higher inputs of protein in their feeds to fuel that level of performance, and a steady diet of interventionist medications in order to survive. Hardly the sort of genetics we're seeking for the sturdy small-scale flock.

Breeders for exhibition do equally brilliant breeding work. I enjoy attending a big poultry show as much as anyone, but I always keep in mind that many if not most birds I see are more works of art, sketched in DNA, than anything I'd want in the rough-and-tumble of my backyard. Many exhibition breeders focus mainly on the fine points of color and pattern, comb and carriage—as opposed to production traits such as rate of growth and egg production, which are not apparent in the show cage. Breeders win ribbons for such refinements, even if their grand champion hens lay only a couple of dozen eggs a year. I have encountered competition breeders as well who have no compunction against crossing in "a little of this and a little of that" to produce their finely honed specimens, even if the result is a muddying of the water in the gene pool.

I hasten to add that my hat is off to competition breeders who conscientiously breed for the *production traits* for which their breeds were originally valued. If you're lucky enough to know such breeders, learn from their breeding practices, and maybe get your hands on some of their stock as well. A good example is Don Schrider, poultry writer and breeder, who manages to win ribbons with his show-quality Light Brown and Dark Brown Leghorns, while breeding to maximize their productive capabilities. Not only did I get hatching eggs from his fine stock this past spring, but I continually pester him for tips about his breeding techniques.

Most of us flocksters depend on the breeding work of commercial hatcheries to stock our flocks with productive birds. Just how much hatcheries emphasize production traits we have no way of knowing, however. What is highly likely is that hatchery birds have been bred in a production environment far removed from the conditions of a working homestead flock. And what is certain is that formerly productive breeds often decline in productivity when mass-bred. Good examples are the Delaware and the New Hampshire, breeds formerly valued both as reliable layers of large brown eggs and as table fowl. Indeed, performance in the latter role was so outstanding that each had its turn as the foundation of the broiler industry, before the Cornish Cross revolution banished all competitor broiler breeds. Until recently, the Delaware declined greatly as a meat fowl and layer and was of interest only on the show circuit. The New Hampshires on offer at most hatcheries should more properly be termed "Production Reds," far removed from Andrew Christie's meaty original, or from the higher-laying strain developed by Clarence Newcomer in the 1940s.[1]

The past decades have seen the same relentless consolidation and centralization in the hatchery industry as in other agricultural enterprises. That means that there are fewer individuals all the time who are making the decisions about how poultry stock is bred. But the news is even worse than that. A few years ago, the American Livestock Breeds Conservancy (ALBC)[2] did a survey of hatcheries and found that, in reality, the sourcing of breed stock was even narrower than the shrinking number of hatcheries would imply, because in many cases multiple hatcheries were actually being supplied with stock for particular breeds from the same breeding flocks.

However well or sloppily poultry breeding is being done, by whatever supplier, one thing is certain: None of them is breeding to *your* specific climate and soil conditions, tailored to *your* goals, using *your* management system. At one time it was common for small farmers and even backyarders to do so—and they were rewarded, like gardeners who save their own seed, with stock that was ever more finely attuned to their own unique situations. Readers who have taken to heart the central theme of this book will not be satisfied with poultry that is *productive* narrowly defined—the payoff in dozens and pounds of eggs and meat. They will wonder as well: How can our birds most efficiently furnish those eggs and meat while at the same time hustling up a lot of their own grub? What useful work can they accomplish if we provide them a life as close as we can manage to that of *Gallus gallus*? How do we ensure that our flock will be more healthy and more suited to our own circumstances with each generation?

One reason I breed Old English Games has to do with their *productivity*. Now that statement would get me laughed out of the room in most poultry circles—fast-growing meat hybrids grow far faster and make larger carcasses, and superlayer hybrids lay more than twice as many eggs. But it is no accident that the Old English Game has a thousand-year history of being valued as a utilitarian farm fowl. Imagine that readily available purchased inputs—feed, medications, new chick stock—became unavailable or much more expensive tomorrow, an eventuality I believe readers should seriously consider. In the changed circumstances the productivity of the Old English hen would become apparent—her ability to hustle up most of her own grub, maintain her health without purchased supports, and furnish her grateful owner with at least one clutch of chicks a year, and probably more.

It is disturbing to realize that traditional and historic breeds are being lost—as in *forever*. Breeds are not like precious coins that can be put in a vault for safekeeping and brought out when desired; they can be conserved only as living animals repeating the round of growth and death and reproduction. But in the long run they will not be conserved by well-intentioned efforts to save them as zoo specimens or a library of DNA. The key to the conservation work encouraged by American Breeds Livestock Conservancy is its recognition that breeds will be conserved only to the extent the *economic* traits for which they were originally developed are valued and used. A sense of altruism about saving heritage breeds may inspire us to take up their breeding, but only a self-interested *What's in it for me?* will carry the work into the future. And remember that even those who do not choose to breed their own stock can help in the work of conservation and breed improvement—by seeking out and buying from breeders dedicated to those goals, rather than from the usual mass-market sources.

If it's unsettling to recognize that there is no guarantee that anyone out there is breeding to keep the breeds we work with at their historic peaks of productivity, and if it's a certainty that no one is breeding stock to fit best our own homestead or farm—doesn't that suggest an obvious strategy: *to breed our own*? And if we do so, wouldn't it feel good to know that we are helping conserve the priceless genetics of the historic breeds that have come down to us? We never know when we'll need them.[3]

But if your head spins when you hear about alleles, genes, autosomes, homologous chromosomes—I assure you that mine does too. Fortunately, we don't need a degree in genetics to do sound poultry breeding. As in many things agricultural, the best direction for progress could be a giant step backward. Savvy farmers of an earlier time discovered and used techniques for improving their livestock and poultry, and we can still learn from them today. That's what the ALBC did in 2006 and 2007: Based on principles well established in a number of texts of the first half of the twentieth century (though largely ignored since, as more and more of us have relied on hatcheries), the ALBC carried out a breed-improvement project with the Buckeye—at one time an outstanding dual-purpose breed developed in Ohio at the end of the 1890s, but subsequently badly neglected as a production breed. Indeed, by the time the ALBC's breeding project began, the Buckeye was almost extinct—there could hardly have been a better test case of improvement breeding. Guided by those earlier selection and breeding criteria, the project, under breeder Don Schrider's direction, achieved in just three years an average increase of 1 pound live weight and a decrease from twenty weeks to reach slaughter weight, down to sixteen. Impressive gains indeed, and an illustration of what is possible in a serious breed-improvement program.

More recently, Will Morrow of Whitmore Farm in Maryland initiated restoration breeding of the much-neglected Delaware. Using the principles of the ALBC Buckeye program to identify and select his best performers, he brought his Delawares back up to their standard weights in two generations (see figure 25.1).[4]

While the bad news is the sad neglect of many production breeds, the good news is that the field is

Fig. 25.1 A pair of Will Morrow's Delawares. In just a few generations of judicious selection, Will has bred his Delawares back to their previous high standards for both meat and egg production.
PHOTO COURTESY OF WILL MORROW

thus wide open for homestead and small-farm breeders to restore some of the traditional breeds to serious levels of utility—that is, to make them productive enough to efficiently supply the family or small local markets. Within neglected traditional breeds, there is a *lot* of leeway for big gains in productivity in a relatively short time. For that reason progress can be surprisingly rapid, *if* we understand how to select for the traits we want.

Note especially, for example, that in both Will Morrow's work with the Delaware and in Don Schrider's with the Buckeye, weight gains did not require sacrifice of good egg production levels. To be sure, there are genetic limits on possible gains; too, at a certain point in a production breeding project, continued gains in meat qualities will come at the expense of laying capacity, and vice versa. In breeds whose productive traits have declined, however, gains in both may be possible.

Setting Goals for Your Breeding Project

Improvement breeding is of course about breeding superior performers. But isn't it empowering and liberating to realize that it's not some anonymous expert who determines what *superior* means? However rudimentary your knowledge at the beginning, *you* are the expert about selection criteria and breeding goals for your flock.

You might decide you'd like to improve laying performance in one of the traditional breeds such as Leghorns, Anconas, and Hamburgs that have been developed to maximize egg production, with some sacrifice of meat characteristics. If the aim is a good meat bird, however, it still makes sense to strive to improve egg production as well as meat qualities. After all, a breeding program with a lower average rate of lay requires more hens to provide hatching eggs, resulting in lower efficiency in the breeding project. This incentive toward improving *both* laying and meat qualities may explain historically why so many preferred farm breeds have been dual purpose. It may be that production breeding is about putting the *dual* back into *dual purpose*.

If you're breeding for meat qualities, which is superior—a bird that grows faster to slaughter weight or one with better flavor? If you're growing to sell, pricing constraints in a market that demands plump carcasses and is less discriminating about flavor may dictate the former choice. In that case breeder Matt John of Shady Lane Farm in Indiana advises that the best bet might be to work with one of four traditional breeds that have been neglected since the Cornish Cross became the standard broiler choice, but which retain the genetics from earlier breeding for efficient meat production: Naked Neck, New Hampshire, Plymouth Rock (White or Barred), and Delaware.

Matt advises seeking out well-maintained strains of these breeds, since there have come to be so many knockoffs offered by hatcheries, often drastically below their previous productivity levels. The point is perfectly illustrated by a lesson Don Schrider received on a North Carolina farm in 2007 from noted poultry breeder Frank Reese Jr. Frank asked Don to close his eyes and handle a hatchery Plymouth Rock hen. Don exclaimed in surprise: "Her body is shaped just like a Leghorn—and she is just as small!"

If you're not constrained to make maximum slaughter weight your first priority, dedication to producing truly gourmet poultry might incline you to work with breeds that are slower growing but have traditionally been valued as superior table fowl—Dorking, La Flèche, Crèvecoeur, Houdan, and Old English Game.

As for layers, which is superior in terms of your own goals and needs—a hen with early onset of lay or one with higher production long-term? One with better annual production, or who produces fewer but larger eggs, or who holds her production better in the winter? If you are taken with the beauty of very dark-shelled or speckled eggs, are you willing to give up some overall production if necessary to favor that trait?

Improvement Breeding Principles

Many of us are apt to think of breeding as hopelessly complex or arcane. Once we gain some experience with natural management of our flock, however, most of the key concepts are quite commonsense. Here are some of the principles underlying the ALBC Buckeye improvement project and its offshoots.

Breed the All-Around Bird

Don't focus exclusively on production narrowly defined—rate of growth, fleshing, egg production—since productive capacity depends on a high level of health and vigor in the bird overall. We are not breeding drumsticks or cartons of eggs, but birds we want to have a high state of fitness in every aspect. Traits such as rate of growth, disease resistance, early onset of lay, fertility, and hatch rate are controlled by interrelated sets of genes. It is better to choose breeders with a good balance of such traits, in preference to those who excel at a single production trait alone.

It's especially important to eliminate as a breeder any bird who has a weak immune system, as demonstrated by suffering illness of any sort. The chicken's immune response is naturally robust—our breeding efforts should help keep it so.

As long as breeders of whatever age, male or female, are meeting the requirements of the breeding program, they may be retained as breeders, to foster longevity and more extended productivity in future generations. Genes for longevity are associated with genes for vigor and health, so their presence among our selection criteria is advisable.

Maintain Genetic Diversity

Whatever breeding scheme you use, as much as you can manage, ensure that it maximizes *genetic diversity* among the offspring. It is intuitively obvious that we get the best results—offspring—if we breed our best individuals. There is an important caveat, however: Mating "best to best," with no further refinement, may prove more a recipe for decline in productive traits than enhancement. If you select your "best" from the entire season's hatch, it could well be they are all from a particularly favorable clutch. Mating them will mean mating brothers to sisters, which leads to *inbreeding depression* (decline in vigor and productivity) more quickly than any other breeding pattern. Even more important, choosing breeders who happen all to be from the same superior clutch means you are *not* breeding individuals from the other clutches of the season; their genes will be lost in future breedings. Maintaining genetic diversity is at least as important as selecting the best breeders, since only genetic diversity will keep the breed resilient and capable of evolving with changed circumstances.

Cull Rigorously

Selecting the outstanding few inevitably means selecting against the average many—another way of saying that *high culling rates are the key to selection.* For example, Will Morrow started his Delaware breeding project with 250 hatching eggs from a large hatchery, then culled in the first year to the top 10 percent of the resulting birds to serve as his breeders the following year. Since that time, he hatches about two hundred Delaware chicks each year as potential breeders and selects the top 5 percent of cockerels, since fewer cocks are needed for breeding, and the top 20 percent of pullets.

Though there may be variations in actual practice, it's a good idea to stick pretty closely to "the 10 percent rule": *Keep only the top 10 percent as breeders each year.* Combined with the rule above, however—about maintaining genetic diversity—this really means: Keep the top 10 percent of each *mating* during the season.

To clarify, imagine there are three groups of breeders in your breeding program—however selected and using whatever mating system you have chosen. You hatch 40 chicks from each group, for a total of 120 for the breeding season. The 10 percent rule suggests that you should retain only the best twelve as breeders. But the need to maintain maximum genetic diversity suggests that you not simply select the top performers from the 120 offspring, but instead select the top performers from each *mating*; that is, 4 each from Group A, Group B, and Group C.

In a recovery project in which only small numbers of individuals are available, you won't have the luxury of starting with a large pool of potential breeders. In this case as well, however, the key is heavy culling, keeping at most about one chick for every ten hatched.

If high culling rates seem drastic to you, remember that nature is herself a hard culler—she ruthlessly winnows individuals with flaws or weaknesses of any sort from the gene pool. While the process may seem cruel from our point of view, the result is strains of animals near perfect health and in almost perfect alignment with their environment. And don't hesitate to imitate Mother Nature's culling methods by deliberately pushing breeders with environmental extremes. A major reason for the hardiness of the New Hampshire is the fact that Andrew Christie, when developing this breed out of the Rhode Island Red in the 1920s, selected his breeders based on how well they did in minimal shelters exposed on pasture to his harsh New Hampshire winters.

Culling rates of 90 percent or so may seem logistically difficult in a small-scale setting. But remember that the culls from the breeding project can serve as the family's meat chicken. Another possibility is to share a breeding project with a poultry friend nearby: Manage the birds as one flock in terms of selection and breeding, but split their actual care between you.

Other Guidelines

While selecting breeders for good size and rate of growth, don't choose as breeders birds who are inordinately fast growing or larger than their peers—such birds often have a tendency toward immune system weakness.

Remember that selection of breeders is followed by selection of eggs. It is good to set large eggs, since genes for large egg size are highly heritable—and larger chicks make a more robust and vigorous beginning. For this reason don't set the smaller eggs of pullets when they start laying, at about eighteen to twenty-two weeks—wait until they are twenty-six to twenty-eight weeks old and their eggs are larger. Note, however, that abnormally large eggs should not be selected, since they may have problems with hatchability. Don't set eggs of odd shape or surface texture—ridged, lopsided, chalky shell, calcium bumps—these traits are highly heritable. Even if the flaw is only a temporary glitch in the system, hatchability of such eggs is drastically reduced.

Breeding should focus on improving production traits in the early stages. Once the flock's productivity is making good progress, secondary traits such as desired feather and eggshell color can be selected for as well.

Minimum standards for retention for breeding should become more stringent each year. In the ideal case the standard this year will be what the best achieved last year.

Improvement Breeding in Practice

Enough theory. How is the breeding project actually managed? How do we keep track of who's who in the flock? How do we select the chosen few for breeding? How will we organize the mating of our breeders, season after season, to replace the usual free-for-all matings in the flock?

Identification of Breeders

Even the simplest of breeding schemes will require that you keep records about breeders, if not as individuals, then at least by group—family, year of hatch, whatever. There are a number of ways to assign identities to your chickens, both temporary and permanent.

You can start when chicks are quite young—say, up to two or three weeks old—with *toe punching*. Using a special punch, similar in design to a ticket punch, only smaller, make one or more holes in the webbing between the toes. Make sure that the hole is clean—that is, that the punched-out tissue is not still attached on one side (like the infamous hanging chads)—otherwise, the hole is more likely to grow back in. To further guard against regrowing, repunch the holes after a couple of weeks. If you take the trouble to repunch, growing back is unlikely; the bird at maturity will clearly show a hole in the webbing, an eighth to a quarter inch in diameter.[5]

It is possible to avoid the problem of regrown toe webbing by notching it with several overlapping punches—or snipping with a small surgical shears—resulting in visibly shorter webbing at that position that will never regrow. As well, there is no hole to catch or fill with dirt as the chicken scratches.

Identification rings and bands of various sorts are available from poultry supply houses. All have their frustrations. Numbered metal bands make it possible to identify specific individuals, but I have found them clumsy to put on and difficult to remove, creating stress for the bird in either case. As well, you would only want to apply them after the bird has finished growing and the shank is full-sized.

Plastic spiral bands or *spiralettes* allow for coding both by color and by the leg on which it is placed. Snap-on plastic *bandettes* are both colored and numbered, allowing for coding individual birds. The numbers wear off over time or get badly obscured by grime.

An inherent problem with both these plastic leg bands is that they have to be changed as the bird gets larger—if ignored, they cut into the flesh of the growing leg. Monitoring and changing bands makes them rather high-maintenance with a sizable flock. Even if placed only on adults, the spiralettes and plastic bands sometimes work free of the leg or, worse, move up over the hock, constricting the thigh just above the hock, resulting in crippling and pain if not noticed. (I've never had either problem with metal bands applied once the bird had matured.)

Numbered wing bands are fastened into the

Toe Punching

Note two things about the toe punch chart in figure 25.2. First, there is nothing official about the coding positions—the one shown here is just for illustration. So long as you are consistent, you can pattern your codes as you wish.

Second, the chart shows there are sixteen codes possible, but only if leaving both feet without punches is also an ID code. However, I avoid "both feet blank" as a possible code. If punched webbings of some birds regrow, as occasionally happens, they will seem to be part of the "both feet blank" group. If "both feet blank" is not allowed, a bird who has regrown its webbing is an obvious unknown—it will not be misidentified with another group. Even if you have solved the regrowth problem by notching, there's always the possibility that you missed a chick when marking groups for identification.

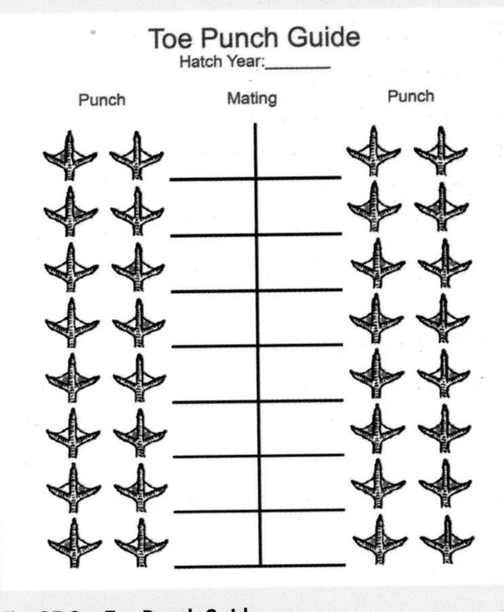

Fig. 25.2 Toe Punch Guide.

webbing of the wing with a special pliers. Except in some of the hard-feathered breeds, the band is hidden from sight when the wing is folded. Don Schrider prefers wing banding over all other options when it is necessary to identify individuals. (For assigning ID to groups, he prefers toe punching.)

Selection of Breeders

There are a number of mating systems. Whichever one you choose, however, improvement breeding starts with *selection* of breeders. The following are some of the selection practices used in previous, more hands-on farming eras, which ALBC's Buckeye project and its offshoots emphasize.

Tracking Weight

Both rate of growth and mature size are key production factors, so both must be tracked prior to selection. Note that the two are not always positively correlated, however. Both are important, so it is desirable to use as breeders both those individuals with a good rate of growth and those who attain a larger adult weight.

In order to measure progress between generations, it is critical to weigh at exactly the same age when evaluating prospective breeders. (You can cull for such obvious defects as crooked keel, crossed beak, or malformed feet anytime.) ALBC recommends weighing at eight weeks and again at sixteen. Will

Fig. 25.3 If you hood the bird's eyes with your hand, not quite touching, it will remain acquiescent long enough for you to read the scale.

Morrow has found that individuals with superior weight performance at earlier stages tend to be superior in later stages as well. He weighs at twelve weeks only, an acceptable slaughter age for his Delawares, and thus combines breeder selection with production of broilers for his market.

Figure 25.3 illustrates my technique for keeping a bird quiet while I weigh it in the scoop of a hanging scale. Don Schrider places the bird to be weighed in a square plastic bucket set on a digital scale. The shape inhibits the bird's getting its legs into position to make a getaway long enough for Don to get his reading.

Selecting for Egg Production

Certain traits related to high egg production can only be determined through hand selection: space between the pubic bones, distance between the pubic bones and the tip of the keel, a soft pliable abdomen, and other structural indicators of large internal capacity for the ovaries and other reproductive organs.[6]

Attention to molting characteristics aids selection of high-producing hens. Hens who wait until later in the fall or early winter to begin molting are generally those who have laid best in the laying season. Those who molt quickly and get back to the business of laying will produce more eggs in a year's time.

The more the performance of individual hens can be tracked, the more accurate the selection of breeders. Only the use of *trap nests* (see appendix A) yields specific production data for individual hens. Nest trapping may or may not fit into a given breeding project, since it is time consuming, but it does enable more targeted selection of high-laying breeders. If maintenance of winter production is important in the breeding program, for example, tracking of laying in winter by individual hens is necessary for breeder selection.

Remember that the male transmits genetics for egg production as well: A cock with a productive dam will tend to pass the trait to his daughters. Cocks should be hand selected using the same conformation criteria used for selecting productive hens.

Older hens who maintain egg production well are especially valuable for use as breeders.

Selecting for Meat Production

Judging the capacity for meat production is not just a matter of weighing the growing birds at a given age or checking fleshing of the breast. Certain *structural* characteristics are necessary to realize a bird's potential for putting on muscle. Heart girth; flatness, length, and breadth of back; depth of body; straightness and length of keel—all are related to allowance of room for a large digestive system and hence capacity for growth. For producing good table fowl, therefore, hand selection is also essential when choosing breeders.

A couple of other less intuitive traits correlate to good meat qualities: The width of the skull is related to the structural traits mentioned above and is an indicator of productive capacity. Also, birds with wider feathers tend to grow faster: Wider feathers retain body heat better than narrow ones, thus more energy from the bird's food is available for growth.[7]

Selecting for Foraging Skills

Though measurable production traits are important, breeding more self-reliant farm and homestead flocks means selecting for high levels of foraging skills. If you notice that a particular individual is a "real go-getter" when it comes to hustling its own grub, you might favor it as a breeder even if he isn't quite as heavy, or even if she doesn't lay quite as many eggs, as individuals more inclined to hang around the feed trough.

Selecting for Temperament

Remember to cull for temperament as well as other traits, to ensure that your birds will be even more pleasant to work with in future generations. Only you can define ideal temperament in general, but *all* of us would do well to cull manfighters, any birds—especially cocks—displaying unprovoked aggression toward humans.

Mating Systems

Breeding is the act of deciding which birds should be mated to which birds. The purpose of breeding is to produce the next generation of productive birds while maintaining genetic diversity within the flock. We need to consider diversity in order to limit breeding birds that are too closely related—which results in the loss of productivity over a few generations. We can classify breeding into two roles: improvement breeding and conservation breeding. Since both of these roles of breeding are of high importance, and will influence our choice of breeder birds and breeding methods, let me address them separately.

Good improvement breeding has the ability to influence the future productivity of our flock for generations to come. In improvement breeding, decisions about which-birds-to-mate-to-which-birds to produce the best offspring need only be guided by two concepts: offsetting faults and emphasizing good traits. Simply put, to rid a flock of a fault, always try to mate together birds that do not share that fault. To ensure that a good trait occurs more frequently within a flock, always try to mate birds together that share that good trait. Most of us tend to focus heavily on improvement breeding when we first start, and this is a mistake—simply offset faults, emphasize good traits, and place more focus on selection and maintaining diversity.

Conservation breeding should be a part of every breeding project, whatever its intent—it is the aspect of breeding management that has to do with maintenance of genetic diversity. We most often think of conservation breeding in connection with saving rare breeds of livestock and poultry. In fact, conservation breeding sets about saving rare breeds by first maintaining all available bloodlines or families, then by using breeding systems that minimize inbreeding in order to maintain genetic diversity. For the purpose of a small-scale flock, conservation breeding is about managing the relations of the flock such that we can produce healthy replacements for decades.

There is a difference, however, between line breeding and inbreeding. While both are the result of mating related individuals, think of line breeding as mating individuals of the same line and inbreeding as mating closely related individuals within the same line. Phil Sponenberg of Virginia Tech often says, "When it works, it is line breeding; when it doesn't it is inbreeding." Line breeding can have the benefit of locking in desirable qualities when used judiciously. Problems arise when we indiscriminately mate closely related individuals or continue to do so for too long, locking in bad traits, and resulting in a loss of reproductive function—for example, lower fertility in males, fewer eggs from females, and reduced rate of growth. Line breeding does offer great results in improved and consistent performance, and it need not be completely avoided. In fact, with the loss of genetic diversity that has occurred in most breeds over the last one hundred years, we can hardly avoid line breeding to some extent.

What we do need to avoid are those matings that represent close relationships: brother to sister; half brother to sister; sire to offspring for multiple generations; dam to offspring for multiple generations. Brother to sister is the most intense form of inbreeding and can result in decreases in size and performance in the offspring. In brother-to-sister matings, there is always a chance that the birds share many identical genes. In sire-to-offspring, dam-to-offspring, and half-brother-to-sister matings, the birds usually have some genetic differences but will produce offspring most like their common ancestor. Breeding closely related individuals for multiple generations increases the likelihood of similar genetics and thus hastens inbreeding effects—inbreeding depression—within your flock.

There are several different breeding systems you can use to mate your flock in order to manage the genetic diversity it contains. They are: Out and Out Breeding; Flock Sourcing; Flock Mating; Rolling Matings; and Spiral Matings. Each has good points and bad, and some are less sustainable than others.

Out and Out Breeding is the act of bringing in new roosters each season from different sources. In this system there is no known inbreeding going on, and vigor is high, but it is almost impossible to make improvements in production or adapt the flock to your region. One hundred years ago this system was used with mongrel flocks: New roosters of a different breed were brought in each season. The result was inconsistent production, and the system fell into decline. Even when using this system with a pure breed, while new males are not closely related to farm females, diversity of the breed you are working with is left to a handful of breeders who are successful at selling males.

Flock Sourcing is a similar system in which you find one good source and purchase new males from that source each season. Vigor is usually good, production is good, and quality improves—as long as your source is a skilled breeder. But you are relying on that breeder to do all the improvement work, and he may not select for traits beneficial to your farm or homestead. This system also does nothing to help maintain the diversity of a breed—since genetically, your flock will be simply an extension of the primary breeder's flock without uniqueness.

Flock Mating is the system of keeping the flock together as one breeding unit. In this system when there are multiple roosters, the birds manage the choice of mates with no regard for possible relation. Large hatcheries often manage large populations as one flock and cull the adults each year—keeping only young birds for the next season. Such wholesale culling only serves economic interests—reduced feed costs—while increasing the likelihood of brother-to-sister mating, half-brother-to-sister mating, and eventual inbreeding depression. When multiple generations of individuals within the flock are used, this system can work more effectively—but a population size of about two hundred birds, 10 percent of which are male, is usually considered the minimum size. On a homestead, featuring a much smaller number of breeders, you might mate the flock as one unit, using males only one season, using their sons the next, and bringing in new males every two or three years to reduce inbreeding.

Rolling Matings is a system that has also been known under the name "Old Farmer's Method." In this system pullets are mated to old cocks and hens are mated to cockerels each year. It requires the separation of the flock into two breeding groups, but no record keeping is required as

Continued on following page

Continued from previous page

long as you can tell the young birds from the old birds. After the breeding season is over, you mix the two flocks back into one, culling the females down to the number you can accommodate, comparing young to old—and culling all the males down to the number you are able to keep, comparing young to old again. In the fall the pullets are mated to these "old" males, and cockerels are used with these "old" hens. In this system the closest relationships are avoided, like brother to sister, though some line breeding does occur. Rolling Matings is a reasonably good system that manages diversity fairly well, even more so when care is given to ensure offspring of both breeding groups are retained each year. With Rolling Matings improvements in production can be realized through good selection over the generations.

Spiral Matings is a system that is used in both poultry and other livestock and is called "Clan" matings by some. In this system three or more families are set up—each with a unique toe punch, which is used year after year to identify that family. Note that both male and female offspring are toe punched with their mother's family mark. Families can be named by color band, like Blue, Red, and Pink; they can be named A, B, and C; or they can be named after their source, such as Ussery, Schrider, or Sand Hill. Once this system is up and running, Blue family sons would be mated only to, say, Red family hens and pullets, while Red family sons are mated only to Pink family hens and pullets, and Pink family sons are mated only to Blue family hens and pullets. Males are used only one or two seasons and, no matter what generation you are in, are always mated to females of the next family in rotation, such as Blue to Red. Females remain in their own family and never rotate—and can be retained as long as they are productive and there is room.

Regardless of the system you choose, you will find reasons to bring in new blood from time to time. So how do you handle this? In both Out-and-Out Breeding and Flock Sourcing, you need to bring in new males each season; this can be done by purchasing chicks or adult males. In Flock Mating, when you have a small flock, you will need new male(s) every three or so years. In all these systems you must use care when choosing the source for the new males to avoid bringing diseases into your flock.

Selecting for (or against) the Broody Trait

An important selection criterion is the brooding instinct. Clearly the trait is inheritable—some breeds retain it to a high degree, others almost not at all—even if the genes for this complex behavior are unknown. Most commercial breeding seeks to eliminate broodiness among hens. You may well decide you want to encourage the broody trait instead, to achieve more self-reliance in your homestead or farm flock. On the other hand, a flock featuring broodies exclusively would leave us with no egg production during the entire spring brooding season. An elegant solution might be the breeding of both a world-class broody breed such as the Old English Game to furnish a reliable working-broody subflock and a layer breed such as Light Brown Leghorn to keep egg production up. Expression of broodiness would then be equally crucial for selecting breeders in both breeds: Among the Games any hen who failed as a broody would be culled; among the Leghorns broodiness would be selected against by culling broody hens.

When selecting among broody hens, further refinements are possible.[8] I hatch eggs using natural mothers exclusively and prefer starting all my new stock in one big wave in the spring. My selec-

In Rolling Matings, new blood can be brought in as males or females. Females can be integrated with the same age group breeding flock—for example, pullets are added to the pullet breeding flock. New males can be added also and likewise would be used based on age—old cock mated to pullets or cockerel mated to hens. But when you bring in new males, you can expect fights that can result in severe damage to one or more males in the flock. So new males are best used to replace an entire age group of males (young or old) or used in a third mating on the side—in the latter case their young need not be identified separately from the other cockerels or pullets. Rolling Matings is a good system that can run, depending on the number of breeder birds used, for ten or more years without the introduction of new blood.

In Spiral Matings we can integrate new blood by inserting both males and females as a new family, or by replacing males or females of one family. The beauty of this system is that within a few generations the new genes will be amalgamated into your flock and you will soon have consistency in the offspring. If you choose to breed your flock using Spiral Matings, you will have little need for new blood. In fact, depending on the size of the flock, you can go twenty years or more when using three families and eighty years or more when using five families.

Spiral Matings offers a system that allows you to maintain genetic diversity within your flock, while selection and culling within each family allow production to improve. This is conservation breeding (maintaining the genetic base) combined with improvement breeding (keeping your "best") and is what we who want productive animals should now consider the best direction.

My final thought is something that master Leghorn breeder Francis LeAnna of Wisconsin once said to me: "One of the great joys in breeding, no matter which system you choose, is knowing that you are responsible for the results." So choose the system that will work best for your homestead or farm, and know that you can do this!

—Don Schrider, poultry breeder and writer for *Backyard Poultry* and other poultry publications[9]

tion, therefore, favors hens who go broody early, and who do not return to broodiness after their spring clutch. You might prefer a more random expression of broodiness, to spread rearing of new chicks over the green season, or favor hens who brood two or even three clutches of eggs in the season, if you have a lot of hatching to do. As for winter broodiness, it makes no biological sense to me—I usually cull a hen who goes broody in winter.

After Selection

The previous observations have focused on the selection process for a successful breed-improvement project. How the breeder proceeds after selection is the next challenge. There are a number of mating systems from which to choose. Some require complex tracking of individual breeders through multiple generations. As with use of trap nests, the enhanced specificity may or may not be judged worth the additional time and effort. Fortunately, more simple breeding management systems that minimize record keeping are available.

In the past I have used Rolling Matings for its greater simplicity of record keeping. In the coming season I plan to begin using Spiral Matings, sometimes known as family or clan matings, as I begin more serious improvement breeding of Old English Games; as well as a new flock of New Hampshires, Newcomer strain.

Something New Under the Sun

I referred above to the critical need for conservation breeding as a way of keeping alive our heritage of useful poultry breeds from the past. But Joel Salatin, in his recent book *The Sheer Ecstasy of Being a Lunatic Farmer*,[10] makes the excellent point that we shouldn't be concerned only about saving the heritage breeds, many of which were bred in Europe. We should instead be breeding our own uniquely American, region-specific strains and breeds. Should you care to take up Joel's challenge, know that the task certainly can be done. Nettie Metcalf developed her Buckeye, suited to her harsh Ohio winters, out of crosses among Barred Plymouth Rocks, Buff Cochins, Rhode Island Reds, and black-breasted red games. Other American breeds, developed out of assorted crosses to produce breeds attuned to their local environment, include Plymouth Rocks, Dominiques, Wyandottes, Rhode Island Reds, Jersey Giants, and more.

If there are a couple of breeds you like a lot, but you would really prefer one that combines the virtues of both, seems to me you're almost ready for takeoff. If you cross the two, the first (or F1) generation may well reward you with higher average performance than that of the parents—it's called hybrid vigor. The next (F2) generation, however, is likely to be all over the map genetically, with as many underperformers as stars. But you can choose from them the individuals who come closest to your ideal. Proceeding by selection and breeding as suggested in this chapter, you should eventually end up with your own unique breed—your gift to the world. And *eventually* needn't mean so very long. I believe Nettie Metcalf's development of her Buckeye took seven or eight years. My friend Tim Shell developed his Corndel, a cross between Delaware and a proprietary strain of Cornish Cross, into a reasonably stable genetic package for use as a pastured broiler in about three years.

26 | MANAGING THE BREEDING SEASON

Please don't let anything said in the previous chapter make you think that breeding your own stock need be a big, complicated project. It's pretty straightforward for the cock and the hen, that's for sure. This chapter will discuss a few of the logistics of organizing the breeding season, however casual or refined your breeding project. Information about how I manage breeding here is intended as illustrative only—you may well find other ways to manage your own breeding season.

The Mysteries of Fertilization

We are familiar enough with the biological intricacies of reproduction in our own species but may be a bit clueless about the ways in which avian reproduction differs. Yet an understanding of these fundamentals is essential if we are to avoid some common mistakes producing the eggs we want to hatch.

The female of a wild bird species starts laying eggs only after she has found and mated with a male. Laying eggs for her is always and exclusively about reproducing her species—she will never lay eggs that are not fertile.

In the case of the chicken, however, selective breeding has resulted in hens whose egg production is no longer connected strictly to a seasonal reproductive cycle. For the domestic hen ovulation is more analogous to that of a woman, who ovulates periodically whether or not she has a mate. The hen lays an egg every day or so, regardless of whether there is a cock in the flock whose attentions would fertilize the eggs. It is worth emphasizing this point, since I am so often asked, "Do I have to have a rooster in my flock?" If you prefer, for whatever reasons, to have hens only in your flock, be assured they will still lay eggs—though, of course, the eggs will not be fertile. For fertile hatching eggs we have to be sure the hens have sufficient exposure to an active, virile cock.

What constitutes *sufficient*? Generally, it is recommended that from eight to twelve hens per cock will virtually guarantee 100 percent fertility in the eggs—the lower end of the range might be more appropriate for heavier breeds; the higher end, for lighter ones. But a ratio of up to twenty-five hens per cock will supply eggs with a high fertility rate. Thus the hen-to-cock ratio you choose for the flock can be quite variable, depending on how important it is to ensure 100 percent fertility in the eggs.

An interesting difference between human and avian reproduction is the term of viability of sperm in the female's reproductive organs. In humans the viability of sperm inside a woman's body—and thus the window of opportunity for fertilization—is more limited, as little as one day, rarely more than three to five. But in the hen, the cock's sperm live much longer—and continue to fertilize eggs the hen is growing and laying over the course of many days.

Just how long do sperm survive inside the hen? There can be a good deal of variation—it is said that the period for sperm viability can be longer in some breeds than in others. But imagine we have a number of hens who have been frequently mated. We isolate them, with no further exposure to a cock, and incubate a batch of eggs each day thereafter, tracking the resulting hatches on a graph. The first clutches of eggs would likely hatch at a rate approaching 100 percent. At a week to ten days we would see a falling curve on the graph, representing the decrease in fertilization in the eggs as sperm inside the hens died off. At some point, maybe about three weeks, the hatch rate would fall to zero—indicating the death of the last sperm in the hens' oviducts and complete absence of fertilization.

This is not merely a theoretical exercise. Suppose we want to be completely certain of the parentage of the chicks we are hatching—that is, certain that fertilization comes *only* from the cocks we have isolated with our hens, and not from more free-for-all matings that occurred in the general flock *before* we isolated the breeders. How long do we wait after isolation to be certain that sperm from previous matings have now died, and that fertilization can only come from the parent birds in our breeding pens? Again, it depends on how important absolute purity is in the breeding program. Imagine isolating Wyandotte hens who have been mating regularly with Rhode Island Red cocks but now, after being removed to the breeding pens, are being mated solely by Wyandotte cocks. The longer the time since the last breeding by the Rhode Island Red cocks, the fewer and less active will be the sperm from those encounters, while there will be vastly greater numbers of more vigorous sperm from the more recent matings with the Wyandotte cocks. It becomes more and more likely that in the competition to find and fertilize the germ cells in the developing egg yolks, the Wyandotte sperm will win out. The ratio of pure Wyandotte to Rhode Island Red x Wyandotte chicks would track as a rising curve on a graph, until the former represent 100 percent of every hatch.

I usually wait about ten days to two weeks after isolating my breeders before collecting hatching eggs.

Isolating the Breeders

I prefer to keep the flock together for as much of the year as I can, avoiding the additional maintenance of servicing separate pens. As the breeding season approaches in late winter, I select my breeders and isolate them in the Breeders Annex.

The exact period of isolation depends on several factors. Since I rely on broody hens (natural mothers) to do my hatching, there is little point in making the separation much more than a couple of weeks—the desired withdrawal period discussed above—before my first mother hens are apt to enter broodiness, perhaps in late March. Those using incubators might prefer an earlier start on the breeding season.

On the other hand, as late winter edges into early spring, the testosterone starts to rise in the cocks. Up until that point, cocks who have accepted subordinate positions in the hierarchy with good grace may be more inclined to mount serious challenges for access to the hens. And these guys play for keeps: On a few occasions cocks in my flock have fought to the death in this high-stakes period. As a practical matter, therefore, I isolate the breeding cocks as soon as I begin seeing conflict among them that goes beyond routine sparring.

When we built our new poultry house, I converted the existing chicken coop—which came with our house when we moved in—to serve as isolation quarters for breeders. I am not a breeder who would find it acceptable to squeeze my breeders into tiny cages. I divided the Annex into five pens, each with plenty of room for three or four adult chickens—up to seven in the largest. With the five Annex pens, my breeding program can easily accommodate seven breeding cocks: five in the Annex and one each in the greenhouse pens and in the main poultry house. I have occasionally stretched that to nine, making subdivisions with wire partitions in the latter.

The pens are simply chicken wire partitions on light framing, allowing maximum light and ventilation. Each has a wire-on-frame door opening onto the central passageway, which can be latched open or closed as needed. Each is fitted with a roost or two, a nestbox mounted on the wall above floor level, and a feeder and a waterer made of recycled plastic containers, hung on the wall.

In the excess space above the pens, I installed broody boxes for isolating setting hens—if one of the breeder hens goes broody, I simply move her to one of the isolation boxes above in preparation for her work.

The Annex door opens onto a yard defined by a single 164-foot roll of electric net fencing, each end anchored on a corner of the coop—a total area of about 1,700 square feet. Close confinement of my birds is no more an option in the breeding season than any other time of year. On the other hand, release of all the breeders onto the yard would result not only in fighting among the breeding cocks but in the very cross-breeding I have gone to such pains to prevent. My solution is to allow each breeding group access to the yard in turn, typically from at least three pens per day in rotation. In the yard the breeders enjoy all the benefits of exercise, fresh air, and sunshine, just as they do when with the main flock. They also have access to large piles of compost inside the yard, which they shred and turn throughout the breeding season, while garnering the sorts of live, nutrient dense feeds—pill bugs, slugs, grubs, and earthworms—likely to ensure maximum vitality in the hatching eggs.

In some cases the same cock will remain with the same group of two or three (up to six) hens. In other breeding situations, I might rotate cocks through more than one pen, increasing genetic diversity in the offspring.

Collecting Hatching Eggs

A common misunderstanding is that a broody hen only sets her own eggs. A wild mother bird, of course, sets a nest containing her eggs only. In contrast, the chicken broody will happily set any clutch of eggs you choose to put under her—I have even set chicken broodies on duck eggs. Indeed, one of our bantam hens once hatched a goose egg! The eggs you choose to set can be from another source—from a fellow flockster, from a commercial hatchery—or from any birds in your flock from whom you want progeny.

Bear in mind that a wild mother bird lays an egg a day in her nest, where they remain viable for many days, at ambient temperatures. That is our key to proper handling of hatching eggs: Store them in an egg carton at room temperature—not beside a heat register, nor in the full light of the sun. Store gallinaceous eggs—those of guineas, turkeys, and chickens—narrow-end down, since the air cell inside the egg is at the other end, which is where we want it for now. Orientation of waterfowl eggs is not critical for storage up to a week. If holding longer than that, store them on their sides and rotate them a half turn daily—clockwise one day, counterclockwise the next.

You can write on the shells in pencil if you wish—date laid, parentage code—but do not use a marker. I don't usually mark my eggs. I place them into labeled cartons, one for each of my five breeding pens. When a carton is full, I rotate out the oldest eggs each day. Do note that hatching eggs rotated out in this way are still perfectly edible. Contra the assumption that eggs must be refrigerated, a fresh egg will keep as long as it would if laid by a broody hen who had hidden a nest—easily a couple of weeks. We never refrigerate eggs, and you do not need to either, if you eat them within a week or so.

Using this strategy, I always have on hand a minimum of a dozen freshest eggs—collected over a number of days—from each breeding pen, ready to be set without delay, as soon as the next hen goes broody.

If an egg is dirty, I do not keep it for hatching. If there's just a trace of smear or stain, in my experience that is no problem for the development of the egg, and I just set such an egg as is. I'm making this point because it is sometimes recommended to wash hatching eggs with a special solution. I never wash hatching eggs. The wet coating on an egg when it is expelled by the hen—called the bloom—not only helps lubricate the egg to ease its passage but provides a protective barrier against bacteria. Washing eggs can actually reduce the time they can be safely stored, and I believe may decrease protection of the developing embryo.

How long can you store eggs and be sure they're likely to retain hatchability? At least as long as a hen who has hidden a nest would take to assemble a clutch of eggs—up to two weeks. Conservatively, if you store hatching eggs under the conditions described above, they should all remain hatchable for ten days. After that time there will be a falling-off of hatchability.

However, if older or less-than-ideal eggs are for some reason the only ones available to you, know that many a clutch of eggs three or even four weeks old have yielded surprisingly good results. A friend of mine had sold all her eggs when suddenly her prize cock was killed by a fox. In desperation she went to her customers, some of whom had some of her eggs remaining in their refrigerators to donate to the cause. Hatch rates turned out to be excellent and saved my friend's breeding line.

At the End of the Season

As said, I have kept up to nine cocks in preparation for the breeding season, ensuring greater genetic diversity in the matings. My new breeding plans imply that number will go up to twelve.[1] Once the breeding season is over in mid- to late spring, however, and I am no longer separating the breeders, life in the general flock tends to be pretty unsettled with that number of cocks. So the end of the breeding season is the time when I cull cocks not needed for the next breeding season. Typically I cull to a single mature cock per mating group that I will use once more for breeding. There are plenty of replacements coming on as the season progresses, and the younger cocks accept the dominance of the older as they mature. This strategy provides as many younger cocks as needed from this year's hatch, while almost eliminating serious aggression between them and older cocks until it is time once again to separate my breeding groups in the Breeders Annex, in preparation for the breeding season.

Mature cocks are even tougher than old stew hens, and at our place they are reserved for the same stockpot.

27 | WORKING WITH BROODY HENS

A regular part of the green season at our place is seeing a Carolina wren faithfully executing the intricate duties of motherhood—building her nest, incubating the eggs, and caring for her chicks until they are ready to fly off into the joys and hazards of the wide world, beginning the cycle anew. Seeing her at this great work is always a special joy.

When we need to raise new chicks—for dressed poultry for the table or for replacement layers—why not let our hens imitate the mother wren and hatch out their own young? There are great advantages to recruiting a mother hen for this work, perhaps the most important being that she is a lot smarter at the task than we are. No incubator manual, no instructions for brooding chicks received through the mail in a homemade brooder will ever put you in the same league of expertise as a mother hen. There are few joys on the homestead to match seeing a hen with her clutch of chicks. As the current economic and energy crises deepen, we may need to rethink our dependence on outside inputs. A mother hen can save us the multiple expenses of incubators, heating elements, and thermostats. She will continue to ensure the viability of the eggs and chicks during prolonged outages in an increasingly stressed electric grid, to say nothing of the rising cost of electricity itself. The chicks she hatches are essentially free, in contrast with chicks sent through the mail whose purchase price may become more and more a limiting factor. And she reduces our dependence on purchased feeds by finding for her babies natural, nutrient-dense foods of a quality superior to anything we can buy in a bag.

I have found, over many years advising flock owners about working with natural mothers, that the most common mistakes arise almost exclusively from the same source—a lack of understanding of the details of avian reproductive biology. So let's begin with the mother wren's spring rituals—mating, nesting, caring for the young—to understand the natural behaviors in play. Then we may better shape realistic strategies for working with broody hens and broody females of other domestic fowl species.

The Mother Wren

Please understand that the following discussion of brooding behaviors in this wild species—the Carolina wren—is greatly generalized and does not address the many variations in avian reproductive behaviors. I am not writing a zoological treatise, just trying to develop some general comparisons to better understand broodiness in chickens and other domesticated fowl.

Seasonality, Mating, and Building the Nest

The time of year when the wren's mind turns to love and motherhood is not haphazard—it is strongly keyed to the opening of the green season, when chicks can be fed from the abundant insect life. A female may hatch only a single spring brood, or may hatch a second in the summer, but she will never nest in any other season of the year.

Increasing day length and warmer temperatures trigger nesting and mating behaviors. The female finds a mate—her eggs will not be fertile without the earnest attentions of a male—who usually has already started building a nest. Note that the nest is small and well hidden, the wrens' best provision for thwarting hungry predators. Though some bird species nest communally, the wren wants to be secluded, secure from intrusion by other nesting females, who would also like a place to hatch some babies.

Laying a Clutch of Eggs

Mating and completion of the nest usher in hormonal changes that bring on the production of eggs in the female's oviduct. The wren typically lays one egg per day, until she has assembled her clutch. It is extremely important to understand three things about the egg-laying period: First, unlike in our own species—in which sperm remain alive only a limited time in the female's body—in avian species the male's sperm remain viable and mobile for much longer. Thus *the sperm from a single mating continue to fertilize multiple eggs*, as they are grown sequentially in the oviduct and laid over the course of many days.

Second, the wren lays *only the number of eggs she can completely cover and keep warm* and no more—she has the instinctual wisdom to know that embryos in eggs not adequately covered against the chill will die. Once she senses that she has assembled as large a clutch of eggs as she can effectively incubate, hormonal changes shut down further egg production.

The final critical point is this: During the egg-laying period, the wren comes to the nest, lays her egg, and flies away—she does not remain on the eggs, which thus remain at the season's ambient temperatures. Inside, the fertilized cells from which the embryos will grow are poised and waiting but dormant. The brief warming of eggs already in the nest, while she's laying the next egg, does not trigger incubation. *It is only the prolonged and sustained warming of the eggs to the mother's body temperature that will do so.*

Incubation

Only after the clutch is fully assembled, and egg production has ceased, does the mother wren settle into the nest and remain on it full-time. Only at this time does the uninterrupted body heat of the mother cause the fertilized cells in the egg yolks, regardless of which day the egg was laid, to start to grow into embryos. All at the same time. All on the same schedule for hatch.

It is worth emphasizing: *No additional eggs are laid after the mother begins incubation.* Embryos in eggs subsequently laid in the clutch would be on a different schedule for hatch—a disaster.

Once the embryos begin to grow, they cannot revert to dormancy—they must remain at incubation temperature and grow rapidly, or die if they are too long without the mother's body heat. The mother makes brief outings only from the nest, for food or water. She also leaves to poop, to avoid fouling the nest. Such brief forays from the nest do not chill the embryos sufficiently to kill them. Indeed, among some avian species—waterfowl, for example—a brief daily cooling of the eggs as the female takes a nest break is apparently necessary for proper development of the embryos.

Once the mother wren is deeply settled into incubation, traumatic disturbance—say, by something she sees as a predator—is likely to break up her nesting behaviors, wrenching her out of the broody state of mind and causing her to abandon the nest. It makes more sense, perhaps, to leave the compromised nest and start the entire process elsewhere, rather than risk return of the predator and loss of the current clutch after yet more time and effort.

Hatch

Hatch occurs after about fifteen days. The sounds and movements under the mother as the chicks heroically break out of the shell call the wren's mind into the next phase: nurture of the chicks. She will not wait indefinitely on a final egg that for some reason fails to hatch, nor will she make any efforts for a chick that is weak and struggling after hatch. She focuses all her energies on the vigorous chicks.

Nurturing the Young

During the following couple of weeks, the mother serves three vital roles for her rapidly growing young. Assisted by the male, she feeds them an abundance of rich natural foods, mostly insects. She ensures they stay warm enough, since the flimsy down with which they hatch does not provide sufficient insulation for the chicks to maintain body heat. They do not chill dangerously on routine flights by the mother to catch insects, but the mother covers the chicks in the nest at night, as well as during wet, windy, or especially chilly conditions, when the chicks need some on-the-spot warming. Finally, any mother bird does her best to protect her babies from predators.

Defensive behaviors vary among avian species. I don't think wrens actually attack threatening predators—their main defense is the cleverness with which they hide their nests. Some mother birds, especially ground-nesting species such as grouse, put on a distraction display—floundering away from the nest, faking a broken wing—to draw a nosy predator away. Some species will attack a predator outright—we have observed an entire local community of barn swallows gang up on crows and blue jays flying too close to a nest with swallow nestlings. A mother Canada goose will fearlessly attack a fox who is trying to get at her goslings.

The mother works incredibly hard providing high-payoff insect foods for an amazingly fast rate of growth in the chicks. The season is short, and the chicks are vulnerable to predators in the nest—snakes, rodents, other bird species—so they have to grow up and start living on their own in a hurry. By the time they are well feathered enough to retain body heat on their own, and are strong enough to fly, their hardworking mother has enabled them to do something truly extraordinary: In one swoop they start flying, feeding themselves, and roosting in the wide world, without retreat to a nest. The mother has made a heroic effort getting them to this point, but she now breaks the maternal bonds and resumes a life focused on her own needs.

The Mother Hen

Understanding natural brooding behaviors among wild birds prepares us to anticipate challenges and provide solutions when working with broody hens. The important thing to appreciate from the beginning is the degree to which chickens, and other domesticated fowl, have moved away from strictly natural behaviors. In some ways, as an avian species, chicken nesting behaviors parallel those of the wren—and in some ways they have changed considerably in the alliance with *Homo sapiens*. Actually, both behaviors that are still natural and those that depart significantly therefrom can be used to advantage when working with broodies, so long as we are clear about the nature of a given behavior. The following discussion will focus on comparisons and contrasts with non-domesticated avian nesting behaviors, in order to reveal effective management strategies for broody hens.

The Great Forgetting

The first and biggest question about the natural behavior of the hen you want to use for hatching chicks is this: Does she "remember" the intricate lore of incubating, hatching, and nurturing chicks? The question I am asked most frequently about using natural mothers is: "But my hens don't go broody—how can I make them do so?" The simple fact is that *the reproductive behavior of most modern breeds has been deliberately and dramatically shifted away from the natural.*

In the era of mass production of chicks using artificial incubation, broodiness is considered not merely an unnecessary nuisance but an economic calamity. After all, a hen who has gone broody and is determined to be a mother—like our broody wren above—ceases laying eggs. Thus a major component of modern breeding has been a rigorous selection against the broody trait—that is, the elimination from breeding flocks of hens with a strong mothering instinct, in favor of hens who happily lay their egg per day, having forgotten that doing so has any relation to reproducing the species.

Shouldn't it *bother* us that we consider the ideal chicken one who has forgotten how to reproduce her kind? In any case, the effort to eliminate broodiness as an inherited trait has been extraordinarily, though hardly surprisingly, successful: If we make going broody a capital offense, it doesn't take the chickens long to get the point.

Hens of most modern breeds, therefore, cannot be relied on as mothers. It is true that hens of some breeds—Cochins, Brahmas, and Buff Orpingtons come to mind—have a reputation for being more likely to brood than others. But trying to find good broodies among such breeds is hit-or-miss at best. Even when the trait is present, it may be more or less intact. Some hens, for example, will go broody in the nest, set[1] with great determination for a couple of weeks, then abandon the enterprise and return to daily activity in the flock. Or perhaps a hen will complete incubation, then prove somewhat clueless about caring for the chicks.

To be assured of hens with sound mothering instincts, it is best to revert to the older, historic breeds of chickens, in whom the broody trait is more the norm than the exception. My favorite breed for a working-broody subflock is the Old English Game. Other game breed hens—Asils, Malays, Shamos, Kraienkoppes—are also likely to have strong instincts to be mothers. Dorkings are an ancient breed, going all the way back to Roman times. Not all Dorking hens will brood for you, but many will. Bantams as a class are likely to retain the broody trait, though because of their small size, they are more limited in the number of eggs you can set per hen, in contrast with standard breed hens. Sadly, even among these petite breeds, the prejudice against broodiness is strong and is being selected against by some breeders.

The Brooding Season

As said, there is a pronounced seasonality to nesting behavior in the female wren: She goes broody in the spring, at the same time that all the other wrens in the area make nests and incubate eggs. A particular female may or may not brood an additional clutch later in the summer, but she will never do so in the fall or winter.

Seasonality is one of the ways in which chickens' mothering instincts differ most sharply from strictly natural behaviors. Hens with the inclination will begin going broody as late winter turns into early spring. But there is no set time for onset of broodiness—some hens may wait a month or even longer before making it obvious they want to be mothers.

Another trait with a big range of variability is whether the hen returns to broodiness later in the season—many will not, though some will hatch a second or even a third clutch of chicks. Folks who raise Silkies tell me they want to do nothing but set and will return to broodiness as soon as the earlier clutch of chicks are ready to make it on their own.

A hen may even take it into her head to set a clutch of eggs in the winter. Going broody in winter is rare, but you will occasionally find that a hen is setting a clutch of eggs in an out-of-the-way corner, indifferent to the bleakness of the winter landscape outside—a testament, I suppose, to the degree to which the domestic chicken has loosed her moorings to a more natural seasonality. Whether you decide to let a hen who goes broody in winter hatch her clutch is up to you. My own sense is that brooding so much at variance with the nature of the season—fragile, highly vulnerable chicks simply have no place in the iron grip of winter, however dedicated the mother—violates biological boundaries. At the least, I break up a winter broody, and I might even cull her outright.

Broody Breeds

If you want to find absolutely reliable broody hens to hatch chicks for you, don't play guessing games with hens of hit-or-miss breeds like Buff Orpington or Brahma. While such hens usually make excellent mothers if they do decide to brood, you can't count on their doing so. Here are three options if you want broodies almost guaranteed to work for you, starting with my personal favorite.

Though Old English Games have a thousand-year history as a utilitarian farm fowl, it is true that cocks of this breed have been exploited in the cockfighting pits. That may explain why hatcheries typically do not offer this breed for sale[2] (though many of them do offer bantam versions of Old English): They may be nervous about being tainted by an assumed support of cockfighting if they do. As well, breeding management of any game breed would be complicated by the heightened level of aggression among cocks. To locate breeders of Old English Games, join the Society for the Preservation of Poultry Antiquities (SPPA). Membership puts you in touch with breeders of all the older breeds, all over the country, from whom you may be able to get stock.

There are a couple of other options you might consider, if smaller broody hens are adequate to your needs. The Nankin is one of the most ancient breeds of bantams, widely used in England from sometime earlier than 1500. Before the era of the artificial incubator, it was favored as a broody for hatching eggs not only of farm chickens but also of pheasants and other game birds.

Silkies are another bantam breed favored as mother hens. Their feathers lack barbs or quills—someone suggested they "look and feel like Persian cats." Silkie hens have a reputation as extremely devoted broodies: "They don't want to do anything but set," I've been told by those who use them. A Silkie hen may assemble her next clutch of eggs even as she is completing the rearing of her current clutch of chicks.

Breeding for the Broody Trait

If you can't find hens who will reliably brood for you, remember an experiment cited by Charles Darwin in his *The Variation of Animals and Plants Under Domestication* (1875): Two decidedly nonsetting "Mediterranean class" breeds were crossed, and all the daughter offspring of the cross expressed the broody trait. I find it comforting that, despite the intensity of our manipulations, our domesticated fowl retain just under the surface their native wisdom, ready to reassert itself with a single roll of the genetic dice.

While it is generally a good idea to maintain the purity of breeds of poultry that have been historically useful, here is a case where deliberate breeding of "mongrels" may be a good thing. For a number of years I have been experimenting with my "Boxwood Broody" cross: the crossing of Old English Game cocks onto proven broody hens of larger breeds. The goal is to produce broody hens who can cover more eggs per clutch than the smaller Old English but with the same level of broody skills and motherly devotion. My Boxwood Broodies are a motley assortment regarding color, pattern, comb, and shank—selection is based not on any particular conformation but on how well the hen performs as a broody. Always, a high level of broody skills trumps body size.

Spotting Broodiness

"But how do I know a hen has gone broody?" Unlike the wren, the chicken broody does not usually make a new nest for incubating her eggs. There can be exceptions—as when a hen who is completely free-ranging hides a nest and assembles a clutch of eggs, with no intervention from you. But typically, *the hen will go broody in the same nest where she—along with her sisters in the flock—normally lays her eggs.*

Thus a further contrast with natural behavior: The chicken broody will enter true broodiness when the requisite hormonal changes occur, irrespective of whether she's actually setting any eggs, which the flockster is gathering every day. (I like to leave a fake egg or two in all my nests, about which more below, to give a hen something to settle on if she's tending toward broodiness.)

The key to spotting broodiness is when she decides to set the nest full-time. Don't assume that a hen is broody simply because she is spending a lot of time on the nest—some hens linger a long time when tending to the business of egg laying. But the broody hen does not leave the nest to go to roost with her sisters at night. She takes on as well a deeply settled, Zen-like intensity that is impossible to describe but highly symptomatic once you have some experience of it. An exploratory hand will draw forth a flattening of the hen in the nest, a fierce raising of the hackle feathers, and an outraged *Skraaaawk!* That hand may receive a peck, usually a mere token of vigilance against the predator—though on a couple of occasions an especially indignant peck has drawn blood from my own hand, especially in the case of guinea hens.

Once you have determined a hen is indeed fully broody, *you must remove her from the egg-laying nest.* Far too many beginners try to leave the broody hen in the regular egg nest she has chosen as the ideal place to hatch her chicks. But the hen may leave the nest to relieve herself, then return to the adjacent nestbox by mistake, leaving her incubating eggs to chill and die. Or other hens may enter the nest to lay their own eggs. Even if you have marked the eggs in the clutch so you can distinguish them from eggs laid by other hens, sooner or later all that jostling will break an egg, coating the clutch with goo. Since incubating eggs "breathe"—via essential oxygen exchange right through the shell, which is porous at the microscopic level—a coating of broken egg can smother the growing embryo. Trust me on this one: I've been there. It doesn't work. Don't do it.

Isolating the Broody

Let's be clear about an essential point: Truly natural behavior—as with the nesting wren—would mean that the broody hen would not tolerate being moved into a different space without breaking up her broody inclination. But chicken broodies have diverged sufficiently from natural norms that they are likely to remain broody despite being moved. This is not to say that every broody hen will tolerate it: You will occasionally encounter a hen who will not settle anywhere else than in the egg nest she has chosen for incubation. But because chicken broodies as a group will tolerate this level of management intrusiveness—I *insist* on it; that is, I define a good broody as one who works with me in a mutual endeavor to hatch her chicks.

Fig. 27.1 One of six broody boxes installed above the breeding pens in my Breeders Annex. The broody has feed and water free-choice, and enough space to get off the nest to relieve herself. The floor of the broody box is half-inch hardware cloth for easy cleaning.

Do note that broody females of other domestic fowl species retain the more natural behavior in this regard: *They cannot be moved, after thoroughly settling, without breaking up broodiness.* With such broodies it is necessary to plan ahead and make sure there is a nest for each female. Once an individual becomes broody, add some sort of physical barrier to prevent other females from using her nest and disturbing her.

When you move the broody hen, you must do so with as little intrusiveness as possible. *Move her at night only*, when she is in the trancelike state that constitutes chicken sleep. Move her to an isolated place of her own, physically separated from the other flock members, into a previously prepared nest. And set her initially on some fake eggs on which she can settle in the new nest: You can purchase plastic or, less commonly, wooden eggs. You could also use golf balls, perhaps even smooth stones—anything loosely suggestive of "egg" in shape and size.

As for the isolation area or broody box—any out-of-the-way corner will do. Give the hen enough space for a nest, feed, water, and room to stretch and poop. Like the mother wren, a good broody has the instinct not to foul the nest if she has enough space. Use cardboard, scrap plywood, wire screens—whatever is convenient or in the recycle heap to establish a physical barrier that excludes other hens.

If you decide to do a lot of hatching with broody hens, you will want to install permanent broody boxes. I prefer mounting them on the wall, so no floor space is lost for the rest of the flock. I have a set of six so mounted, made of leftover plywood from a friend's construction project. Each box should be at least 24 by 30 inches and 16 inches high. By all means use wire mesh—half-inch hardware cloth is best—rather than solid floors. The wire increases ventilation through the unit and is easy to clean—just use a scraper to push the poops through. Wire also doesn't accumulate an inch of dust in the off-season like a solid floor (the cleaning chore from hell).

As for the nest, you can use cut-down cardboard boxes—which get mulched at the end of the breeding season—or scrap lumber or plywood to make sides about 4 inches high to retain the nest materials: burlap, straw, wood shavings, and the like.

It is not unusual for a hen to be somewhat agitated the day after being transferred to the broody box, moving about restlessly inside, as if trying to find her way back to her chosen nest, perhaps squawking indignantly. Most often, however, she will settle on the fake eggs in the new nest by the end of the day. If she fails to do so, you might leave her in the broody box an additional day. If she still fails to settle, she is unlikely to do so. Then you have to decide what to do about her. You might return her to the main flock and see if she returns to broodiness—especially in the case of a first-timer, she may settle with less resistance on a second attempt.

Because a hen who goes broody is not laying eggs for me, I take seriously her offer to work as a mother. If she fails for any reason to carry through, I'm inclined to cull her to the stew pot. Such a strategy may seem drastic, but it leaves me with a subflock of dependable working broodies. On the other hand, I *never* cull a good working broody. As long as she continues to perform strongly as a mother each season, she retains an honored place in our flock, until she dies a natural death—or I pass her on to someone else who wants to work with mother hens.

Breaking Up a Broody Hen

A key goal in the above section was avoiding breaking up the broody hen with our interventions. But suppose a hen enters broodiness at a time not convenient to the demands of the season or our breeding plans. You will be amazed at how persistently a broody will set the nest—sometimes for weeks on end. Is there a way to encourage her to get over it and return to egg laying and normal life in the flock?

An excellent way to break up a broody hen is to do exactly as you would if you wanted her to incubate: Isolate her in a broody box with feed and water and a wire floor—but in this case without a shred of nesting material. Keep her there—typically a week or so—until she lays an egg. Then return her to the main flock.

Another effective strategy is to isolate her with an active young cock, whose constant attentions get her mind running in less motherly directions.

If I am maintaining two separate flocks—as I often do—I find that the simplest method for breaking up unwanted broodiness is to remove the hen from her chosen nest and put her into the other flock. Generally this transition—not only into a different physical space but into an unaccustomed social hierarchy—breaks up the hen immediately. Should she try to brood in the new flock quarters, I immediately return her to the other flock. Theoretically, I would keep bouncing her back and forth until she forgets about incubation, but in practice I've never needed more than one additional bounce to accomplish the task.

Setting the Clutch of Eggs

Once the broody hen is thoroughly settled, you are ready to set the hatching eggs from your chosen breeders, collected as discussed in the last chapter. *Working only at night*, remove the fake eggs from under the setting hen and substitute the eggs you've reserved. Here's another big contrast with natural behaviors: The hen is not like the broody wren, who has laid her own eggs in a nest she built herself from scratch—she's in a nest you've provided and placed her in, on eggs you choose to put under her.

How many eggs do you set? The actual number is strictly a function of the body size of the hen, the size of the eggs you are setting, and, to a lesser extent, the point in the season. Remember what was said above about the mother wren's instinctual understanding when "enough is enough" regarding clutch size: *She will lay only the number of eggs she can completely cover and keep warm.* You must develop the same sort of instinct for clutch size. If you are setting standard-breed chicken eggs, you might set five or six under a bantam broody, nine or ten under an Old English Game, up to fifteen under a big matronly Wyandotte, Brahma, or Buff Orpington. Perhaps twenty guinea eggs could be set under a standard chicken broody, perhaps only three duck eggs under a bantam. If in doubt, err on the side of setting slightly fewer. If it is still early in the season and nights are cold, set fewer—the hen's body heat must offset lower ambient temperatures in order to maintain incubation temperature in the clutch.

I cannot stress enough the importance of not going for broke when determining clutch size, assuming you might as well risk a few eggs that don't make it, for the sake of maximizing the number of eggs that do hatch. Not so fast—that strategy is more likely to *minimize* the hatch. During the course of incubation, the hen continually moves the eggs around in the clutch. If she is setting more eggs than she can completely cover, there is a chill zone around the edges of the clutch in which embryos fail to maintain incubation temperature, and die. When the hen rotates those eggs into the center of the clutch, and eggs with viable embryos from the center into the overexposed edges, those embryos may also chill and die. The result is a potential loss in the clutch greater than that of the inadequately covered eggs alone.

Once you put the eggs to be hatched under the hen, all the embryos start growing at the same time, thus are on the same schedule for hatch. Obviously it would be a mistake to add any more eggs after the initial setting of the clutch, since those embryos would be on a delayed hatch schedule.

When you set the clutch of eggs under the hen, you should mark on your calendar the expected hatch date. But I'm going to give you a tip that to my knowledge you won't see anywhere else: Any source of information on the subject I know will tell you that incubation of chicken eggs takes twenty-one days—as indeed it does, in an artificial incubator. In my experience, however, hatching of eggs under natural mothers is as likely to occur in twenty days as in twenty-one. Make it standard practice to staple on the door of every broody box a note that specifies: the broody inside (name, flock number in your records, or other identification); how many eggs she is setting and of what breed, cross, or parentage; and earliest possible hatch date, twenty days out. (On rare occasions I've seen a clutch take a day or two *more than* twenty-one, which would not happen in an incubator.)

Keep in mind, however, that the hen is not tracking incubation period with a calendar. Her setting behavior will end not after a given number of days but when the sounds and movements of chicks hatching under her signal the end of incubation. Only then will she begin the next phase, mothering the new chicks. This fact is particularly useful if you use chicken broodies to hatch out eggs of species with a longer incubation period (guinea, turkey, and mallard-type duck, twenty-eight days; goose, twenty-nine to thirty-one days; Muscovy, thirty-five to thirty-seven days)—a good chicken broody will make no objection at all to remaining on the nest for the additional time. Similarly, if you need to schedule a hatch to your own convenience, you can keep a settled broody fixed on a clutch of fake eggs a week, maybe up to two, before you set hatching eggs under her—she will set the additional three weeks with marvelous patience.

Managing the Broody Hen

Once you have set your hatching eggs under the broody, she will do the rest. Check on her unobtrusively each day—she will not be disturbed if your movements are quiet and gentle. Refill her waterer as needed, and provide feed free-choice. Broodies usually eat some feed each day, but it is not unusual for a broody to have little interest in food during this period. Do not be concerned. The broody may in any case lose up to a third of her weight during incubation, so be sure all hens you set are in good condition and well fleshed. As for feed, I've found that if I change to a leaner feed for broodies—coarsely cracked corn and small grains, for example—there is less chance they will have loose, diarrhea-like poops, and the broody boxes remain cleaner.

Some broodies like to leave the broody box occasionally, while others never do so even if given the chance. If a hen makes it obvious she would like to leave the box for a quick outing, I generally allow her to do so. She will typically emit an explosive poop of an odd, distinctive smell, then maybe take a quick dust bath and return on her own to the broody box, since she instinctively knows the eggs must not cool too much. If she fails to return—say, by mistakenly getting into one of the egg nests to continue setting—the embryos in the cooling eggs will die. If I allow a broody off the nest, it is only when I am caring for the general flock, and I make certain the broody is back on her nest when I leave the area.

Candling the Eggs

It is a good idea to candle the eggs midway through the incubation period. Work at night, in full darkness, right beside the broody's nest. Remove the eggs from the nest, and working quickly, shine a strong light through the egg. You can buy candling lights, or make one, though I just use a strong flashlight. At about day ten, a growing embryo will show as a small pulsing mass at the center of a spiderweb of red supply veins. Keep examining eggs until you are sure you recognize a living embryo with its support system. Then it will be obvious when you find a nonliving egg—one with only a yolk showing, or a dark mass. Such eggs should be discarded immediately.

It is tempting to skip the chore of candling—and admittedly, I sometimes do—on the assumption that "it'll all come out in the wash" come hatch day. And frankly, you can usually get away without candling in a typical clutch. But remember, a non-viable egg is a rotten egg, and the putrefaction in it generates gases that can sometimes cause it to explode. Not only is the resultant smell not to be believed, but the remaining eggs get covered with a thick coating of goo, which may block the necessary gas exchanges through the shell, in effect smothering the embryos. The contents of the exploded egg carry a heavy load of nasty bacteria as well, which can penetrate the pores of the other shells. You should candle instead.

Hatch Day

Plan ahead for the hatch. Remember that chicks become active an amazingly short time after hatch. If the nest has sides that might prevent a chick who has fallen out from getting back in, place a little straw around as a ramp on which to climb back in. A chick who cannot get back under its mother will chill and die.

Check progress on the expected hatch day without being too intrusive. With most broodies you can slip a hand gently under the hen and feel the eggs. If you feel a crack in one of them, pull it out and examine it. The first stage of hatching is pipping—the chick cracks open a little hole from the inside. (At this point, if you hold the egg up to an ear and tap with a fingernail, you hear the chick peeping inside. Kids *love* this.) Later the first crack extends around the entire shell, which breaks open into two halves, the wet, exhausted chick sprawled between. After an hour, the chick will be dry and fluffy and as said surprisingly active (see figure 27.2). During the day you can remove the empty eggshells from the nest as more chicks hatch.

Remember that the embryos all started development at the same time. However, their rate of growth varies sufficiently that the first chick may be out of the shell sixteen hours or so earlier than its slowest sibling. The hen has the wisdom to know that she must not leave the nest early, and waits patiently for the last chick to hatch. The early arrivals hatch with the last of the yolk material in their systems and are thus able to wait awhile without feed or water.

Fig. 27.2 Hatch day.

Awhile means easily forty-eight hours, and up to seventy-two. I've never seen a clutch of chicks take anywhere near the latter, though I once had a clutch of Muscovy ducklings take two full days to hatch. The fact that the hatchlings are well enough furnished with internal supplies of water and nutrients to wait so long before eating or drinking explains why new hatchlings can be sent through the mail. Do note, however, that this "pause" period in the young hatchlings' life ends as soon as they have their first water and feed. From that point they must have regular access to feed and water—there is no going back to the grace period when they could do without.

In practice I typically wait until the following morning for the last chicks to hatch. Any egg showing no sign of pipping at this point is unlikely to hatch. If you shake it, you may hear a liquid gurgle inside—proof of a non-viable egg. Even if there is pipping that has not progressed, if you tap on the egg and hear no peep, it is clear the embryo has died attempting to hatch. Such failed eggs should be removed from the nest and the hen encouraged to leave and start caring for her chicks.

Sometimes a chick is unable to break free of the shell on its own. While it's tempting to intervene and help it out, this apparent kindness is ill advised. Breaking out of its shell is difficult for the chick, but that difficulty itself is nature's first challenge for the new life. If it isn't strong enough to meet the challenge, and you give it a boost it would not otherwise have had, it is likely to start life weak and struggling. Perhaps it's lacking in vigor, a trait you would not want to pass on to offspring. Better to let it make that first big step or fall on its own. Like the hen, you should focus your efforts on the vigorous chicks in the clutch.

Broody Hens as Foster Mothers

If my description of the advantages of a broody hen over the artificial brooder sounds good to you, you might conclude that it would be a good idea to give purchased day-old chicks to a broody hen to mother. Will she accept such an offer? Maybe. Most of my attempts to graft purchased day-old chicks onto a

Hope, Supermama Extraordinaire

My most spectacular success grafting chicks occurred with a White Jersey Giant hen named Hope, who came off the nest with only six chicks of her own. Shortly after hatch, I received forty-two day-old chicks in the mail and put them in an adjacent section of the poultry house, set up as a brooder. Believe me when I say that Hope *asked* to be mother to those new chicks, and when her request finally penetrated my thick skull and I opened the door between the sections, Hope rushed in and began busily mothering them.

But that is not the end of the story. That night I tried grafting ten purchased goslings onto a broody goose. The goose was willing, but the goslings wouldn't fix on her (wouldn't recognize her as mother), and kept wandering off through the dew-wet grass. By morning goslings were going down like dominoes. In desperation I scooped up the remaining seven and prayerfully offered them to Hope. She didn't even *blink*.

For weeks thereafter the center of the action in the henhouse was Hope's kaleidoscopic clutch—a tumble of colors, sizes, and shapes during the day; a tight happy clump spilling out from under Hope in a corner at night. And Hope was the happiest hen alive.

Hatching Chicks in an Incubator

An elderly neighbor Ellen and I got to know soon after we moved to Boxwood told us that, when she was growing up, her mother used to provide a custom hatching service in the village. People wanting a batch of chicks would furnish hatching eggs from their hens, and pay Mrs. Green a fee to hatch them out. She set them behind the woodstove and fiddled with them as needed to keep the temperature in the right range. If she could pull it off with a woodstove, you should be able to do so with an incubator—and tips from my friend Don Schrider, who uses two of them to hatch his prizewinning Light Brown and Dark Leghorns.

Over the centuries various methods have been successfully used to incubate the eggs of fowl. The Egyptians built what amounted to ovens, attended to by crawling inside and adjusting the temperature by feel. In America early breeders used broody hens to hatch the eggs of nonsitting breeds of chicken. Later, kerosene and electric incubators were developed. I have even known someone who tossed his eggs into his compost pile only to find chicks springing forth three weeks later. The point is, you should understand that nature has designed the egg such that it will hatch even when conditions are not exactly perfect.

Successful incubation actually starts prior to the egg being laid: Be sure the hatching eggs you use are from healthy, vigorous stock that have been fed well. Remember that the egg is the sole nutrition for the development of the chick, as well as its food for the first three days after hatch. Commercial layer mash is hardly suitable to producing eggs with sufficient nutrition to produce vigorous chicks.

Proper handling and storage of hatching eggs is important. Think of a fertile egg as being pregnant; it is growing ever so slowly at room temperature, and rough treatment and temperature extremes will damage it. Keep hatching eggs in a location with a fairly constant humidity and temperature, ideally about 60 degrees Fahrenheit, where the eggs will not be disturbed—a basement works well, as long as it is not overly damp, cold, or drafty, as does a pantry or a closet so that sunlight does not heat the eggs up during the day. Monroe Babcock, a famous commercial Leghorn breeder of the past, recommended storing the eggs in an enclosed container with little or no ventilation—the idea being to prevent environmental changes and prevent development in the fertilized eggs before it's time to place them in the incubator.

I like to store my eggs in egg cartons in a portable cooler. I also place a scrap piece of 2-by-4 under one end, changing ends daily, to rotate the eggs—this prevents the germ cell from sticking to one side of the egg and as we'll see is even more important during incubation.

Select the eggs to go into the incubator with care. First, heavily soiled eggs should not be used. Remember, an incubator is an ideal environment to grow things, including bacteria—let's not introduce contaminants to this environment. You can remove small manure spots by lightly sanding just the soiled area of an egg. You can also wash the eggs in a disinfectant bath, but consider: Doing so removes the "bloom"—a protective antibiotic coating applied when the egg is laid. If you do choose to wash the eggs, do so just before incubation; use water that is warmer than the eggs so that disinfectant and germs are not drawn into the egg as its temperature changes; and be sure to use a minimal amount of disinfectant so that you do not kill the embryo (Pine-Sol or chlorine bleach, about one teaspoon to a gallon of water; commercial eggwash solution, as directed). More environmentally friendly cleaning solutions include quaternary ammonia, ethyl alcohol, hydrogen peroxide, and peracetic acid, a mix of acetic acid (vinegar) and peroxide. Second, eggs that are odd in shape, rough in texture, too big, too small, or poorly colored should not be used. Such egg characteristics are highly heritable, so select eggs for the incubator that look like what you want to see in the egg basket when your future hens start laying.

Let's talk about incubator types. There are a variety of incubator designs: from those meant to hatch three eggs to big commercial units that will hatch tens of thousands per batch. Be on the lookout for great deals on old and used incubators. To decide what kind of incubator you need, let's think about how many chicks you want to produce and how often. If you use one incubator, you will only be able to set one batch every three weeks. A Styrofoam incubator with an egg turner installed, usually holding forty-two eggs, is fairly

Fig. 27.3 Don Schrider's two incubators, set up in a closet to help keep the temperature even. His GQF incubator on the right is the one in which he incubates the eggs for two weeks. He completes the hatch in the Brower incubator on the left, confining the mess to it and cleaning between batches. PHOTO COURTESY OF DON SCHRIDER

inexpensive but has many nooks and crannies and is thus hard to disinfect between batches. Brinsea makes an all-plastic incubator, which disassembles, and the parts can be put in the dishwasher for cleaning.

GQF Manufacturing makes a four-hundred-egg incubator with a tray at the bottom that will allow you to hatch up to 135 chicks per week while continuing incubation of other batches in trays above. When operated in this way, however, the dust produced by the hatching chicks' down gets on the subsequent batches, and thus cleaning is hard to do without cooling the other eggs too much. I like to hatch a few very large batches of chicks, so I use a GQF incubator of this capacity, which performs that all-important function of turning the eggs twice a day, as my main incubator. I solve the GQF's cleaning problems by using as well an old Brower incubator made of

Continued on following page

Continued from previous page

redwood—I use it as the hatching incubator for the last week of incubation. This system allows me to clean the GQF only once per season and confine the mess to the Brower, which I clean between batches.

Incubator operation and location are important for success. Incubators come with instructions for operation, including adjusting the temperature and humidity for best results. Most incubators will have fans installed to move the warm air around the eggs. They usually operate at a temperature of 99 degrees Fahrenheit and humidity at 55 to 65 percent. Location of the incubator is of prime importance. Choose a location with a constant temperature and humidity, out of the sun, and where it will not be disturbed: basements, guest rooms, and closets all make good choices. I once had a friend that put his incubator in his garage but had bad results—the incubator could not keep up with the changes in temperature around it. Another friend also used his garage but built an insulated closet for the incubator. The best location I have used was a basement—humidity and temperature were constant, and hatches were excellent.

Let me end by giving a number of tips I have found useful over the years:

- I use either Pine-Sol or bleach to disinfect my incubators. Add just a teaspoon of Pine-Sol to a gallon of the water used to maintain humidity, and the incubator will stay clean and fresh smelling. Bleach can also be used, at about half this rate, but it may cause galvanized parts to corrode. Use your nose! If an incubator smells bad or "off," it needs cleaning. If the disinfectant burns your nose or eyes then it is way too strong for the chicks too. (Again, more environmentally friendly cleaning solutions include quaternary ammonia, ethyl alcohol, hydrogen peroxide, and peracetic acid.)
- Save eggs for hatching for up to two weeks. You'll get best results with eggs saved for up to ten days, but eggs two weeks old will still hatch well. After two weeks hatchability drops drastically. Eggs stored in a refrigerator can hatch, even if they do get a little too chilled, so if your family has accidentally put your hatching eggs in the refrigerator, all is not lost.
- Candle eggs at 10 to 14 days. Working at night with the lights off, I set each egg on the lens of a strong flashlight to shine its light through the egg. Remove all eggs that have stopped growing, as they will overheat the other eggs in the incubator.
- If a good egg shows a slight crack when you candle, you can still usually save it. Simply light a candle, and drip the wax over the crack. Use as little as possible so that the chick inside the egg can breathe.
- If the power goes out, wrap a blanket around the incubator to help hold in the heat. Eggs can stand a few hours with the heat off as long as they do not chill. Don't despair; they will simply hatch a little later.
- If hatching eggs have been mailed to you, let them sit at room temperature overnight for a better hatch. This allows the eggs to re-form: The egg's structures—yolk, chalazae, albumen—have been jarred and shaken in transport and need time to "resettle." Also, the eggs gradually

Fig. 27.4 Don's GQF incubator turns the eggs automatically, an essential task that he would otherwise have to do by hand. PHOTO COURTESY OF DON SCHRIDER

come up to room temperature. Giving them a "rest" this way avoids shocking the embryo at the start of incubation and yields much better results than eggs placed immediately in the incubator.

- Don't open the incubator until 24 hours after the chicks are due to start hatching. Chicks hatch with a three-day supply of food and fluid, absorbed with the last of the yolk, so they can stand the wait while their brothers and sisters hatch. But opening the door will cause a drop in humidity, with the result that chicks cannot break the membrane holding the shell together.
- Hatching chicks is one of the wonders of having a home poultry flock. Good luck.

—Don Schrider

Fig. 27.5 Hatch day in Don's Brower incubator. PHOTO COURTESY OF DON SCHRIDER

Fig. 27.6 I like to get the new family out on pasture from day one. PHOTO BY BONNIE LONG

broody in this way have been successful. However, never assume that success is certain, and be prepared to brood the chicks yourself if the hen doesn't cooperate. The hen should have been on the nest a couple of weeks—she is unlikely to accept a graft if she has been broody only a few days. But you can hold a willing broody on her nest with fake eggs for four or even five weeks, until your purchased chicks come in. The hen is not counting off days on a mental calendar—she moves on to the next phase when she hears live chicks under her.

To make a graft you should again work only at night. Remove the fake eggs and slip the chicks, who have been kept quiet in their shipping carton through the day, under the hen. Check on them later that night, and again at first light. Chances are excellent the hen will be delighted to welcome "her" new babies into the world. I have only a couple of times had a hen reject grafted chicks—but in the worst case, the hen killed a few of the "intruders." Monitor closely, and be prepared to intervene.[3]

The New Family

My practice is to move the new family directly from the hatching nest to the pasture. If that seems drastic to you, consider this: The first time I did so, the mother, one of my Boxwood Broody cross hens, went onto the pasture in the last week of March with nine chicks right out of the nest. Daytime temperatures ranged from 40 to 45 degrees, with sharp winds of 20 to 25 mph. Water froze in the waterers at night. In contrast with pampered chicks in a brooder, these chicks were out foraging in those challenging conditions from day one, enjoying superior natural foods found by the hen, enjoying a just-in-time warming session with Mama whenever needed (see figure 27.7). She didn't lose a one.

A good broody is completely fearless in defense of her young. One summer, when I had a lot of young growing chickens in the flock, a Cooper's hawk began making hits, becoming more brazen with each successful meal. As I was feeding one morning, the Cooper's made a dive for a chick belonging to one of my Old English Game hens. Just before contact, the mother hit the hawk—a shrieking flurry of wing and stabbing beak and outrage. The Cooper's suddenly concluded that dinner was not nearly as important as pressing business somewhere else—*anywhere* else. The hen maintained her attack as the hawk made a desperate scramble over the fence and finally managed

Fig. 27.7 Just-in-time warming session. PHOTO BY BONNIE LONG

to break free and land on the branch of a tree, shaking and stunned.

That same fierceness is directed at any member of the flock so foolish as to show a lack of respect for a broody's babies—ensuring that she and her chicks can run with the main flock on the pasture from day one, if that is your preference.

A problem may arise when two mothers come off the nest at the same time and begin fighting each other. My solution is a "halfway house" for new families: a low pasture shelter divided into two sections by a wire mesh partition of half-inch hardware cloth (see figure 27.8). If there is only one new family, I leave the doors open—they shelter in it at night, or during the day if there's rain. If there are two new families coming onto the pasture at the same time, I place each in a section of the halfway shelter. The two hens can see and interact with each other through the mesh divider but cannot fight. After a few days, I release both groups to the pasture. Usually the hens now accept each other's presence and get along peaceably. On rare occasions there will still be a determined fight, with neither hen willing to back down. If it becomes obvious that a continued standoff will result in serious injury—loss of an eye, for example—I return one of them to the main flock and allow the other to adopt her chicks. I hate denying the hen who has put so much effort into making a family the opportunity to carry through to completion, but sometimes that is the best solution. I've had a few occasions when an especially aggressive mother viciously attacked the *chicks* of another. Such behavior goes beyond tolerable levels of aggression, and I cull such a broody.

Fig. 27.8 Hen with chicks in a special pasture shelter I call a halfway house. PHOTO BY BONNIE LONG

I don't want to overemphasize the possibility that mother hens may fight. Non-game-type broodies are inclined to be more laid back. Indeed, we've had numerous cases in which two mothers have formed little communes, the chicks running to either mother indiscriminately for help finding something to eat or for a warming session, and all settling down for the night in one cozy clump.

If new chicks are on the pasture with the general flock, there are special considerations for feeding. Since young growing chicks should never be fed commercial laying feed, I make a compromise feed that everybody can eat, omitting any heavy boosting of calcium; I offer crushed oyster shell for the laying hens as a free-choice supplement, which the chicks ignore.

The feed itself is 16 percent protein, which any poultry book will tell you is not enough for young chicks. However, the mother hen works diligently finding live food for her brood—earthworms, insects, green plants, and wild seeds. It is my belief that the presence of live food in the diet—food of a quality superior to anything I can offer—is the main reason my chicks are so healthy, and why I have never had a single case of pasting up in a chick with a mother hen on pasture.

I also use a second low pasture shelter set up as

Fig. 27.9 Creep feeder when needed, additional pasture shelter at night.

a creep feeder, with doors made of slats 2⅝ inches apart, so that the young birds can enter between the slats but the adults are excluded (see figure 27.9). Note that this is a rather narrow opening, designed to exclude my rather small Old English Game adults. If you have larger breeds exclusively, you could increase the spacing somewhat. I give high-protein supplemental feedings for the chicks—crushed hard-boiled eggs, earthworms from the vermicomposting project, Japanese beetles, soldier grubs—inside the creep feeder. I open the creep feeder at night as an additional shelter.

More recently I've made a couple of easier and cheaper creep feeders: circles of welded or woven wire, stiffened with a wired-on top, made of the same fencing wire. I use either one with 2-by-4 mesh or one with 4-by-4, depending on what size birds are to be given a pass to the inside of the circle. Of course, such a creep feeder doesn't serve double duty as a shelter.

As said earlier, when the mother wren has readied her brood for the wide world, she breaks her bond with them and goes her own way. Mother hens break the bond with their young as well, when they are ready to thrive on their own. But the exact point at which a mother will break the bond varies enormously from one hen to another. Some go their own way early on, roosting with the other hens at night and foraging for themselves. Others maintain the bond much longer—in extreme cases until the chicks are half grown. Chicks whose own mothers

Does the Cock Take Part in Parenting?

Anyone who sees a family of geese is impressed with how devotedly the gander helps with parenting the goslings. Does the cock take any part in raising the chicks? The answer is largely no—it is the hen exclusively who sees to the nurture of chicks. However, over the years we have seen a couple of partial modifications to that pattern, endearing because they are so unusual.

When I've had a single pair of Old English Games in a pen in the Breeders Annex—one hen and one cock—I've seen on several occasions that, when the hen went broody, the cock settled down in the straw facing her, an air about him of reverent solicitude.

Last spring one of the young cocks developed an apparently strong family relationship with one of the broodies and her chicks: Wherever they foraged on the pasture, he was right there with them, a behavior I had never seen before.

Even if a cock exhibits no parental interest in the chicks, he is respectful of them. I have many times seen a hen flare at another hen's chicks—but I have never seen a cock do so.

Fig. 27.10 This young Old English Game cock became part of the family after the hen moved out to pasture with her chicks. PHOTO BY BONNIE LONG

have returned to their usual place in the flock, but who still feel the need for a little loving, join the broods of more motherly hens for a while. So long as a broody does not abandon her young before they are ready—fully feathered, increasingly robust, and able to forage for themselves—I don't care when she chooses to break the bond.

I hope you have the chance to give a willing broody hen a try—for a joy and intimacy experienced more intensely than in any other part of husbandry. If you have children, it is especially gratifying to be able to share with them this miracle of new life and mother love.

PART SEVEN
Poultry for the Table

PHOTO BY MIKE FOCAZIO

28 | BUTCHERING POULTRY

Butchering skills are essential for anyone serious about keeping an ongoing homestead flock. Whether you hatch your own stock or buy in day-olds straight run, you soon have a large surplus of males for whom there is no long-term place in the flock. Even if you keep layer hens only, they will cease or greatly decline in egg production long before the end of their natural lives. Maintaining them "on welfare" is a fine option if you are keeping pet chickens but hardly a practical choice for those whose flock is part of a productive homestead or farm.

Flock keepers often try to find someone else to do their culling for them. I can't tell you the number of times I've received a call out of the blue, a beginner flockster on the other end of the line who started his first flock blithely assuming that *of course* it would be easy to find someone to butcher his birds for an appropriate fee. Reality sets in as I assure him that, while dressing out birds for my own table is good work, doing so for someone else is drudgery—and I don't do drudgery, at any price. "Of course, you're welcome to bring your birds and join me the next time I slaughter," I always assure them. "I'll be glad to give you some pointers."

I expect that with the current growth in the numbers of backyard flocks as a serious part of the home economy, there may be a growing number of small custom butchering services in the future—perhaps based on mobile processing units that arrive on-site for the day's work, leaving neat packages for the freezer behind. But at present most flocksters find that individuals interested in providing such a service are few and far between, and they are stuck with doing their own butchering.

It is easiest to learn butchering skills working with someone more experienced. In lieu of that better option, I hope you will find the following guide useful. Read it thoroughly, study the pictures, and don't be discouraged if there are points that do not make sense. When you have your bird on the worktable, you will recognize key anatomical features from the pictures, and obscure points from the text will become clear. It can be helpful to work with a partner who assists with point-by-point reference to the guide, while you do the hands-on work. Good luck!

Culling Strategies

The idea of slaughtering their own birds often makes beginning flocksters nervous, so it is important to think of culling as an essential tool for effective flock management. One year, for example, we had a large group of Silver Penciled Wyandotte cockerels that caused such an uproar Ellen started calling them "the Mafia"—they were constantly sparring, with *everybody*, and mounting the poor pullets and hens nonstop. I took them to the slaughter table en masse, and the chicken run was the Peaceable Kingdom once more.

I like to hatch with natural mothers, but

occasionally a hen who has gone broody fails to carry through. Such a hen is failing me as a layer, failing me as a mama, so instead she serves quite nicely in the stew pot.

Many flocksters like to raise big batches of meat birds for the freezer. I still sometimes do that myself, but more and more I rely on the needed culling through the season to put meat chicken on the table—a strategy for reducing the amount of plastic packaging waste I generate, as well as expensive energy to store dressed birds in the freezer. If you're culling on a larger scale for market, see chapter 30 for more details on butchering on a larger scale.

Progressive culling of birds of all ages opens up culinary adventures unavailable to those dependent on supermarket chicken: extremely young, tender birds just past the fully feathered stage (sometimes called *poussin*, from the French word for "chick"); larger but still tender fryer or broiler size (what I call "spring chicken"), excellent for sautéing, baking, or broiling; older cockerels, though less than a year old, for braised dishes such as *coq au vin*; old cull hens, with meat that is flavorful if cooked slowly on low

A Note About Slaughter

Most of us are uneasy over the prospect of slaughtering for the table poultry that we have so intimately nurtured. As we should be—I cannot imagine being complacent about the killing of a beautiful animal for food. As with any moral issue, however, the answers we should distrust most are the easy ones. We need to consider the question of slaughtering in its context. That context at its broadest is how we eat, and as a consequence "how the world is used."

It is good to reflect first of all on the origins and nature of *domestication*. The common notion that domestication was initiated and controlled by humans from the beginning is probably a complete misunderstanding. In a sense certain opportunist species *chose* domestication by moving into ecological niches opened up by human activities, most especially the practice of agriculture. The accommodation that emerged was one based on *mutual* benefit. The result has been that domesticated species have been wildly successful, in terms of reproductive success, in comparison with their wild cousins, and that most domesticated species are dependent on their partnership with humans and would likely not survive outside it.[1]

If we accept that domestication is a relationship of mutual benefits, however, certain consequences inevitably follow. Of greatest significance for poultry husbandry is that domestic avian species are like humans in that they produce about equal numbers of males and females. In a domesticated flock, however, keeping all the males that hatch is unmanageable—culling down to just a few is a necessity. You may choose to keep hens only in your flock—but remember that your purchase of pullets-only chicks implicates you directly in the "euthanizing" of excess cockerel chicks by the hundreds of thousands. You may choose never to cull old, no-longer-productive hens in your flock, but those who supply you with chicks will most certainly not make that choice—that is not an option for them economically. Again, you are implicated in the killing of their retired breeders if you buy their chicks.

heat but whose main payoff is the fabulous broth they make. Mature cull cocks are also more appropriate to making broth.

Getting Ready

Some elements of preparation, such as the level of tooling up, will vary according to the frequency you slaughter and the number of birds you usually process. Other preparation is the same whatever the scale of the operation.

- **Starving the birds:** I strongly recommend isolating birds to be slaughtered overnight without feed. Do provide water free-choice. The brief starving of the birds clears the gastrointestinal tract, making for easier, more sanitary butchering.
- **Chill:** Do think of freshly dressed poultry as highly perishable, and be prepared to chill the carcasses as you complete them. You can ice them down in a cooler, or pop them into the refrigerator after each bird is finished. I arrange space in the kitchen refrigerator ahead

To my mind the conclusion is obvious: We can choose either to keep a domestic flock or not as we wish. But if we do so, there is no escaping our implication in the unavoidable killing of chickens—the entire enterprise, keeping domesticated chickens in any fashion, is not a possibility without it.

But killing to eat is unavoidable, even if we eat nothing but tofu. In our current agricultural practice, the growing of grains and soybeans, supposedly the basis of a "cruelty-free" diet, causes enormous suffering of living beings of all sorts—from soil organisms killed by excessive tillage or agricultural chemicals, to birds and small mammals and amphibians whose habitats are destroyed by plowing to the ditch banks.

It seems intuitively obvious to me that eating is an intensely moral act. And I respect the eating choices others make. But I emphatically *do not* cede the moral high ground to anyone eating a plants-only diet because that supposedly means less suffering for living beings. I have made clear that I despise the industrial CAFO model (confined animal feeding operation) as a moral abomination, with its reckless disregard for the quality of life of billions of living beings and for the well-being not only of creatures with whom we share the ecology but of ourselves as well. I take my cue from the fact that natural ecologies the world over involve complex mixed communities of both plants *and* animals. Keeping both in the homestead and small farm should be the key to a regenerative and sustainable (which is to say moral) agriculture.[2]

My respect and gratitude to my chickens is as profound as toward the microbes that create soil fertility in my garden, the bees that pollinate my crops, the decomposer organisms that keep my world clean and sweet, rather than a wasteland of putrid corpses. In addition, they contribute to my nourishment and maintenance of my health. My response can only be a sense of *personal* indebtedness to the chicken on my plate, which leads to a sense of gratitude and duty not only to my flock but to the totality of the Creation of which it is a Gift.

of time and chill the birds, covered loosely with waxed paper, until I'm ready to complete processing.

- **Butchering waterfowl:** This guide focuses on butchering chickens. Butchering geese and ducks is basically the same anatomically, but there are thousands more feathers in these species. You pay your dues when you dress waterfowl!
- **Use it all:** You honor the bird who has made such a contribution to your homestead by utilizing it to the maximum extent possible, minimizing the parts you define as "waste." Learn to make stock from what I call the spare parts. Learn to love liver.

Setup and Equipment

The key operations are: killing the bird, scalding, plucking, and eviscerating. Setup requires, at a minimum, a scalder of some sort, a worktable, cutting tools, and running water. You can work indoors if you like, but I prefer to work outside, setting my worktable in the sun if the day is cool and in the shade of a big white oak if it is hot. I use a 15-gallon fiberglass scalder heated by a thermostatically controlled electric element—like that in an electric water heater. I encase my scalder in 2-inch foam insulation for greater efficiency. You could use instead a large enameled canner, like the one in figure 28.9, on the stovetop or a portable burner. I also use a mechanical plucker featuring a drum driven by an electric motor, into which are set many stiff rubber "fingers" that slap the feathers off the bird as the drum rotates. The expense of a mechanical plucker—or the effort and time to make one—may only be justified if you process a lot of birds, but a plucker does speed up the operation considerably. You will be more efficient working at a table of a comfortable height, which can be made entirely from scrap or recycled materials. A stainless-steel work surface is best, for sanitation and ease of cleaning. My basic setup is shown in figure 28.1.

Fig. 28.1 My basic butchering equipment. PHOTO BY MIKE FOCAZIO

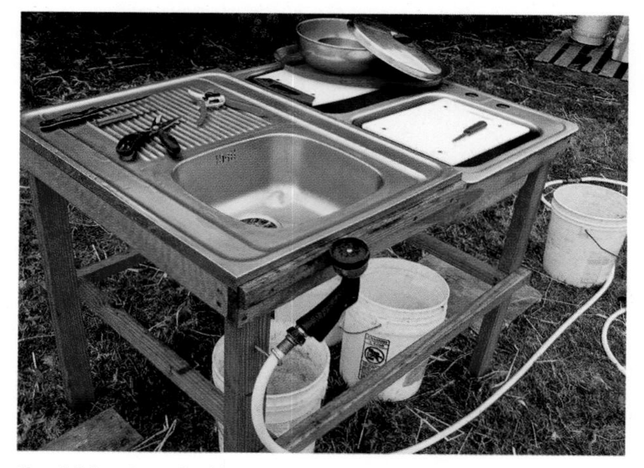

Fig. 28.2 A worktable with essential accessories is the key to efficiency. PHOTO BY MIKE FOCAZIO

There is more information on tooling up to dress broilers for market in chapter 30.

My father and I made my worktable from scrap lumber, a double stainless-steel sink from a friend's kitchen renovation, and a single sink with drain board I picked up at a junkyard for four bucks. To accommodate friends who join me for slaughter day, I made insets for the sink wells to provide additional work surfaces: small synthetic cutting boards screwed onto wooden feet (see figure 28.2).

Note the supply hose hooked onto the leg of the table. The pistol-grip sprayer is just the thing for that quick splash of water whenever needed. I spray frequently—for better sanitation, to prevent drying of the skin, and to keep flies and yellow jackets from getting too interested.

I like to have on hand a tray for carrying dressed carcasses into the house and a stainless-steel bowl with a lid to hold the usable innards until I get them inside. And of course, the homestead revolves around 5-gallon buckets—I use them under the table to catch the rinse water, so the area doesn't become a muddy mess, and position them on either side of each workstation, to catch feathers and offal.

About Cutting Tools and Knife Technique

Do yourself a favor and invest in good cutting tools—yes, the good ones are expensive. One of the greatest frustrations in my butchering workshops is the wretched cutlery many beginners bring to the work. Knives that are badly designed, or that will not take or hold a keen edge, are clumsy and fatiguing to use—and of course, a dull tool is always more dangerous than a sharp one, because of the greater force required to cut with it.

You will discover your own preferences for cutting tools. I prefer two knives—one with a thin, flexible, 3-inch blade for more delicate cuts, and the other with a stiff, heavier, 6-inch blade for more hefty, resistant cuts. I also recommend a good pair of shears, for cutting off the neck at the worktable, and for cutting up carcasses in the kitchen. Poultry shears vary tremendously in quality, and I have broken at least half a dozen over the years. After eventually breaking the spring on the best model I ever found (the black-handled "Soft Touch" shears in figure 28.3), I lost patience and bought a Felco No. 2 pruner (red handles). I don't expect I'll ever break that one.

Let me emphasize two points about knife technique, based on the most common mistakes I see beginners make. *Never use the point of the knife when cutting.* All the cuts you need to make are *slicing* cuts, some of them rather delicate—to avoid piercing the entrails—so always use the edge of the blade. And keep it sharp!

You may have wondered about my preference for a stainless-steel work surface, having expected a chopping-board surface designed for contact with the blade. Using my methods, *the blade need never contact the work surface.* Either make downward cuts from above, so the carcass itself prevents contact of blade and work surface, or pull on the part to be cut, using the weight of the carcass to create tension, and make your cut against that tension—rather than sawing or chopping down onto the work surface, as when using a chopping board.

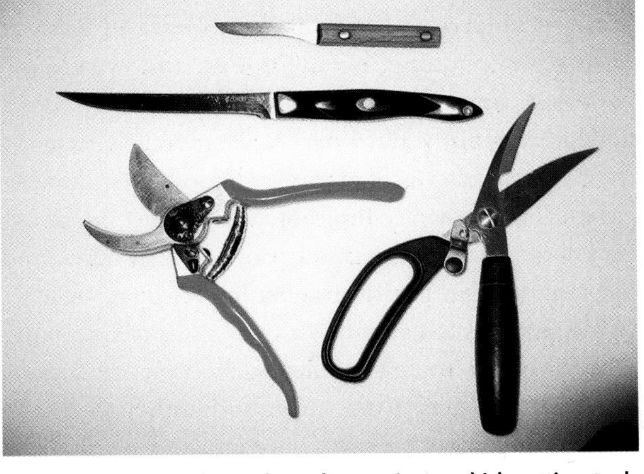

Fig. 28.3 You will learn through experience which cutting tools work best for you. Don't skimp on quality–good tools are a joy to use, and mediocre ones are anything but. PHOTO BY MIKE FOCAZIO

Killing the Bird

Emotionally and psychologically, the killing of the bird is typically the most difficult part of the process. Naturally you will want to do the job as quickly and humanely as possible. I use the three methods described below, depending on the age, size, and species of the bird I'm butchering.

Note that, whatever method you choose, *it is essential that the bird bleed out completely.* The dressed bird will not keep as well, nor taste as good, if the blood remains in the muscle tissue. And there is no need to let the blood go to waste. Traditionally, blood from slaughtering larger animals was caught in a vessel and used in cooking. When I lived in Norway, I enjoyed fresh blood pudding, blood sausage, and even blood pancakes from a couple of pigs slaughtered on a neighboring farm. But I never met anyone who made the same culinary uses of poultry blood, and I haven't either. There are a couple of possibilities, however. You could catch it in a bucket of water or straw and use it as a liquid fertilizer or mulch for heavy feeders in the garden such as corn, or trees and shrubs, giving some thought to forestalling such curiosity seekers as dogs and foxes in the neighborhood. I catch the blood and pass it on to my composting worms and soldier grubs. Don't tell the contagion police, but I have even experimented with feeding the coagulated blood to my chickens, who ate this nutrient-rich food with apparently only beneficial effect.

The Chopping Block

A solid, stable round from a log makes a good chopping block (see figure 28.4). Drive a couple of large nails into the block, spaced to grip the bird's head as you pull on the feet. Under such restraint, the bird will not continue struggling. Don't be rushed—pull on the feet, stretching out the neck, take a breath, and steady yourself for a decisive blow of the hatchet that chops off the bird's head with one whack. Hold the flapping bird over your collection vessel, so the blood will drain without spattering.

Fig. 28.4 The chopping block. PHOTO BY MIKE FOCAZIO

The Killing Cone

A killing cone made of sheet metal—in different sizes, depending on the species to be slaughtered—is a useful accessory. Hang it on the side of an outbuilding or a tree. Insert the head through the hole at the bottom of the cone, pulling it to stretch the neck and draw the wings and legs more tightly into the confinement of the cone. Use your sharpest knife to make a quick, decisive cut just to the side of the "jaw." Note that you sever the jugular vein *only*, not the windpipe, resulting in less stress for the bird. Keep your grip on the head as the bird bleeds out thoroughly, guarding against a final spasm that might flip it out of the cone (see figure 28.5). Homemade versions of killing cones

Fig. 28.5 The killing cone. PHOTO BY MIKE FOCAZIO

include burlap sacks or plastic buckets with holes cut in a corner of the bottom through which to pass the bird's head.

English Method

You can kill the bird by what is sometimes referred to as the English method, if it is young enough for you to break the head off the neck. I find chickens at the fryer-broiler stage, and most old hens, easy to kill with this method. I cannot break the necks of mature cocks, ducks, or geese, so for those birds I use the cone or the chopping block instead.

Holding the head of the bird firmly in your strong hand, and the feet in the other, brace the bird over one thigh, and imagine you are going to pull it apart. At the right point of tension—which you can only learn by experience—give a sharp twist-snap downward-outward, and the head will separate completely from the neck. Hold the bird away from your body until its spasms subside (see figure 28.6).

You will find this method difficult the first time you try it and may mistakenly conclude you are not strong enough to make it work. Trust me: It is not a matter of brute strength but of technique—that is, the right degree of tension and proper action in the wrist. The first time the head comes off—so easily, really, with a sort of liquid giving-way—will be something of a surprise.

An advantage of this method is that it is less messy—the bird does bleed out, but the blood is retained inside the skin of the neck. Note, however, *it is essential that the head actually break away from the neck for this method to work*—when that happens, the jugular vein is severed as well, and the bird bleeds out properly. It is possible to kill the bird simply through trauma to the spinal cord but without breaking the jugular, resulting in a dead bird that has not properly bled out. Squeeze the skin between the head and the stump of the neck: If you don't feel a completely flaccid, empty-balloon space at least big enough to insert three fingers, you have not properly broken off the head.

Fig. 28.6 The English method. Note use of mask to prevent inhalation of dander. PHOTO BY MIKE FOCAZIO

About Using a Mask

The reflex thrashing of the bird after killing kicks loose a good deal of poultry dander. If you find that your sinuses are heavily congested after slaughtering, especially during the night, use a good dust mask during this phase—and later in the plucking phase as well, if you are using a mechanical plucker. The one I use is the Respro Sportsta Mask, extremely effective and comfortable enough to wear for all dusty chores on the homestead.[3]

The Naked Fowl

Do not be discouraged if the work at the table is at first painfully slow—with practice the hands quickly learn the required efficiencies, and processing becomes faster and more fluid.

PHOTOS 28.7–28.35 BY MIKE FOCAZIO

Fig. 28.7 Proper scalding is the key to a clean, easy pluck.

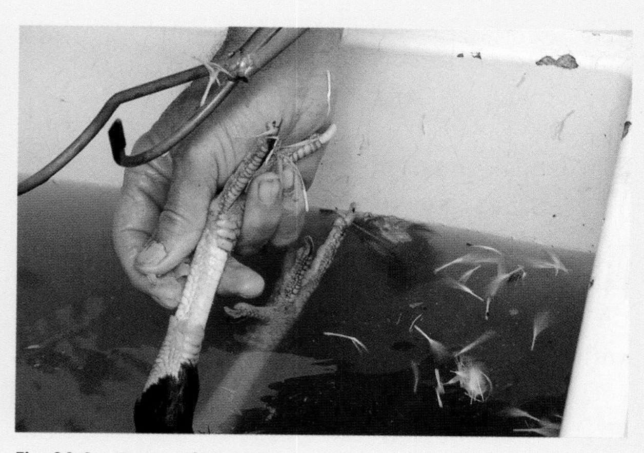

Fig. 28.8 Testing the scald.

Fig. 28.9 Dunking.

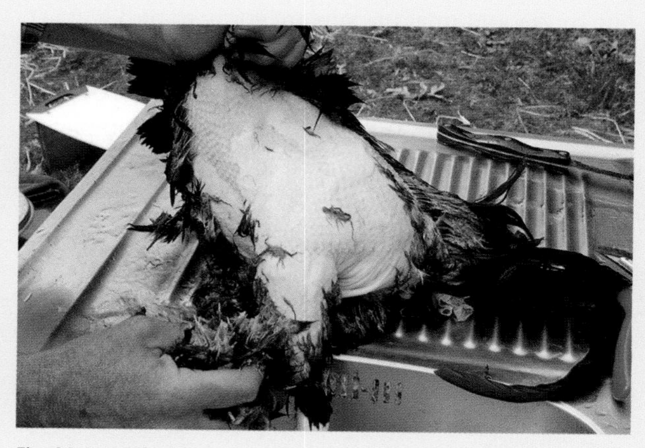

Fig. 28.10 Plucking manually is easy if you have scalded properly.

Fig. 28.11 A mechanical plucker speeds up the work.

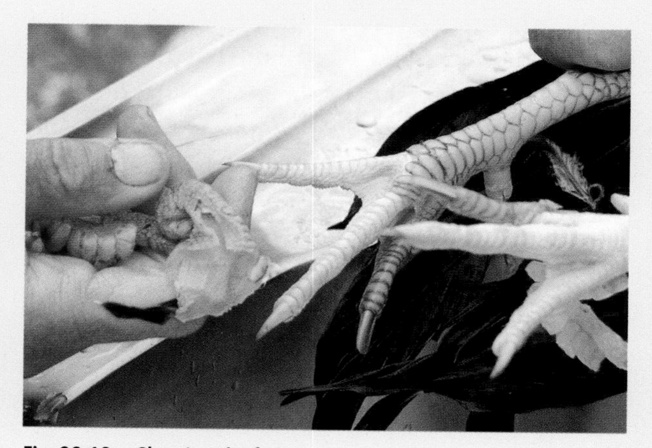

Fig. 28.12 Cleaning the feet.

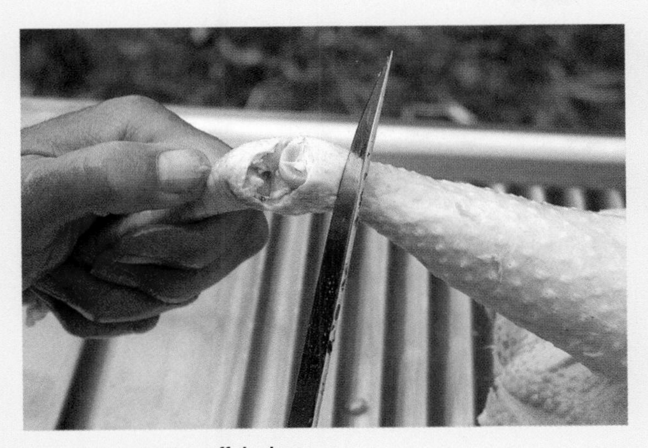

Fig. 28.13 Cutting off the leg.

Fig. 28.14 Starting the evisceration.

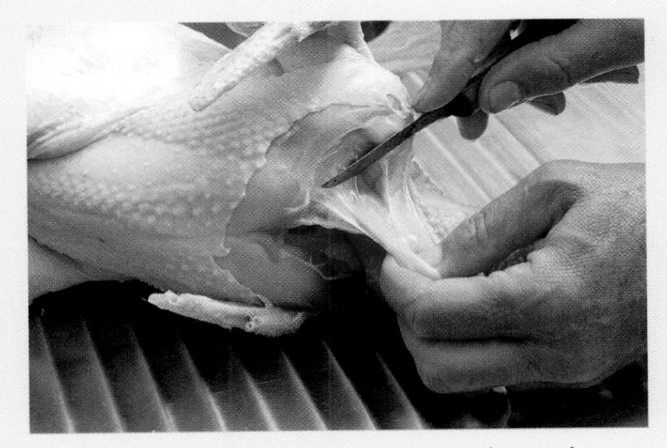

Fig. 28.15 The crop–empty if the bird was properly starved.

Fig. 28.16 If the bird was *not* starved, the full crop may spill its contents.

Fig. 28.17 Leading with the thumb, separate crop, esophagus, windpipe, and skin from the neck.

Fig. 28.18 Cut all that tubular stuff off as close as you can to where it enters the body cavity.

Fig. 28.19 Cutting off the neck.

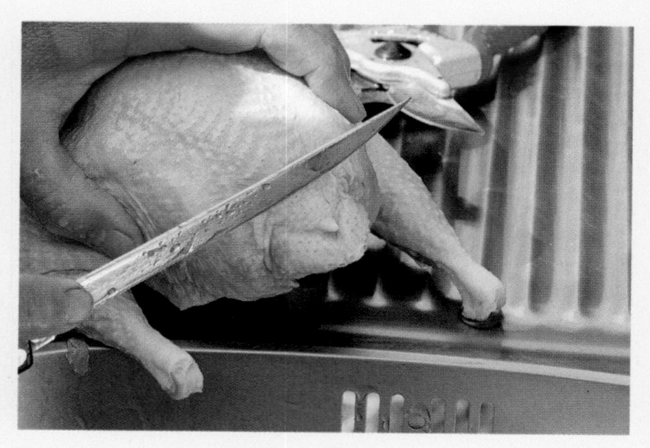

Fig. 28.20 Removing the oil gland. Part 1.

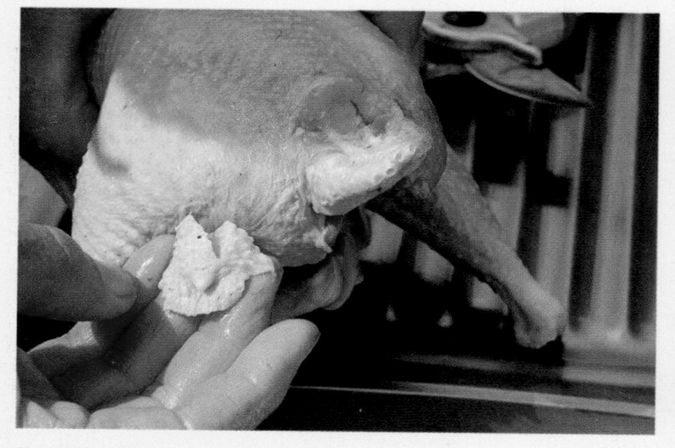

Fig. 28.21 Removing the oil gland. Part 2.

Fig. 28.22 Make a small opening into the abdomen, slicing through skin and fascia only.

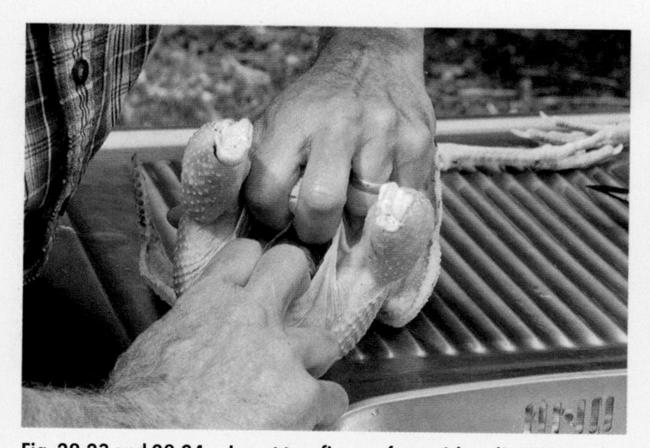

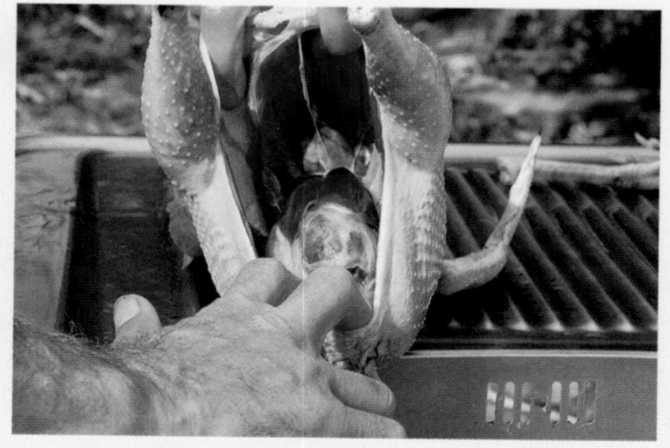

Fig. 28.23 and 28.24 Insert two fingers from either direction and give a stout pull to open the body cavity. Note that in an unusually tough old bird, you may have to cut through the skin and muscle on either side of the body cavity to give full access.

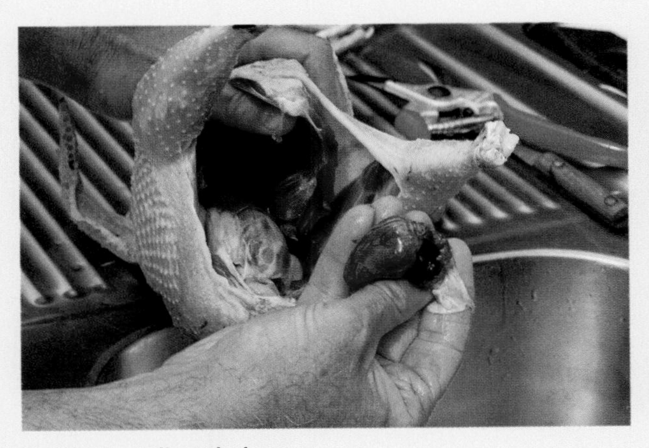

Fig. 28.25 Pull out the heart.

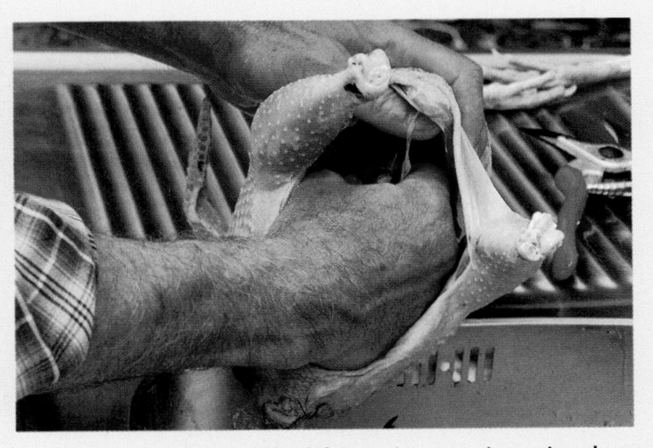

Fig. 28.26 The fingernails lead the way in a scooping action along the rib cage and spine, underneath the viscera.

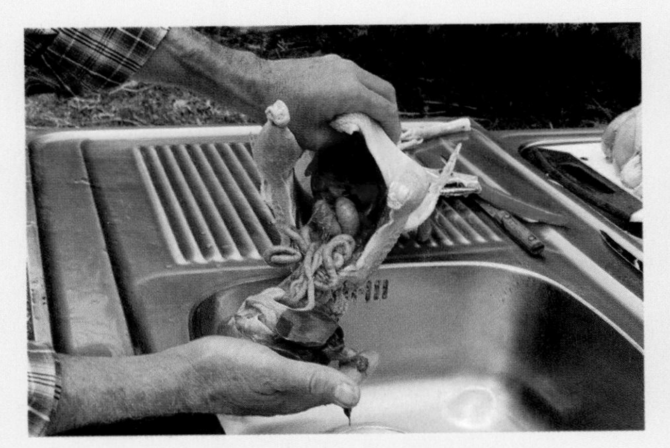

Fig. 28.27 Pulling out the package of entrails.

Fig. 28.28 Cutting the liver from the bile sack.

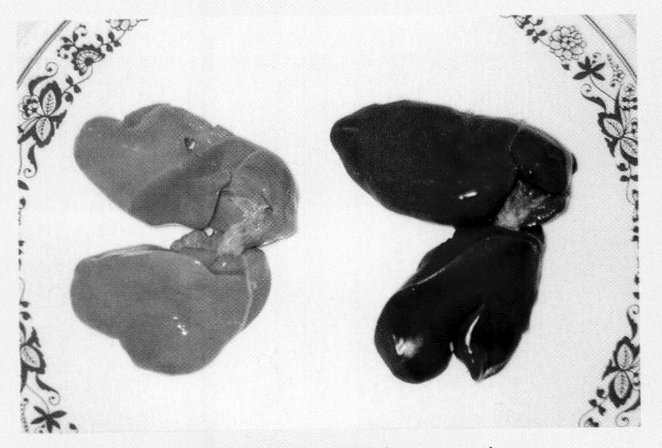

Fig. 28.29 Livers of old and young birds compared.

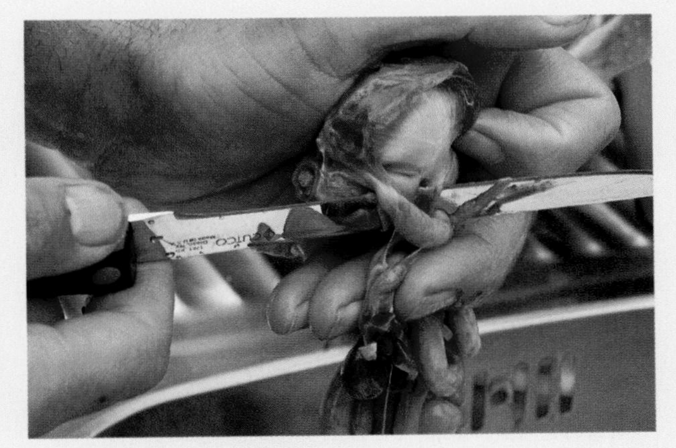

Fig. 28.30 Cutting away the gizzard.

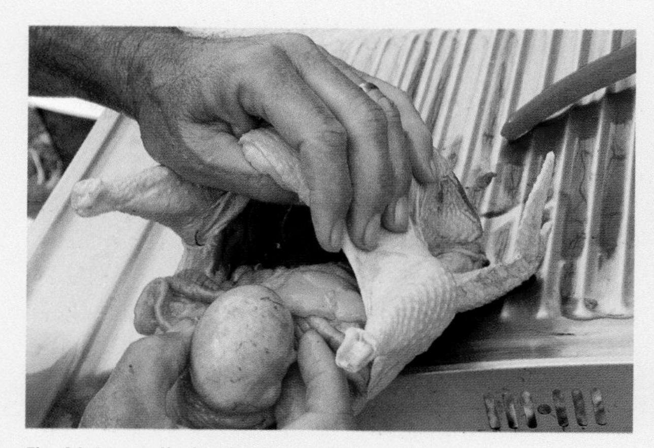

Fig. 28.31 Fully developed egg in the oviduct.

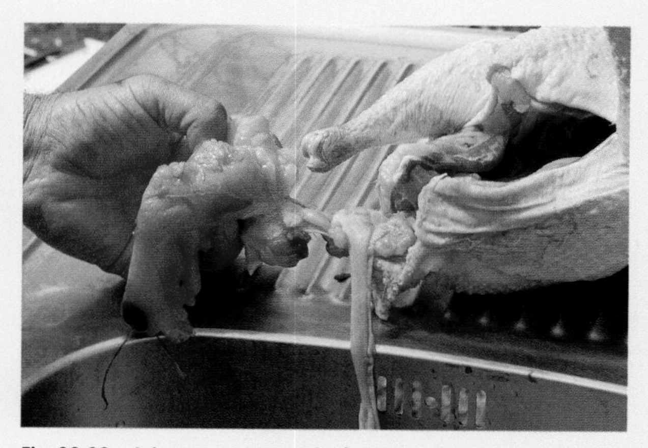

Fig. 28.32 It is easy to remove the fat reserve around the end of the intestine, and easy to render it into high-quality cooking fat.

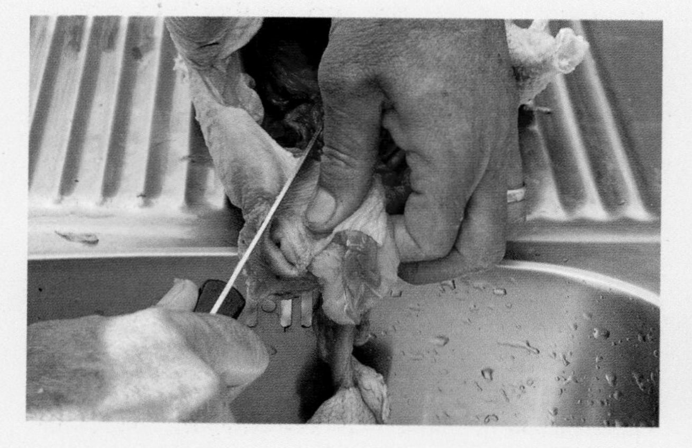

Fig. 28.33 To finish removing the entrails intact, start with cuts on either side of the vent . . .

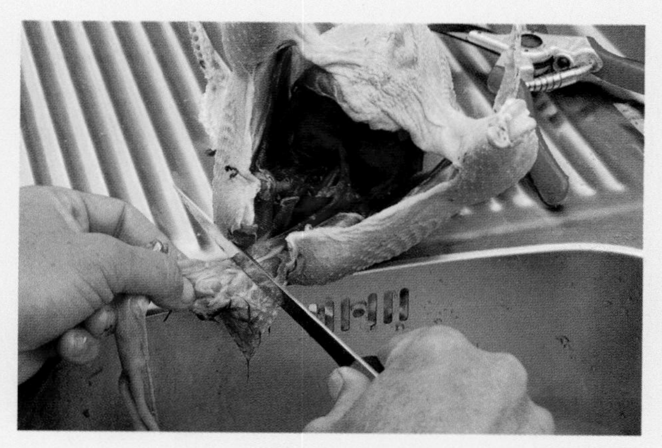

Fig. 28.34 . . . and end with a slicing cut underneath both cloaca and vent.

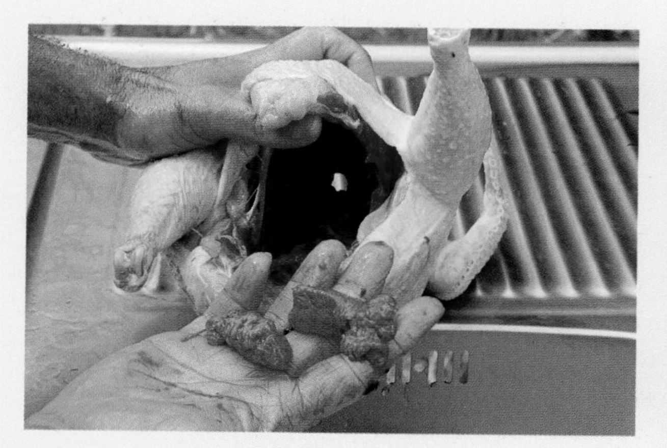

Fig. 28.35 Popping out the lungs can be a bit tricky.

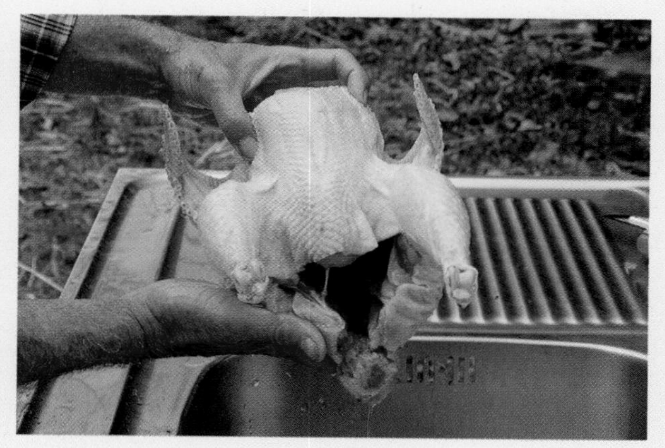

Fig. 28.36 Don't forget to say *Thank you.*

Scalding

See figures 28.7, 28.8, and 28.9.

The key to a clean pluck is a good scald. Note that *scalding temperature is nowhere near the boiling point.* I set my thermostatically controlled scalder to 145 degrees Fahrenheit, and that is a good temperature to aim for if your scalding container is over a stovetop or burner. However, it isn't really necessary to use a thermometer to measure the temperature. Just stick in a finger. Can you immerse the finger without getting a burn, but not for more than a second without burning? That's proper scalding temperature.

Note that overscalding—either through too high a temperature or scalding too long—starts to cook the skin, which then tears when you pluck. Underscalding, on the other hand, fails to loosen the feathers sufficiently, and they are difficult to pluck. No exact formula can be given for scald time—how long to scald depends on the age of the bird, the species, the point in the plumage cycle, probably the phase of the moon. You will learn only through experience when enough is enough in the scalder.

Add a few drops of liquid soap to the scalder, to break the surface tension of the water and increase penetration to the skin—especially important with waterfowl. Put the bird into the hot water, and use some sort of poke to agitate it up and down. After a few agitations use your poke—I use an old three-prong cultivator missing one tine—to snag a leg. Pinch-squeeze the scaly covering of the shank: When that covering easily breaks loose from the skin of the leg, remove the bird from the scald and dunk in cold water, to stop the skin from overheating from the residual heat in the water under the feathers.

Plucking

See figures 28.10 and 28.11.

If you've scalded properly, it will be easy to remove the feathers in handfuls. It's better to start with the largest feathers, on the wings and tail, since they are the ones that start resisting pulling out first as the follicles cool. As said, you may find a mechanical plucker a good investment.

Alternatives to Scalding

You may see references to two alternatives to the above approach to plucking: *waxing* and *brain piercing*. You can dip the bird to be plucked—especially ducks—into melted wax, let the wax solidify, then peel it off, feathers and all. Or so it is said. I've tried waxing a couple of times, but the wax granulated into a thousand tiny fragments that had to be individually removed. It may be that I was not using the right kind of wax, so if you want to try this method, be sure to procure a wax designed specifically for removing feathers.

You can also drive the point of a stiff-bladed knife through the roof of the bird's mouth and into its brain. If done correctly, this method supposedly loosens the feather follicles, allowing easy dry plucking without the necessity of a scald. Again, my attempts at this method turned rather nightmarish, and failed to yield easy dry plucking. Good luck if you try it.

There is one alternative to scalding that I can assure you works just fine: skinning the bird, just as you would skin a rabbit. I prefer to keep the skin on a dressed bird, so I always scald and pluck if I'm butchering a larger number of birds. But if I need to dress out a couple of birds on the run, I usually skin them, in lieu of heating a large amount of water in the scalder. To skin, simply make a slice through the skin of the back large enough to hook a couple of fingers under the edge, and pull. Most of the skin pulls away easily. The skin at the base of the large feathers on the second joint of the wing takes a bit more effort; and I just cut away the pinion—the last joint—rather than trying to get its feathers off. Young growing birds are easy to skin; old stew hens and ducks require more effort but are not too difficult; but old mature cocks—forget it! They're *really* tough.

Cleaning the Feet

See figure 28.12.

Always save the feet. They are a valuable addition to the stockpot, yielding not only essential minerals but also collagen, which is beneficial for the entire digestive tract. Remember how you used the scaly covering of the shank to test the scald? Simply continue with the same pinching-pulling action to pull the covering off leg and toes like a glove. Be sure to pinch tightly and pull the toenails as well, and the cuticles will pop off easily. If the cuticle resists popping off, use your shears to snip off the entire toenail. The result is a pristine foot you will be glad to have in your stockpot.

Legs

See figure 28.13.

This is the first point to take care to avoid sharp edges of cut bone, which could puncture the wrapping you use for freezing your bird. (If you're cooking it right away, of course, the point is moot, and you can cut leg or neck any way you like.) Therefore, you should not use your shears to cut through the joint at the hock to remove the leg.

Instead, lifting by the foot, pretend you are going to break the leg sideways at the hock with the weight of the suspended bird. The resulting tension on the joint makes it easy to slice through the skin and find the cartilage-padded interstice between this ball-and-socket joint. Once the edge of your blade has found that space, it is easy to continue cutting through skin, connective tissue, tendon—anything but bone—until the leg is cut away.

Head

Unless you chopped off the head, at this point you will need to remove it. If you used the English method, cut through that empty-balloon segment of neck skin with a knife. If you used a killing cone the head will still be attached, so cut it off with your shears.

The head can also be reserved for the stockpot if you like. Pull off the feathers, rub the coating off the skin of comb and wattles, and pinch hard on the beak—the horny cuticle will pop off, like the toenail cuticles earlier.

Should You Singe?

The work thus far has produced a completely naked fowl, ready to eviscerate. But if you look closely, you may see fine "hairs" sticking here and there out of the surface of the skin. Some folks are repelled by these hairs—more properly called filoplumes—and feel compelled to singe them off to produce an absolutely pristine carcass, using the flame of a gas stove, or even a propane torch. Concern about a few (almost invisible) filoplumes is a good deal too fussy for my taste—when the bird comes out of oven or pan, nobody is going to notice or care. I have never yet taken a blowtorch to a dressed chicken.

If you do choose to singe, sweep the carcass lightly with the flame—you do not want to heat the skin, and the filoplumes burn away with the merest touch of flame. Wash thoroughly afterward, to prevent carbon residues remaining on the skin.

Evisceration

For convenience, the following descriptions assume you are right-handed. Lefties will need to switch hands and directions as appropriate.

Crop, Windpipe, and Esophagus

See figures 28.14 through 28.18.

Lay the plucked bird on its back, its long dimension perpendicular to you, its neck to the left. Start the evisceration at the skin where neck joins breast. Using your small knife, slice through the skin on the far side, and continue slicing through skin (only) to make a full circle around the neck (see figure 28.14). As you slice you expose the *crop*, a semi-translucent membranous pouch—to the bird's right, close beside its neck, on the side nearest you in this position—in which the bird stores its food for pre-processing, before passing it on to the gizzard. Because you wisely starved your slaughter birds overnight, the crop is empty and this step is not messy (see figure 28.15). If the crop is full, it may rupture, spilling the fermented contents over the top of the carcass and the work surface (see figure 28.16). This is no disaster—a thorough rinse will whisk them away.

Force a thumb between the neck and everything attached to it: the crop—in the process pulling it free from the top of the breast meat—the esophagus, and the windpipe. Pull all that, along with the skin, free of the neck (see figure 28.17). Separate the neck skin from the three tube-like elements, and reserve it for the stockpot as well. Then pull on the tube-y things, and cut them off as close as you can to where they enter the body cavity (see figure 28.18).

Cutting the Neck

See figure 28.19.

Do not leave a stub when you cut off the neck, exposing edges of sheared bone that could poke through your wrapping later. Instead, position the blades of your shears on either side of the base of the neck, but well up between the shoulders of the wings. The resulting cut keeps the sheared edges tucked up between the shoulders, and out of contact with the packaging.

Removing the Oil Gland

See figures 28.20 and 28.21.

If it remains on the carcass, the oil gland, which secretes oil the bird uses to preen its feathers, may affect the flavor of your cooked bird. To remove it, place the bird breast down, tail end to the right. To the left of the gland's nipple are two little mounds—the lobes of the gland. Just to the left of the lobes, slice down vertically until you hit bone. Then turn the edge of the blade right, toward the end of the tail, and make a scooping cut to slice under the two fatty lobes and the nipple. If you plan not to keep the tail but to cut it off when cutting away the entrails, you don't have to bother with the oil gland in this step.

Opening the Body Cavity

See figures 28.22, 28.23, and 28.24.

Lay the bird on its back, perpendicular to you, neck end to the left. Make a shallow slicing cut into the abdomen, just to the right of the end of the keel or breastbone. Cut through skin and fascia, the translucent membrane surrounding the inner organs, *only*. Remember good knife technique: Avoid poking at the skin with the point of the knife—the bird's intestines are immediately below (see figure 28.22).

Sooner or later, you will have a slip of the knife and nick the intestine, spilling some of its contents into the body cavity. Do not despair—even with such a mishap, your home butchering is vastly more sanitary than that of commercial poultry. After eviscerating, simply do a more thorough rinse of the interior cavity than usual.

Make the cut into the abdomen just big enough to hook two fingers from either direction, and give a stout pull, tearing a larger hole in the skin and fully opening the interior cavity. Note that, if the bird is old and unusually tough, you may have to slice through additional skin and muscle on either side of the initial cut to make it easier to pull open the body cavity (see figures 28.23 and 28.24).

Please note that the pressure exerted on the intestinal tract at this point can cause what I call a "poop attack"—the forcing out of some residual fecal material from the cloaca. A poop attack is unlikely if the bird was properly starved in preparation for slaughter, but you can guard against it anytime by hanging the vent end of the bird over the drain when you pull open the carcass. If there is an expulsion, rinse it away from the vent, being careful to avoid backwash into the body cavity.

Heart

See figure 28.25.

Reach as far as you can into the body cavity to find an organ that feels like a large grape—the heart. Hook two fingers around it, and pull it out. Squeeze out the remaining coagulated blood, rinse, and set aside—I use the steel mixing bowl with lid that I mentioned—along with the neck and feet and other good things you will be saving.

Liver

See figures 28.26 through 28.29.

Reach again into the body cavity, fingers and hand encircling the gastrointestinal tract. The fingernails lead the way, tight to the rib cage, finding the seam between the chest wall and the ropy tubes of the tract and other organs, the gizzard filling your grasp like a slippery apple (see figure 28.26). Grip all and pull. The tract and the organs pull free in one mass—connected only at the base of the abdomen—which you allow to hang over your drain (see figure 28.27).

The liver is the large, dark red organ beside the gizzard. The clear tissue connecting it to the other organs is easily torn away with the fingers to leave it attached at one point only—to a small, dark green sack the shape and size of a caterpillar, the bile sack. The bile it contains is essential for the bird's digestion of fats but is extremely bitter. Sacrifice a bit of the liver as you pare it away from its connection to the bile sack with your small knife (see figure 28.28). If on occasion you do have a spillage of bile, rinse with more than usual thoroughness and proceed.

Note the picture of this cut as an example of proper knife technique: The weight of the hanging gizzard is used to put a little tension on the connection between liver and bile sack, allowing the cut to be made against that tension, rather than against the steel work surface. Of course, the blade must be sharp, to enable cutting against such a tiny bit of tension.

The liver of a young, healthy bird—on the right in figure 28.29—is plump, dark red, and glistening. By all means save it—it is extremely nutritious. The liver of an old bird, on the left in the picture—equally healthy, equally well fed—is usually pale brown, indicating longer service as the bird's major metabolic filter. While I honor this liver for the good work it has done, I do not eat such livers.

Gizzard

See figure 28.30.

The gizzard is a large, muscular organ with a tough interior pouch, filled with bits of rock. In lieu of chewing its food, the chicken processes it inside the gizzard, using the small grinding stones and digestive enzymes. One tube goes in, one comes out. Cut off both, flush with the surface of the gizzard, and add to the other "goodies."

Testes, Eggs, and Fat Deposits

See figures 28.31 and 28.32.

If the bird you are butchering is a cockerel, the testicles—light yellowish, football-shaped, and tucked up against the spine at about its midpoint (visible in figure 28.27)—may be the size of kernels of corn. If you're eviscerating a mature cock, you may be astounded: In proportion to his body, they are enormous. Pluck out the testicles. If from a mature male, add to the goodies bowl if inclined to eat them. Toss them into the gut bucket if disinclined.

If you are butchering a hen, you may well find inside a number of egg yolks of various sizes, or even a fully developed egg. The one in figure 28.31 is a finished egg ready to be laid, still inside the oviduct. Keep these eggs, and yolks the size of a pea or larger.

Especially in a hen, and even more so in hens butchered in the fall, you will find a deposit of glistening, yellow fat at the lower end of the body cavity. If you want to render it into high-quality cooking fat, it's easy to pull it out with your fingers.

Cutting Away the Entrails

See figures 28.33 and 28.34.

You've now drawn out all the interior organs without spilling stuff-we-don't-want-on-our-meat, and they're hanging from the vent in one long, intact package. Keep them intact as you complete the evisceration. Hang all that ropy stuff to one side of the vent, and position your blade on the other side, between the vent and the sharp point of the pubic bone. Simply slice down until the blade hits bone.

Now move the ropy stuff to the side of the vent where you just made your cut, and repeat the cut on the opposite side (see figure 28.33). Pull on the entrails and slice under the vent itself, leaving it intact along with its connection to the end of the intestinal tract (the cloaca; see figure 28.34).

If you chose not to remove the oil gland earlier, modify the above in order to remove the entire tail: After cutting through the skin between pubic bone and vent on both sides, grasp the tail with attached viscera in your left hand, and cut it off the end of the spine using your heavier knife.

Lungs

See figure 28.35.

Reach into the cavity one final time and remove the lungs. Again, lead with the fingernails as you follow the curve of the rib cage, finding the seam between rib and lung, and pop the spongy lung tissue free. This step takes some practice and can be a bit tricky—sometimes the lungs shred and resist coming free easily. Don't worry—leaving a little lung tissue is no problem. I like to remove them to make a neater carcass.

Kidneys

The bird's kidneys are lumpy masses the color of the liver, tucked either side of the backbone, near the vent end. Though removed in commercial processing, I find them quite tasty, so I leave them in place to be cooked with the bird.

Give a final rinse, inside and out, and admire the creature who has made so generous a contribution to your homestead, now ready to grace your table. Don't forget to say "Thank you!"

Cleanup

You will end your slaughter day with a pile of wet feathers and a bucket or two of entrails. I throw the feathers onto the deep litter in the poultry house. The birds eat some of them—feathers are almost pure protein—while the rest get buried in the litter, where they quickly decompose. The entrails can be composted or buried. If you compost, take care to assemble a pile that heats up rapidly and does not become an attractant to dogs and other curiosity seekers.

Sometimes I place the entrails out in the edge of our woods, as an offering to friends—Fox, Raccoon, Possum—who live in the neighborhood and who need to eat as well. It's my way of honoring our kinship and saying "Thank you" for coexisting peacefully while my flocks stay—mostly—intact. When I return to the site next day, the entrails have invariably been entirely removed—this gift is not creating a nuisance.[4]

29 | POULTRY IN THE KITCHEN

The kitchen is where all the effort producing our own eggs and dressed poultry pays off. And part of the payoff is the assurance is that, with the application of the plain common sense our grandmothers used in their kitchens, the foods from the flock are not a threat to our family's safety.

This chapter discusses handling and storage of both eggs and dressed poultry. It also offers a few of our time-tested recipes for using these gifts from the flock.

Handling and Storing Eggs

Imagine a packet of nutrients that, under the right conditions, can turn into a baby chicken—in twenty-one days. That packet would necessarily contain a powerhouse of nutrients, don't you think? It's certainly something you'd want to see frequently on your table. The name of that packet—one of the most perfect of all foods—is "egg."

Eggs contain all eight of the essential amino acids from which the body builds proteins, and high-quality fat that assists the absorption of not only the yolk's own fat-soluble vitamins (A, K, E, and D) but fat-soluble vitamins from other dietary sources as well. Eggs also provide B-complex vitamins and the essential minerals iron, phosphorus, potassium, and calcium. Importantly, they are a valuable source of cholesterol—a fat-soluble nutrient essential in ensuring integrity of the walls of every cell of the body, but especially of those in the nervous system and brain[1]—and of choline, a fatty substance that is a major component of the brain and is essential to proper functioning of the liver.

Happy, well-fed chickens—both standards and bantams—ducks, and guineas grace the table with high-quality eggs with a lot of easy culinary possibilities. Handling and storing are simple.

- **Is it necessary to wash eggs?** If you keep your nestboxes lined with plenty of clean straw, frequently renewed, most of your eggs will come from the nest perfectly clean. It is *not* necessary to wash such eggs. Indeed, I advise against it: The hen's just-laid egg is wet with a coating that quickly dries (the bloom)—a coating that not only eases passage for the hen but leaves an anti-bacterial residue on the surface of the egg to help protect it from contamination. Washing it off may actually decrease keeping time for the egg. On the other hand, if an egg comes from the nest with even the smallest trace of mud or poop, I do indeed wash it. We have always used paper towel dipped in a half-and-half mix of water and white vinegar, with excellent results.
- **Can I use a cracked egg?** An egg fresh from the nest with a small crack in the shell is okay to use right away. Do not store such an egg, and do not use one at all if the membrane inside the shell has ruptured, exposing its

Eggs: The True Truth

Contributed by Ellen Ussery

Eggs are a superfood, though they have been badly maligned over the years. Let's set the record straight.

We have been told that eating eggs will give us heart disease because they contain cholesterol, that the fat will go directly to our arteries and clog them up just like it would clog your drain if you poured fat into the sink. This is the diet heart hypothesis and it has been shown to be as untrue as it is simplistic.

An extensive review of the scientific literature conducted at the University of Connecticut clearly indicates that egg consumption has no discernible impact on blood cholesterol levels in 70 percent of the population. In the other 30 percent, eggs do increase both circulating LDL and HDL cholesterol.[2]

You've probably been conditioned to believe that anything that raises LDL cholesterol (so-called bad cholesterol) should be avoided like the plague. But even the medical mainstream has come to recognize that not all LDL cholesterol is the same. It's true that small, dense LDL particles have been linked to heart disease. This is primarily due to the fact that they are much more susceptible to oxidative damage than normal LDL cholesterol particles.

However, egg consumption increases the proportion of large, buoyant LDL particles that have been shown to be protective against heart disease. Egg consumption also shifts individuals from the LDL pattern B to pattern A. Pattern B indicates a preponderance of small, dense LDL particles (risk factors for heart disease), while pattern A indicates a preponderance of large, buoyant LDL particles.

Thus consuming cholesterol-rich eggs either has no effect on our cholesterol levels or actually *improves* the markers for heart health. So much for the diet heart hypothesis.

To understand more fully why I used the term *markers* in the preceding paragraph, you need to carefully examine what is known as the Lipid Hypothesis. Sorting that out completely is far beyond the scope of this book. Those interested in finding clearheaded, well-balanced, and current information on the subject would do well to start with Chris Masterjohn's website and blog: www.cholesterol-and-health.com. Suffice it to say that cholesterol in the body is a health-promoting substance. It is a critical component of cell membranes, the precursor to all steroid hormones, a precursor to vitamin D, and the limiting factor that brain cells need to make connections with one another called synapses, making it essential to learning and memory.

What is dangerous is *oxidized*, small, dense LDL particles. A major cause of oxidization is consumption of the very polyunsaturated vegetable oils that we have been told to cook with instead of saturated fats. Think about it: Such polyunsaturated fatty acids were not available in more than minuscule quantities until the industrial era. But once they came into widespread use, not only heart disease but other modern degenerative diseases became epidemic. I am not saying that excess PUFAs, as they are not so fondly called, are the only cause of all these ills. But they are certainly a major player, and the only good they do is to increase the bottom line of corporate agriculture and industrial food processors. The

Continued on following page

Continued from previous page

small quantities that we do need are to be found naturally in animal foods and small amounts of properly prepared grains, nuts, and seeds. Please note that olive oil is a MUFA, monounsaturated fatty acid, and is safe to use in moderation.

So ditch all industrial polyunsaturates, and the processed foods containing them, and enjoy your eggs without fear. Cook them with butter, coconut oil, or bacon fat from pastured pigs. Not only will they not harm you, they will contribute to every aspect of vibrant health.

Here is a brief rundown of the nutrients supplied by eggs:

- All eight of the essential amino acids, from which the body builds proteins.
- The fat-soluble vitamins A, K, E, and D, along with high-quality fat that is essential to absorb them.
- The B vitamins—thiamin, riboflavin, niacin, pantothenic acid, B_6, and folate—as well as choline, which is essential to healthy brain and liver function. (Choline is deficient in the standard American diet, and many posit that this lack is a major contributor to the current epidemic of non-alcoholic fatty liver disease.[3])
- The minerals calcium, magnesium, selenium, iodine, phosphorus, potassium, zinc, iron, copper, and manganese.
- The anti-oxidant carotenoids lutein and zeaxanthin, known to support health of the eye.
- The essential fatty acids DHA and AA.

Interestingly, the preponderance of these are to be found in the yolk.

Of course, not all eggs are created the same. Pasture-raised eggs are far superior in many ways. To date, testing has shown that pasture raised eggs contain:[4]

- Two-thirds more vitamin A.
- Seven times more beta-carotene.
- Significantly more folic acid and vitamin B_{12}.
- Three to six times more vitamin D.
- Up to nineteen times more omega-3 fatty acids and thus a smaller proportion of the pro-inflammatory omega-6.
- More vitamin E.
- More DHA. (An essential omega-3 fatty acid needed in abundance in the brain, retina, and neurons. Deficiency is associated with cognitive decline and depression.)

Finally, pasture-raised eggs are full of flavor—boiled, scrambled, or fried. With only a little butter and salt, they will satisfy your palate and sustain you for hours. I do not think this is a coincidence. Our bodies are not mechanistic plumbing systems. They have an exquisitely balanced innate intelligence and can tell us a great deal about what we should eat. We would do well to pay attention.

contents. (I hard-boil such an egg and feed to the chickens.)

- **Is it necessary to refrigerate eggs?** We never refrigerate our eggs. An egg is designed by nature to remain fresh and viable at least as long as it would take the mother bird to assemble a full clutch of eggs or considerably longer. We organize the flow of eggs so that we are always rotating out the older ones—they are never more than a week old when we either eat them or give them away. We would feel completely confident holding them up to ten days unrefrigerated, if they are not sitting in the full light of the sun or by a source of heat. If you need to hold them longer than that, then do refrigerate them—they will hold for weeks in the refrigerator.

- **Is it possible to store eggs long-term?** Fresh eggs are best, but if you need to store a supply for a lean period such as winter, when hens' production declines, there are several options.
  1. For *freezing* whole eggs, stir the whites and the yolks together with a fork, being careful not to incorporate air into the mix. Pour into small freezer containers, or make individual cubes by freezing in an ice cube tray, breaking them out, and storing in a plastic freezer bag—do note that oiling the ice cube tray first is a *must*. You can also separate whites from yolks. Put the whites through a sieve to break up the albumen, then freeze as above. To prevent gumminess in frozen egg yolks, stir in salt (one-quarter teaspoon per eight yolks) or honey or maple syrup (1 tablespoon per eight yolks), depending on whether eventual use of the yolks will be in a savory or a sweet dish.
  2. You might want to experiment with a couple of traditional means of preserving eggs. Early in the twentieth century, a strong solution of *eisenglass* or *water glass* (sodium silicate) was used to cover freshly laid, perfectly clean eggs in the shell in a jar or crock. Even at cool room temperatures, the eggs would keep several months. At temperatures just above freezing, they would keep half a year. I recently corresponded with a fellow APPPA member who a couple of years ago preserved twelve dozen eggs—same-day-fresh and not washed—in water glass. Her family ate the last of them six months later. The only noticeable difference in quality was some thinning of the whites toward the end of storage—but even so the family much preferred them to supermarket eggs.[5]
  3. Another means of preserving eggs has been used in China for a long time. *Thousand-year eggs* start as fresh eggs in the shell smeared with a paste made of very strong tea mixed with clay, quicklime (calcium oxide), wood ash, and salt, stored in baskets or clay jars to cure for several weeks or months. The strongly alkaline coating prevents the contents from spoiling while both yolk and white change texture, taste, and color—the whites brownish or light green, the yolks ranging from dark green or blue to almost black.
  4. Hard-boiled eggs can be preserved by *pickling.* Simply boil and peel the eggs, fill a jar with them, and cover with a hot spiced vinegar. (You can also recycle the pickling solution from pickled cucumbers or beets.) Keep in the refrigerator as long as six months, and use in salads or sandwiches, or as a garnish to almost any dish.
- **Is there a good use for eggshells?** We always save our eggshells. If the egg was soft-boiled and scooped out with a spoon, there will be residues of cooked albumen, which are worth feeding to the chooks—I just crush by hand and toss them out. Shells of eggs that were cracked out before cooking I usually reserve for the garden—finely pulverized and sprinkled around plants, they help deter slugs and snails. The sprinklings are especially valuable around plants that like an extra helping of calcium, such as tomatoes and figs. Of course, shells can always be recycled back through the hens as a calcium source. Disregard nonsense you may read about toasting them in an oven and grinding to a powder—just crush coarsely by hand and toss them out.

The Dalai Lama's Eggplant Parmesan

Presentation of this recipe is dedicated to my friend Maryanne Cristello and to His Holiness the Dalai Lama.

I liked eggplant Parmesan the first time I tried some from an institutional kitchen—a college dining hall—which says something about how difficult it is to get this one wrong. One evening on my way home from work, I got a hankering for it, so I stopped and got the key ingredients—olive oil, eggs, eggplants, canned tomato sauce, and mozzarella and Parmesan cheeses. When I arrived home—a building I shared with friends—my friend Maryanne, a second-generation Italian-American, learning of my intention, asked how I would make the eggplant Parmesan. Based on the only version I knew, I replied, "I'll cut the eggplants half an inch thick, dredge them in seasoned flour, and fry them in olive oil. Then . . ." Maryanne shook her head, her face somewhere between pitying and appalled, and took the sack of ingredients out of my hands the way you'd take a dangerous toy from a child. She then started on the casserole.

As she worked, Maryanne continually bemoaned, "Too bad I have to do this in such a hurry—my mother would spend all day on it." As for me, I was almost in tears from hunger by the time the bubbling casserole made it to the table, about nine thirty. But the first bite told me this simple dish would henceforth be an essential part of my repertoire.

A couple of years later, when Ellen and I were in residence at Dai Bosatsu Zendo—a Zen monastery in the Catskill Mountains of New York State—His Holiness the Dalai Lama paid the monastery a visit. Our *roshi* ("honored teacher") asked me to be head chef for the lunch we would serve and suggested that I make my "eggu-plantu," a great favorite of his. Thus it happened that I made eggplant Parmesan for the Dalai Lama—and sixty other diners.

There was much that was memorable about the Dalai Lama's visit. The most important thing for you to remember when considering this recipe, however: His Holiness had three helpings.

There are three pointers to surefire success, two of which I learned from Maryanne: Slice the eggplant *just as thinly as possible*, and fry the eggplant pieces after dipping in *plain beaten egg*. A third I've learned through experience: *Easy on the mozzarella*.

The eggplant: Your best bet is to grow your eggplants. Otherwise, buy one or two medium to large eggplants, the best you can find. I usually half-peel the eggplant—peel off alternate strips of its skin down its length. You could leave all the skin on, or peel it all off, depending on how much of its bitterness you want to retain in the final dish.

Slice the eggplants crosswise as thin as you can. Reread that last sentence—it is meant to be taken literally. Use a sharp kitchen knife to make slices as thin as you can manage while still getting whole slices—a too-thin slice is one that ends as a half-round. I might get about forty slices from a medium-large eggplant. That's what I mean by *thin*.

Some recipes call for salting the slices to pull excess water from them. Whenever I've done that, the final result was far too salty. I place the slices in single layers between paper towels, put a cookie sheet over all, and weight heavily with a stockpot full of water. After a couple of hours the paper towel has absorbed some of the water from the slices.

Do *not* dredge the slices in flour—dip them in *plain egg*, beaten with a fork, *only*. Then fry them in a single layer in a heavy pan until golden brown, using good olive oil. Set the fried pieces aside as you work—this stage will take a while. And it will take a *lot* of egg—a major reason the dish is so good. For the forty slices mentioned above, I use eight eggs.

The tomato sauce: Summer is the time for tomato sauce. I start by sautéing onions and garlic in olive oil or duck or goose fat in my big, heavy-bottomed stockpot, then add fresh tomatoes—maybe a gallon and a half. I cook until the sauce is as thick as I want it (fairly thick for the eggplant Parmesan), adding salt and fresh herbs—oregano, parsley, thyme, tarragon, basil, any or all—during the last hour or so. Then I put the sauce through a food mill to remove the peels, seeds, and herb stems.

Of course you can buy a prepared tomato sauce. You'll use a lot of it when doing the final assembly—have on hand a couple of quarts for a good-sized casserole.

The cheeses: You'll need a wedge—a couple of ounces or so—of good Parmesan. Grate it finely on one of the punched sides of a kitchen grater. (If you use that pre-grated stuff in cardboard "cans," I don't want to hear about it.) Shred about half a pound of whole-milk mozzarella, using the large holes of the grater.

Assembling the casserole: Ladle sauce into the bottom of an ovenproof glass casserole dish, and cover entirely with eggplant slices in a single layer. Spread over the slices some more sauce, and sprinkle lightly with Parmesan and mozzarella. A light sprinkling is best—if you use too much mozzarella, it will form a tough "sole" in the dish as it cools. Continue with eggplant–sauce–cheese layers until the casserole dish is filled, then top with a final dollop of sauce and sprinkle of Parmesan.

Baking and serving: Bake the casserole in a 325-degree oven for forty-five minutes to an hour—until it is bubbling, not only around the edges but throughout the center as well. Set aside to cool a bit and set—firm up in texture a little. Serve with some crusty bread, red wine or beer, and a fresh garden salad.

This is a foolproof recipe—you'll wow your guests on your first attempt. At its worst, it is *very* good. At its best, it is something like a scrumptious eggplant-tomato custard.

Butchering Day in the Kitchen

I put dressed chickens in the refrigerator as I complete them. As soon as they are well chilled, I am ready to prepare them for storage. Of course, if it has been a long day at the slaughter table, going on to the next step is usually put off until the next day.

Is It Necessary to Age Dressed Poultry?

We don't consider it necessary to age carcasses of young birds before freezing or cooking. Even with older birds, cull hens and mature cocks, there is no need to age if the bird is going to cook long and slow to make broth. We age only those chickens culled from around sixteen weeks to a year. These are not stew birds, but can be a bit on the tough side. They will be more tender—for braised dishes like *coq au vin*—if aged about a week in the refrigerator.

Packaging and Freezing

If promptly chilled, freshly dressed poultry can be held up to a week in the refrigerator. If you do plan to keep it that long, though, *do not keep it tightly wrapped in plastic.* Instead, set it on a plate and loosely cover with waxed paper or freezer paper.

I dry the carcass inside and out before freezing. To package, I have used loose plastic bags, double bagging if the plastic is thin; freezer paper, again, double wrapping; zip-seal plastic freezer bags; and shrink-wrapping. That is the order of my preference as well, from least to most desirable. Whatever method you use, *expel all the air you can.* And make sure the wrapping is thick enough to prevent loss of water vapor from inside the package, through osmosis, to the drier air of the freezer. These two precautions will prevent or delay freezer burn.

When using either a loose or a zip-seal freezer bag, I use a drinking straw to suck the air from inside the package, then seal tightly with twist-ties or the zip seal. If using freezer paper, I wrap and tape as tightly as I can.

Home vacuum heat sealers, widely available where kitchen appliances are sold, come with a range of features and prices but need not be expensive. I have used one for many years now—the resulting shrink-wrapped packages keep better and longer than those I produce using any other method.

The greatest limit with a home vacuum sealer is size. The manufacturers seem not to have anticipated that the homesteader might want to shrink-wrap an entire goose, a roasting chicken, or a turkey. For those use one of the other methods.

Cut or Leave Whole?

Our usual practice is to freeze almost all our dressed poultry whole, then cut as desired after thawing. It may be that the meat keeps better in larger pieces. What is certain is that if you cut before freezing, you create pointed ends of sheared bone that can puncture wrapping. When freezing cut pieces with exposed points of bone, I take care to tuck sharp points inside other cuts with a smooth exterior. It is occasionally necessary to fold a piece of freezer paper several thicknesses to pad an exposed point.

An exception to the leave-whole preference is the processing of duck carcasses. To be sure, we freeze some ducks whole, usually the larger drakes, for special meals of roast duck. Usually, however, I cut up a duck, using the same two knives I use for butchering: the thin, flexible one to fillet the breast off the breastbone in two portions, and the one with the heavier, longer blade to cut wings, thighs, and legs from the back, which we reserve for the stockpot (see figure 29.1).

We love the breast fillets cooked like small steaks, by pan-grilling hot and quick in the fat that renders from the skin as grilling begins. *Duck breast is best served rare.* What I call the bits and pieces—wings, legs, and thighs—are delicious for braised dishes, for example with onion, red cabbage, and apple; or made into *confit*, preserved duck (see appendix D). After a meal like that, we sigh with satisfaction and muse, "Hmmm, wonder what the rich folks are eating tonight!"

Broth Is Beautiful

Recipe contributed by Ellen Ussery

The best use for old hens declining in productivity is the making of broth.

Chicken broth is not only a delicious base upon which to build a flavorful soup or sauce—it is an extremely nourishing food in its own right. Properly prepared, it is an excellent source of minerals, including calcium, magnesium, and potassium. It is rich in gelatin, an extraordinary digestive aid that, although not a complete protein, helps the body more fully utilize protein from other foods—in effect, you do not need to eat as much protein. And modern research has confirmed traditional wisdom: Chicken broth does indeed help prevent and moderate colds and flu.

What follows is not so much a recipe as some guidelines about how I make chicken broth now, based on my own experimentation and requirements. I will also tell you how I used to make it, and why I made the changes. Once you understand the possibilities, experiment with your own chickens in the context of your own lifestyle—if you can find a method that fits comfortably into your schedule, you will be more likely to make it on a regular basis. But let me say up front that it is almost impossible to make a failed chicken broth. Once I added too much water relative to the amount of chicken parts I was using. But even this was still usable as a soup base. I just had to add more flavoring ingredients to the final soup.

Of course, the quality of your chickens is of primary importance—if you raise them yourself, you are off to the best possible start. When slaughtering your chickens, be sure to clean and save the feet, which contribute a lot of collagen, the source of that all-important gelatin—a component as well of the bones, muscles, skin, and tendons. If you don't save them for your dog to eat raw, you can use other spare parts as well—necks, hearts, gizzards, even the heads. If you cut chickens into serving pieces and do not want to eat the backs, reserve them as well for making broth. In addition, the carcasses of cooked birds (from a roast chicken, for example) can be saved. I hold all these ingredients in the freezer wrapped tightly in plastic until ready to make broth.

I usually start with a whole stewing hen, the older the better—an older hen has more collagen and more flavor. I surround it in the pot with as many of the other parts as I have on hand, then cover—just barely—with water. Occasionally I use mostly feet, along with recycled bones of previously cooked birds. But in either case I use a pruning shears to cut open all the long bones, as the enhanced extraction makes a good gel. I then add 2 tablespoons of vinegar (or lemon juice) per gallon of water and let sit for an hour. This soak in the acidulated water helps to extract the minerals. I then bring it to a boil and skim off the scum that rises to the surface. After that I reduce the heat for the gentlest possible simmer, and cook until a poke with a fork indicates that the meat will easily come off the bone. The length of time this takes will vary, depending on the age of your chicken. The older, more flavorful chickens may take six to ten hours or more, whereas a young bird might be ready in an hour or two. At this point I take out the stewing hen, or whatever fresh chicken parts I started with, let them cool slightly, and remove the meat, which I reserve for later use. Note that

Continued on following page

Continued from previous page

you need to be careful that no tiny bones stay with the meat.

At this point I should say that I used to cook the broth in a huge stockpot. But recently I have been using a 2-gallon slow cooker. This way I have no worries about the flame going out on my gas stove during the long cooking. Other benefits are the slow cooker's timer and easier meal preparation when I don't have to contend with a crowded stovetop. If you have a woodstove, that's another great option.

After you have removed the meat from the bones, put them and the skin back into the pot. I used to then simmer the broth another fifteen to twenty hours. However, the firmness of the gelatin, once the broth was fully chilled, varied considerably. Recently I have come to understand that, once gelatin is extracted, too much additional cooking will break it down; so what I do now is just let everything sit in the pot until the next day for a period of passive extraction. Then I bring it to a boil and simmer it for about an hour just to make sure it is sterilized.

To be honest, I am not sure that this passive method actually does extract more minerals, nor do I know that overcooked (broken-down) gelatin is less valuable, nutritionally, than when more firmly gelled. I hope that someday these questions will be experimentally tested in a lab. In the meantime I am happy to speculate and take my clues from traditional wisdom.

Some broth makers are not comfortable with letting the pot sit out at room temperature and therefore put it in the fridge for the passive extraction. Others go through several more cycles of cooking and resting that can last for three days, and still others just stop completely at this point. Frankly, I would err on the side of caution with a bird that I did not know with certainty to be raised with the highest standards and butchered with utmost care.

Whenever you decide to stop cooking, strain out the solids, cool, and refrigerate the broth. I usually put mine into half-gallon canning jars. After a day in the fridge I have a solid gel and a layer of fat on the top. The fat seals the broth and protects it from spoilage, so it is always ready in the fridge. We usually consume broth within the week, but I have kept it this way for six weeks. If you like, you can always freeze it.

I noticed when I started using my passive method that the fat smelled and tasted fresher. When I cooked it for twenty-four hours I used to throw the fat out after I "unsealed" the broth—it did not pass the smell test. Now I'm quite comfortable using it as a cooking fat when I have run out of my precious duck and goose fats.

If you are not getting a firm gel, consider the age of the birds you're using—experiment with shorter cooking times for younger birds and longer times for older ones. The breed of the bird and feeding may also come into play here. There

Stock Parts

Feet, heads, necks, and skin from around the necks can be saved for the stockpot. Dry thoroughly, and freeze in appropriate-sized packages. If you cut up a carcass, the bonier parts such as the back can also be reserved for the stockpot. In addition, we always freeze the bony carcasses from roasted fowl until ready to make stock.

Giblets

If you've never liked chicken livers, this is your chance to appreciate *real* livers, not the dull yellow-brown,

are no hard-and-fast rules for the exact amount of time to cook the perfect broth.

You may have noticed that there is nothing but chicken in this broth. Keeping the initial broth simple gives me total flexibility later. You could, however, add salt and some aromatic vegetables for flavor—carrot, onion, parsley, celery—and some seaweed for more minerals.

Use of the reserved meat depends on how much flavor has been removed. Add it to soup along with rice, pasta, or potatoes and/or carrots plus some green vegetables for a one-dish meal. Make potted chicken by putting it in the food processor along with some seasonings and one-half to three-quarters of its weight in rendered poultry fat or softened butter. This will keep in the fridge for two weeks in a glass jar topped with a good layer of fat. If most of the flavor has been extracted, you could feed it to the dog or back to the chickens.

We often start a meal with a cup of broth. Many times I just heat it up and stir in some flavorful miso, such as the red pepper and garlic miso from South River Miso Company. Otherwise I add salt and drink as is, or simmer with any of the following: a pressed or minced garlic clove and chopped parsley, trimmed shiitake mushroom stems that I have frozen, shredded spinach, or coconut cream concentrate and fresh ginger. The possibilities are endless.

tired-looking livers your mother may have bought at the supermarket. Livers from your younger poultry should be dark red, plump, and glistening. Cook them as a first-day treat, and freeze extra livers in small packets for later use. For a fancy hors d'oeuvre, try Bridget's Chicken Liver Pâté.

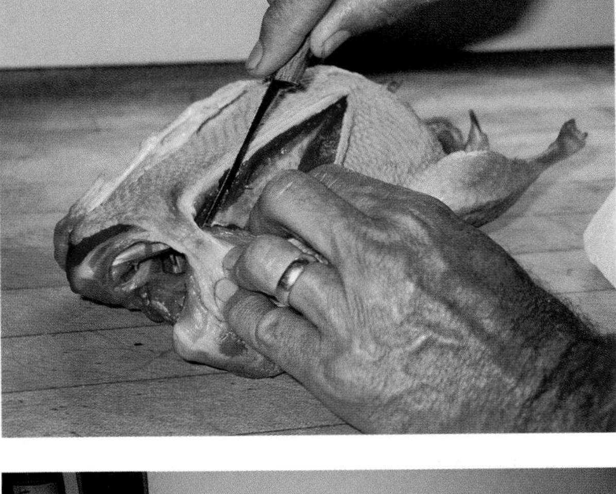

Fig. 29.1 Cutting up ducks provides breast fillets for gourmet meals, bits and pieces for a variety of interesting dishes, plus additions to the stockpot.

The hearts and gizzards can be sautéed and added to soups, sauces, or stir-fries. Or they can be added to the stock parts.

Processing the gizzard is complicated by its structure—a mass of lean muscle around an inner pouch with a tough lining, containing bits of stone the bird uses to grind its food. You can cut through the muscle just to the pouch lining, then peel the muscle back with your fingers to free it from the pouch, leaving a butterfly-shaped cleaned gizzard. It can be tricky peeling off the muscle without rupturing the pouch—sometimes it helps to chill the gizzards a little before peeling them. Another option is to place the gizzard on a cutting board and slice off bite-sized

Chicken Liver Pâté

Recipe contributed by Bridget Chisholm

I've had my feet under Bridget's table many a time, and I can assure you—this is a sophisticated hors d'oeuvre that has even folks who don't like liver coming back for more.

- 1 pound chicken livers
- 1 medium onion, thinly sliced
- 1 clove garlic, crushed
- 2 bay leaves, crushed
- ¼ teaspoon dried thyme (use 1–1½ teaspoons very finely minced if fresh)
- 1 cup water
- 2 teaspoons salt (1 teaspoon for poaching livers, 1 teaspoon for seasoning pâté)
- 1½ cups softened unsalted butter, more or less (see below)
- Salt and fresh-ground pepper
- 2 teaspoons cognac or Scotch whiskey

Place the first seven ingredients in pot (using only 1 teaspoon of the salt), bring to a boil, cover, and cook at a bare simmer for seven minutes. Remove from the heat, and let sit five minutes.

Strain off the liquid, then place the mixture of liver and solids in a food processor. Process, adding the butter a little at a time. You can add a little more than 1½ cups for a milder flavor, or a little less to emphasize the flavor of the liver a bit. After all the butter has been incorporated, add salt and pepper to taste and cognac. Process another two minutes until creamy and smooth.

Pour into small molds and place in the refrigerator to set. Unmold before serving with crackers or thinly sliced chewy bread. The pâté can be frozen for up to six months.

pieces, keeping the blade's edge away from the interior pouch and leaving it intact.

Remember Your Pets

Actually, at our house, all the necks, hearts, and gizzards are reserved for our dog Nyssa. Before dogs started eating that dry crunchy stuff, they ate raw meat, right? And bones. So we decided years ago that Nyssa could only benefit from raw meat and bone as a substantial part of her daily diet. I cut the hearts and gizzards into bite-sized pieces and freeze in mini-packets. I also freeze the necks, wrapped in pairs in plastic wrap. Each day Nyssa gets one or the other, thawed to room temperature. She loves them!

Yes, I know you've been cautioned never to feed chicken bones to dogs. That's excellent advice with reference to long bones (thigh bones, drumstick bones), cooked bones, and bones from commercial chicken generally (which are poorly mineralized, softer than home-raised chicken bones, and apt to break into long dangerous splinters). Raw chunky neck bones have never been the slightest problem for Nyssa. Chewing them helps keep her teeth clean and her gums healthy, and they are unmatched for promoting good bowel function.

We used to offer our cat the same raw tidbits, but she didn't seem much interested. She was a good hunter who regularly ate her kills, so I suppose she considered our offerings a poor second best. I've read, however, of cats who relished such raw tidbits.

Rendering and Storing Fat

I hope you are not among those unfortunates still bamboozled by the rampant superstitions regarding animal fats in the diet.[6] High-quality animal fats are, in reality, extremely beneficial in the diet. Female fowl especially—whether chickens, ducks, or geese, and especially in fall—contain a heavy deposit of body cavity fat in the lower abdomen. While completing evisceration, it's easy to pull it free and store in the refrigerator until you're ready to render it. At that point, heat a heavy cast-iron frying pan or equivalent over low heat, and cut the masses of fat from the birds into small cubes. Add the cubes to the pan and keep on lowest heat to gently melt the fat out of them (see figure 29.2). When the contents of the pan are nothing but golden liquid fat with some crinkly remnants of the original fat tissue, pour the fat through a strainer into small jars. It will keep for months in the refrigerator or indefinitely in the freezer. Spoon the fat from the jar anytime you need a cooking fat for frying sliced potatoes or greens such as mustard or kale; for adding to chopped liver or potted meat; or even as a spread on bread or crackers, in lieu of butter. This fat is also a key ingredient for making confit (see appendix D).

The shriveled tidbits remaining in the strainer are called cracklings. Save them, and use as you would croutons—that is, as garnishes for salads, soups, stir-fries, or baked potatoes, or even as a crunchy snack.

Fig. 29.2 Render fat from ducks and geese (or chickens) over gentle heat.

First-Day Goodies

Butchering a large contingent of fowl is hard work, and at the end of slaughter day we're not much inclined to do any complicated cooking. A couple of the more perishable goodies from butchering make great quick meals.

We prize what Ellen calls the unborn eggs from laying hens. Finished eggs we simply use as we do any other eggs; but the developing egg yolks—there may be up to half a dozen large enough to keep per hen—are both a special treat and a quick meal. We pour hot broth into mugs or bowls, add the small yolks, and wait just long enough for them to warm through. Simple, nutritious, and filling—a fine appetizer, or an entire meal if you add a quick salad.

Supper at the end of butchering day almost always centers on the fresh livers. Sauté some onions in a generous amount of butter or goose or duck fat in a heavy skillet, then sauté the fresh livers hot and quick, deglazing the pan with a little wine or sherry, maybe garnishing with bacon. Livers are rich in fat-soluble vitamins, especially A and D, anti-oxidants, and essential fatty acids—*and are most delicious if cooked quite rare.* Again, the simple addition of some buttered steamed green beans or a green salad makes it a meal.

As said in the last chapter, the testicles of mature male fowl are as edible as those of larger animals such as lambs or calves. Those not put off by "mountain oysters" could give them a try—preparation, texture, and taste are the same, only the size is different. Older versions of *Joy of Cooking* will have recipes for testicles, referred to as "fries"—we do like our euphemisms, don't we? Newer editions have abandoned some of the older culinary lore, so they may or may not still be a guide for this particular dish.

Harvey's Basic Summer Chicken

Rather than freeze all the chickens from a full butchering day, in the summer I almost always cook a couple of them right away. My approach is likely to be the same anytime I'm culling just a couple of birds in a hurry, probably skinning them for simplicity.

Since I have onions, garlic, and tomatoes in quantity at that time, that is usually my starting point. I sauté the alliums in poultry fat; add the cut-up chicken pieces, sometimes browning them first in the fat, sometimes not; then add coarsely sectioned tomatoes right off the vines, enough to cover the chicken. I simmer with no lid—so that I'm simultaneously cooking the chicken and reducing the sauce—until the chicken is as tender as I want. How long that takes depends on the age of the birds, from less than an hour for one a couple of months old to three or so for an old stewing hen.

If the chicken pieces are cooked before the sauce is sufficiently reduced, I remove them and continue cooking down the sauce. When it's thick enough, I put it through a food mill to remove the peels and seeds, then return the chicken pieces.

How I finish the dish from that point depends on the mood I'm in, or what's clamoring to be used in the garden. Adding hot chile peppers and garnishing with chopped cilantro gives it a Mexican twist. Fresh oregano or thyme or tarragon allows me to pretend I'm a hotshot French or Italian chef. If I use a *lot* of additional onion at the beginning, add several tablespoons of good paprika, and stir in cultured cream at the end, I dub the result Chicken Paprikash. And of course, what could be easier than working in some mixed exotic spices for a curry, mild or hot? It's especially good with the addition of fresh green beans.

30 | SERVING SMALL LOCAL MARKETS

Though I do not produce for market, since we started keeping poultry going on three decades ago, I have produced *all* the dressed poultry and eggs we eat, year-round—and we eat a lot of both. I therefore think of myself as working right at the intersection between serious home production and small-farm production dedicated to bringing to local markets "food with a face." This chapter is addressed to the reader who is just a little more ambitious than I—one who, having mastered the inclusive, integrative approaches to poultry husbandry presented in this book to produce all the family's eggs and dressed poultry, dreams of expanding production sufficiently to earn some income—and make a difference—by offering the world's best poultry and eggs in small local markets. As you approach that level of production, I guarantee that guests or friends to whom you make gifts will say, "I've never tasted chicken like that!" or "We haven't been able to find eggs that good anywhere—can we buy from you?" If you decide to try your hand at marketing, I fervently hope you will be motivated by this book's passion for healing and sustainability in a new agriculture—*that's* who I want to be serving local markets for eggs and poultry in the future.

And I hope your whole family will get involved. What kind of sustainability are we practicing if our children do not continue what we're doing on our homesteads and small farms? What better than a poultry marketing project to teach youngsters responsibility, independence, and business acumen?

If you are considering stepping your home flock up to production for small markets, you couldn't do better than to join the American Pastured Poultry Producers Association (APPPA). APPPA's members can give you better guidance—out of a deeper well of hard-won experience, intelligent observation, and constant experimentation—than any other source, bar none.[1] Do note the *pastured* in the organization's name. All the small producers with whom I correspond believe passionately that the place to produce quality eggs and broilers is on good pasture. With the exception of winter, none of them would even consider a confinement model.

Relationship Marketing

Some small producers sell at the farm, others through CSAs (Community Supported Agriculture), and others through farmer's markets. Whatever your choice, see the contact with your customer as an opportunity to establish *relationship*. Foster that relationship first of all by as full an introduction to the quality of your foods as possible. Hand out recipes for dishes using your food products. Offer tasting samples. Garnish your sales table with pictures of happy hens on pasture and other scenes from your farm. Develop a newsletter, e-mailing list, or farm blog.

But the relationship doesn't just center on the food sale of the moment—see it as a potential partnership in the transformation of farming toward more sustainable patterns. Tread softly, however—nobody wants to be lectured. Without being tedious, preachy, or overbearing, see the transaction as an opportunity to *educate*—always remembering that the education opportunity is a two-way street. More customers are getting concerned, for example, about the ubiquity of soy in our modern diet and its deleterious effects. "Do you feed your chickens soy?" is a question you're bound to be asked. When you get such questions, make it a point to describe your foods, as applicable, as: pastured, GMO-free (genetically modified organism), chemical-free, antibiotic- and hormone-free, soy-free—my APPPA correspondents report these are all major selling points with customers well informed about issues of food and health.

Customers who feel they consistently get the

Should You Seek Organic Certification?

An important question as you consider producing for market is whether or not to seek organic certification. Many of my correspondents in the American Pastured Poultry Producers Association choose to do so, perhaps because they get a premium price if their eggs or broilers are certified, because one of their major customers will only buy organically certified products, or because they believe that the organic certification program, even with its shortcomings, is a key part of creating a more benign agriculture. Other members do not choose to become certified, because of the expense, due to the additional management input of maintaining the required paper trail, or simply because they find no advantage to certification in their particular market.

It may be that the degree of anonymity in the transaction is a key to whether it will be advantageous to you. If you sell face to face with your customers, the opportunities to explain directly your production means and to develop customers' trust may make certification irrelevant. It may be far more important to the concerned and informed customer that you are able to explain what *local*, *pastured*, *sustainable*, *organic*, and *GMO-free* mean to you, and to communicate your passion for regenerative farming.

On the other hand, if you sell through outlets that do not give you personal contact with the purchaser, being able to label your eggs or dressed poultry as organic may increase customer trust. If you do seek certification, be aware that some buyers are increasingly sophisticated about the nature and problems of the National Organic Program. One of my APPPA correspondents, who is organically certified, finds that his restaurant buyers are not impressed with certification per se, knowing that the label means little if the certifying agency is too lax, and are quick to ask him which agency certifies his eggs and broilers. They are reassured when he explains that he is certified by Northeast Organic Farming Association, New York chapter, since NOFA-NY has a reputation for strict application of the organic standards.

The same correspondent points out that certification need not be expensive in the majority of states that have certification cost-sharing programs to encourage farmers to produce organically. He finds that maintaining the required paper trail is not especially burdensome, given his need to closely track costs and production in his egg and broiler operations anyway.

brush-off from the commercial food system are frustrated—a major reason they're showing up at your farm or sales table. Be prepared to engage their concerns seriously, and be willing to do some research to resolve their questions. Whether or not you decide to seek organic certification for your products, be prepared to explain why they are beyond organic—that is, even more naturally and sustainably produced than required by the National Organic Standards.

Producers in the APPPA find that interest in their products spikes every time there is a massive recall from the supermarkets, or an outbreak of foodborne illness in the news. Be ready to explain to them why your pasture-raised products are not only tastier but *safer* and better for their health. Setting out take-home brochures from the Weston A. Price Foundation[2] or lists of local foods producers on your display table could be a little like scattering seeds—who knows when and where one will sprout?

At the same time remember Joel Salatin's advice not to waste your time with customers who only whine and complain, or who fail in their responsibilities in distribution arrangements. Customers who simply cannot see the difference between you and Walmart, or who expect you to *be* Walmart—let 'em go to Walmart!

Equipment and Logistics

If you are handy, you can save a lot of money on equipment and infrastructure costs. I've corresponded with small producers who made their own range feeders, pluckers, compressed-air egg washers, ingenious mobile shelters, walk-in coolers, brooders for starting hundreds of chicks at a time, and more. Again, another reason to join APPPA, as members swap information about their latest projects all the time.

Save money on ice by making your own reusable freezer chillers: Make a saturated salt solution—as much salt as you can dissolve in hot water—and fill recycled plastic jugs of any convenient size and shape. Keep them in the freezer until needed. Because of the lower freezing temperature of the salt solution, they will be much colder than ice.

Breed, infrastructure, marketing, management—it all has to fit when thinking through what might work for you. As one of my APPPA correspondents put it:

> Turkeys are the most capital intensive because their equipment and brooder can only be used once a season. Cornish Cross and Pekin ducks aren't much different because the groups get turned over several times during the season. Management level required in the brooder is the highest for starting turkeys, followed by chickens, and then ducks. Transitioning from brooder to pasture is easiest for ducks, turkeys, and then chickens. Ability to forage is best for ducks, followed by turkeys, and then chickens. Ease of processing is chickens, turkeys, and then ducks. Ease of marketing at a profitable level for us is ducks, turkeys, and then chickens. Ease of getting yourself in the door with a new customer: chickens easiest of all.

Feeding Issues

The logistics of feeding change when the flockster starts feeding market flocks, even on a modest scale. It may no longer make sense to buy in 50-pound bags but rather to take delivery in bulk. Bulk ordering has its challenges—storage becomes a bigger part of the equation. But those getting bulk deliveries will probably be in a position as well to get the local feed mill to custom-make feed to their own recipes.

I'm always pushing the idea as well that flocksters—whether serious homesteaders like myself who use several tons of feed a year or small producers for markets—band together to contract with local farmers to grow their feed grains. You buy their feed grains, they buy your chicken and eggs—how's that for keeping it local? That strategy would have a positive ripple effect in the local economy and could help keep a small farm viable. You would also know the life history of your feed grains, as opposed to buying a pig in a poke. Again, however, the big challenge would likely be how to handle storage.

Remember that feed quality can pay off, even though it costs more. An APPPA member recently reported that she started feeding organically raised feed grains and raised several hundred broilers more than she had the previous year—on the same amount of feed.

Regulations–and Regulators

I can see the day coming when even your home garden
Is gonna be against the law.

—Bob Dylan

The biggest difference between producing for the family and producing for even the smallest market is being subject to applicable regulations governing food production and marketing. If we assume that the regulations are intended solely to guarantee the safety of what we put into our mouths, however, we are being naive. The regulatory process is frequently hijacked by the big producers in order to create impossible hurdles for the small producer. Giant, vertically integrated food processors feel threatened by the growing demand for safe, local food and want to stop it in its tracks. If they have their way, any alternative to industrial food whatever "is gonna be against the law." By all means read Joel Salatin's *Everything I Want to Do Is Illegal: War Stories from the Local Food Front*—it serves up just the right mix of boiling-mad outrage and belly-laugh humor to get you in the mood for the byzantine labyrinth of food-police regulations.

The small producer's most important contact with regulations is the regulator, often the inspector in the field. It's important to make the regulator a friend if possible—I've had small producers report that their inspector actually helped them out: "Well, the regulation says you have to do *X*, but you know, you could meet the requirement by doing *Y* [minor change, maybe only a 'paper' one, to bring you technically into compliance]."

On the other hand, if your regulator turns out to be fiend rather than friend, you have to know the regulations *even better* than the regulator. Again, I've had numerous reports of regulators handing out diktats off the top of their heads, with serious disadvantageous implications for the operation—only to be stopped cold by a level-eyed producer quoting the regulations verbatim, chapter and verse.

If you have a problem with regulatory agencies you cannot resolve on your own, by all means call on the Farm-to-Consumer Legal Defense Fund,[3] a leading advocate for small farmers being harassed by agencies of government. Intervention by one of their lawyers on your behalf may more readily sober up an agency acting arbitrarily than your own remonstrances. For example, when a health inspector recently disallowed use of Polyface Farm eggs by a Charlottesville, Virginia, restaurant, Joel Salatin immediately called FTCLDF. Within twenty-four hours of the Legal Defense Fund's intervention, the health department had rescinded its arbitrary ruling.[4]

Pricing

In a market blinded by the illusion that cheap food is a blessing, there is often resistance to paying the higher price of non-subsidized poultry products. The customers the small producer should target are those more concerned with food quality—and safety—than bargain-basement prices. And they're out there: My APPPA correspondents routinely report that they cannot meet growing market demand for "real food."

Remember to pay yourself for your labor in calculating a fair price to charge—it must be fair to you as well as the customer.

Questions about price can also be opportunities

for educating customers. For example, you might tactfully ask customers unhappy with your prices if they're aware that *their taxpayer dollars* pay for the subsidies underlying "cheap chicken" and "cheap eggs" in the supermarket.

Complaints about price can be frustrating. But you might like to know that I've had numerous reports from small-market producers about customers who anxiously worry they're not charging *enough* for their products. Such customers understand the *One dollar, one vote* rule and are willing to pay a worthy price in order to ensure their supplier stays in the market.

One thing beginners should keep in mind is *never* to charge a hobby price when selling in local markets. Even if you are just dabbling with sales of a bit of surplus from home production, undercutting prices of producers who depend on sales for their livelihood is an inexcusable disservice to them. Building a sustainable agriculture that adequately rewards suppliers of quality food is serious business—do your part.

Market Offerings

Whatever you most like to see on your own table from your flock is probably what you should consider first for offering in a market. Your own enthusiasm for the superior poultry foods you provide your family will be infectious among your customers, and their queries about less usual fare may suggest new marketing ventures.

Eggs

Eggs are the easiest part of the typical home poultry project to expand into market sales, if only because they require so little additional in the way of infrastructure and equipment. As such, they may also be an ideal project for children. And I don't mean just gathering the eggs—though that's a fine chore for the wee ones—but organizing and managing the entire care of the egg flock and the market sales.

You will have to check the regulations in your own state as to whether you need to be licensed or inspected and regarding candling, grading, washing, packaging, labeling, and refrigerating eggs.[5] Talk with others serving egg markets in your area as well—in some cases actual practice, for example whether reuse of egg cartons is allowed, is a good deal more casual than is strictly set forth in the regulations. If you sell at a farmer's market, it may have requirements to be met as well.

Regarding breed choice for layer flocks, there are those in the pastured poultry community who wax dogmatic on the hybrid superlayer strains as the only practical and economical choice, because of their higher rate of lay. But many small-market producers base their market egg operations on traditional dual-purpose breeds. If Joel Salatin—as hard-nosed and practical a farmer-entrepreneur as you'll find in the movement—bases his production of hundreds of dozens of eggs a week on Rhode Island Reds, Barred Plymouth Rocks, and Black Australorps (all traditional farm breeds), you certainly don't need to be told that superhybrids are the "obvious" choice for you.

Those who rely on traditional dual-purpose breeds for market eggs find they take fuller advantage of natural forages; are more easily managed; make a better-quality egg; live longer; stay healthier; and are more alert to their environments, thus more likely to escape predation. Remember as well that anyone who enters the market for eggs is after a couple of years going to be looking for a market for stewing hens—many producers have noted how much easier it is to sell the larger carcass of a dual-purpose hen than the much smaller one of a hybrid layer.

For many it is a bonus that the dual-purpose breeds lay large brown eggs. Though they know that egg quality is strictly a matter of diet and lifestyle, some producers despair of trying to educate customers who stubbornly believe that brown eggs are better than white. Use of the traditional brown-egg breeds helps keep such customers coming back. (If you prefer hybrid layers, options such as Golden Comets and Red Sex Links that lay brown eggs are available.)

Some producers keep a mix of white-egg breeds, several brown-egg breeds for multiple shades of brown, even a scattering of Ameraucana hens, finding that a rainbow effect in the display carton helps spark interest. Some producers find that Ameraucana eggs especially, even at the cost of a slight decrease in overall production, may justify including in order to catch customers' eyes—especially those with children—and trigger questions about "those green eggs."

Managing the flock regarding when to retire older, less productive hens and bring along a new flock of layers becomes more complicated than in the home flock, when you have customers depending on your eggs. As said, you'll also need to find buyers for the old stewing hens your operation generates. A couple of possibilities are mentioned below. Some producers sell them as well in a growing market for raw, natural pet food.

You don't have to tell customers how different your eggs are—an attention-getting trick is a display of your pastured egg and a supermarket egg, cracked open in side-by-side saucers. Alternatively, friends of mine hard-boil some of their eggs and some of the supermarket imitations, and slice them in half for display. Differences in interior structure, yolk color, and integrity of the whites are all that you need to tell the tale.

Some producers mark their egg cartons with the laid-on date as a way of highlighting how fresh they are.

I get mixed reports from members of APPPA about getting a decent price for their eggs. Some encounter an appreciation of the obvious superior quality, and a willingness to pay the higher cost. Others find that customers, fixated on $1.29/dozen eggs in the supermarket, stubbornly refuse to pay more than an unfairly low price. Many producers find that initial obstinacy about price is followed by repeated demands for more once customers decide to give their pastured eggs a try, as in this report from the field:

> I had a customer last year at the farmer's market who confronted me about selling pasture raised eggs for $7 doz. I told her to buy them quick because my bookkeeper was putting the price up to $8. I explained that my price has something to do with the fact that we buy organic feed for $745 per ton, pay our crew well, and have workers' comp insurance to pay. Then I advised her that on second thought she actually shouldn't try them, since they were so good, she would never be able to eat any other eggs. She was not impressed. But later that day she came back, still indignant but willing to try half a dozen. I warned her again but to no avail. Next week she came and thrust the money into my hand and picked up a dozen. She's been a regular customer ever since.

Broilers

If you want to supply fresh-dressed broilers to your market, the first decision is what type of bird to grow. Many pastured poultry producers grow Cornish Cross—the foundation of industrial broiler production—as their market bird. Some report that they have learned to work with the Cornish and don't see any reason to change. Others are not happy with the weaknesses of this supercharged hybrid—the leg problems, the heart problems, the laziness as foragers on pasture—but find that customers are so used to the plump, broad breast of the Cornish that they resist buying a broiler with a narrower profile.

Many of my APPPA correspondents prefer meat hybrids more suited to pasture, such as the so-called Freedom Ranger[6] and other meat strains coming out of the French *Label Rouge* system of certified broiler production. A few offer both Cornish Cross and Freedom Ranger broilers, charging an additional 50 cents or more per pound for the latter to compensate for their longer grow-out. The occasional adventurous producer raises traditional dual-purpose breeds such as Barred Rock, Buff Orpington, or Delaware as market broilers, for customers who are okay with a skinny carcass with superb flavor—and willing to pay a higher price commensurate with the even longer grow-out. In today's market, however, traditional-breed broilers are a hard sell.[7]

Regarding regulations, as a broad generalization, most states go with the federal guidelines for poultry production for sale: The grower may process and sell from the farm up to one thousand "bird units" per year with no inspection or regulation. (A broiler is one unit; a turkey, four—any combination is allowed to bring the total to no more than one thousand).

Maintain as low a profile as possible: If you don't have to be inspected or regulated by a given agency, by all means avoid getting on their radar with inquiries—one of my correspondents said that's "like poking a hornet's nest with a stick to see what happens." The same correspondent described calling three different state agencies to inquire about regulations for broiler production. He got three different answers—all of them wrong, as it turned out. If inquiries are necessary, make them anonymously.

Remember that if you choose to fudge existing regulations a bit in your favor, a first-offense action by the powers that be is rarely more than a cease-and-desist warning.

Major questions for setting your price for broilers have to do with whether you distribute them packaged through a CSA or farmer's market, or require customers to pick up at the farm and do their own packaging—and whether you sell them whole or cut up. Most producers sell their broilers whole, but I hear from more and more that they're responding to customer demands for cut-up chicken. If you try the latter, the increase in your own costs results from not only the added labor but also extra time and materials in packaging, weighing, and inventorying. Still, you can make money if customers are willing to pay for the convenience of getting their preferred cuts only. For example, one correspondent reports that she sells whole broilers for \$4 per pound (thus getting \$18 to \$24 each for birds that range from 4½ to 6 pounds), while charging \$13 per pound for boneless and skinless breasts.

In contrast with producing market eggs, broiler production for market requires serious tooling up for more efficient processing—you either make money or give away that bird, not at the feed trough, but at the slaughter table. Market processing likely will require more automated scalding and plucking equipment than that in my backyard operation. Good poultry processing equipment is expensive if purchased new. Again, if you're handy you might make your own.[8] Keep your eye out for farm sales: A friend of mine stumbled on an auction where none of the bidders was interested in an array of stainless-steel pluckers and scalders—and walked away with \$20,000 worth of equipment for \$400.[9]

Turkeys

Many of my APPPA correspondents confirm that turkeys can be very profitable. One reported that she invested \$3,500 converting a mobile construction site office to a state-approved processing facility. She earned back the entire costs for her new facility that fall, as profits on sixty-five turkeys preordered for Thanksgiving. She noted as well that the daily care of those sixty-five turkeys was the responsibility of her six-year-old son entirely, a big boost in his self-esteem from helping earn the family's income.

If you have a lot of room to give a flock of turkeys, they will forage a large percentage of their own feed, reducing costs. Your labor costs are lower when butchering turkeys as well—processing time per bird is pretty close to that of a broiler, but the payoff in salable meat is so much greater.

Most growers of market turkeys stick with the Broad Breasted White, the equivalent to the Cornish Cross for commercial turkey production. But enough raise heritage breeds to merit giving them a try. Though they require more management input and produce a smaller, narrower carcass, you can make them work for you if you find customers willing to pay more for the very best.

Most small producers I know grow only the number of turkeys for which they have preorders and target their turkey grow-out to Thanksgiving and Christmas. However, a young farmer friend of mine is considering

Tooling Up

Efficient processing for any but the smallest broiler market requires automating scalding and plucking and a comfortable place to work furnished with running water. Shelter to work when the weather is not cooperative could be a boon as well.

Mechanical pluckers at the lowest end of the scale might feature a homemade plucker head mounted on a portable hand drill (see figure 30.1).[10] For hands-free plucking of several birds at once, a tub-style plucker is just the thing. Do-it-yourself flocksters might want to make a Whizbang plucker for a few hundred bucks (see figures 30.2 and 30.3). If you want to buy a tub-style plucker, the best among the more affordable options is David Schafer's Featherman Pro (http://featherman.net/pluckers.html#pro). At approximately $1,000, it is a significant investment, but my correspondents in APPPA who use it report that it is an effective and durable unit, even with heavy use serving broiler markets. If your level of production justifies

Fig. 30.2 Mike Rininger's Whizbang plucker, made from a recycled food-grade plastic drum, a donated steel plate custom-cut by a metal sculptor friend, and purchased parts, including an electric motor.

Fig. 30.1 Kate Hunter spent less than $20 to make this do-it-yourself plucker driven by a portable hand drill. Plucker heads are usually inset with purchased stiff rubber fingers. This one uses short rubber bungee cords cut in half instead, set in holes drilled in a PVC end cap. PHOTO COURTESY OF KATE HUNTER

Fig. 30.3 Mike's Whizbang in action. Note that in a tub-style plucker, the stiff rubber fingers are inset into both the rotating bottom plate and the stationary drum sides.

spending $1,365 to $3,395 for top-of-the-line stainless pluckers of proven design, Eli Reiff of Mifflinburg, Pennsylvania, has designed several to fit the needs of pastured poultry producers. See his Poultryman pluckers, widely used among APPPA producers, at http://chickenpickers.com/page10.html or call Eli at 570-966-0769 (see figure 30.4).

Both Dave Schafer and Eli Reiff also offer automated scalders for scalding numerous birds at once (see figure 30.5).

Once you are supplying a steady broiler market, you might want to install an efficient farm processing facility, such as that of my friends Matthew and Ruth Szechenyi, who serve local markets for broilers and turkeys in my area (see figures 30.6 and 30.7).

Fig. 30.4 One of Eli Reiff's stainless-steel pluckers in action.

Fig. 30.5 A scalder made by Eli Reiff, with a rotating shelf that cycles up to a dozen broilers through water kept thermostatically at scalding temperature.

Fig. 30.6 Matthew and Ruth's processing facility–a 12-by-24 hoophouse (made from a FarmTek kit for erecting a tractor shelter) on a 12-by-28 concrete slab.

Fig. 30.7 The interior of the Szechenyi hoop, featuring stainless-steel processing equipment and worktable and drainage to a French drain outside for easy cleaning.

expanding his turkey operation this year. In the past he has had to do some creative marketing of some of his turkeys, mostly "uglies" that ended up with cosmetic blemishes in the field or during processing. He told me the skinless and boneless turkey breasts and the turkey tenders he salvaged out of those less salable carcasses were snapped up by his customers. Many of the legs and thighs he ground and again found eager buyers among his customers. The remaining legs, backs, bones, and wings he sold to his restaurant chefs for making broth. In the process of making these sales, he has discovered that some of his customers would prefer to eat turkey rather than chicken year-round. One bought eight of his Thanksgiving turkeys to freeze, but said she would prefer being able to get them fresh for more of the year. He plans an additional, earlier batch of turkeys this year to test the out-of-season market.

Specialty Poultry

Some of the remaining options for dressed poultry get us into niche-market territory, but all are good eating whose challenges have more to do with marketing than management or table quality. You are well advised to start small with these offerings and try to line up committed orders for the dressed birds before putting in the feed and sweat to ready them for sale.

As observed in the chapter on butchering, processing *waterfowl* is little different from processing chickens—except for the plucking. One correspondent noted that she can process four or five broilers in the time it takes her to do one duck. Thus the price you charge for them should be commensurately higher. Many producers who try to market ducks and geese, however, find it isn't easy to get that higher price.

Still, many producers find that within a narrow segment of their market, supplying *duck* can be a small but profitable sideline. The Pekin is the Cornish Cross of meat ducks, fast growing and clean plucking, and is often the breed of choice for market producers. Remember *Muscovies* when considering ducks—one of my correspondents finds that the combination of busy foraging, health, ease of raising, and the fact that the ducks are excellent mothers means Muscovies are low-input but they generate gourmet-level prices. Another reports she gets $20 per pound for duck breasts from her Muscovy drakes.

Remember ducks as an option for producing specialty eggs as well. Two young friends of mine found eager acceptance of duck eggs in the farmer's markets they serve.

Geese require even more processing time and grunt work at the worktable than ducks. But they can be a center of highly special holiday meals not likely to be available from any other source, for which people are willing to splurge. Again, they might fit in as a small profitable sideline—a friend of mine has no problem selling twenty-five as preorders for the winter holidays—made even more profitable if you raise them mostly on good pasture. The large, all-white Embdens are often the breed of choice, but based on my experience, I'd advise giving fast-growing Africans a try as well. There is no comparison between commercial and pastured goose as a taste sensation.

I hear from a few of my correspondents who try teasing the market with *guineas*, *pheasants*, and *quail*. Don't bet the farm on such exotics, but it could be fun experimenting, especially as an unusual project to share with children.

Finally, I recommend you consider *capons* as a meat-fowl option. In times past "caponizing" cockerels was a common practice, both because it was a profitable use for cull cockerels and because it produced a matchless roasting fowl. The term means "castration"—surgically removing the testicles of young cockerels to make for faster, plumper growth sans testosterone. A well-grown capon can dress out to 12 to 15 pounds—the size of a small turkey. Yet however large it grows, or how long, it remains tender and succulent—and continues accumulating flavor. In addition, it doesn't waste its time and energy duking it out with the other cocks in the flock.

The purpose for caponization, then, is the same as for castration of male livestock such as bull calves. The difference, however, is that mammalian testicles are held externally, while in avian species they are deep inside the body cavity—so removing them is *major* surgery. All the same, a few producers are taking up this option again where they find a demand for the very finest roasting chicken. Caponizing kits containing a few simple surgical implements are available, as well as manuals describing the procedure.[11]

Because caponization is so invasive, you will need more detailed research if you want to give it a try, but here is a brief summary. Caponize young cockerels early, within the first week to three weeks ideally. Make an incision between two ribs, hold it open with surgical retractors, and remove the testicles with forceps or a special cutting scoop. You don't use an anesthetic; you don't sew or bind the incision; and the procedure requires only a high level of cleanliness, not operating-room sterility. The bird's robust immune system, and quiet aftercare for a few days, are the keys to success.[12] The problem is not the complexity of the procedure but the necessity to "screw your courage to the sticking point," as Lady Macbeth urged, and to continue resolutely through your first batch *whatever* happens on the operating table. Failures you will have, but you will learn this skill only if you persist despite them.[13]

I know of some producers of capons who skip the unnerving part and buy in week-old cockerels already caponized by producers specializing in them. Someone skilled in the procedure with an assistant can caponize one cockerel a minute. I even know of one operation that furnishes week-old caponized keets for the production of guinea capons—don't ask me how they sex the males!

Other Income Options

In addition to the offerings from your flock discussed above, there are other ways to turn a homestead or small-farm poultry operation to profit. You will probably think of more, but here are a few you might consider.

Game Birds: A Niche Market Opportunity

If you can find the right market, growing a niche product may be a better option for you than competing with a more standard offering in a market crowded with competitors. My friend and neighbor Denton Baldwin has found a niche market you might consider.

Denton operates Freestate Farms in northern Virginia, based on several leased farms on which he raises beef cattle, pigs, and sometimes a batch of sheep. In line with my point about finding a workable niche rather than butting heads with big established producers of standard products, rather than raising Cornish Cross along with a dozen other nearby small farmers, Denton raises three species of game birds to supply five local top-end restaurants—quail, chukar, and pheasants.

All three game species are in the same order, Galliformes, as domesticated chickens, guineas, and turkeys, with similar feeding and lifestyle needs. Wild relatives include grouse and prairie hen.

Quail raised for market are most typically strains of *Coturnix coturnix japonica*, the domesticated subspecies of the common quail, *Coturnix coturnix*. They are astoundingly productive. Incubation is a mere sixteen or seventeen days. The birds mature in six weeks and begin laying as early as five weeks and are ready for butchering as early as six to eight weeks. Denton gets an average of one egg per hen per day in the final two weeks before slaughter at eight weeks, which increases

Continued on following page

Continued from previous page

profits through sale of quail eggs. Sometimes he retains some of the mature hens for two more weeks, during which egg production is even better. Carcass quality declines somewhat as a result, but that's no problem for creative chefs, who turn such older birds into pâté and fancy potted quail dishes.

Chukar are a type of partridge, *Alectoris chukar,* in the pheasant family, larger in size than the coturnix. Ring-necked pheasant is the collective name for a number of subspecies and hybrids of the common pheasant (*Phasianus colchicus*), native to Russia and the Caucasus but widely introduced elsewhere as a game bird, which readily breed in captivity. Both chukar and pheasant require a sixteen-week grow-out. Denton once had a batch of the latter that required twenty weeks.

Denton prefers buying in hatched chicks of these three game bird species from two sources in the area that he trusts. A recent surge of interest in growing game birds, however, suggest to him that not only the reliability of his supply of chicks but their genetic soundness could be under threat, as there is more demand for hatching eggs and chicks, and inexperienced or unscrupulous producers could rush into the market and push for high numbers alone, at the expense of sound breeding. Denton has reluctantly concluded he may have to start keeping his own breeders for producing hatching eggs to ensure his supply of chicks using an incubator of three hundred to five hundred capacity.

Since coturnix hatching eggs and chicks are available year round, Denton raises successive batches through the four seasons. This year he plans to raise about five thousand. Chukar and pheasant eggs and chicks are available only in the spring, so he has to order at that time as many as he anticipates selling. He raised two hundred pheasant last year, which were so well received by his chefs that he plans to increase production to five hundred this year. He plans to raise four hundred chukar.

Like me, Denton cannot imagine excessively confining creatures in his care. In contrast to the usual models for raising game birds in high confinement—low all-wire cages in the case of coturnix—he raises all three species in the numerous horse stalls (each up to 12 by 18 feet) in the barns of one of the farms he leases. Though he may start up to two hundred coturnix in each stall, as they grow he splits them into groups of no more than one hundred per stall. At that spacing they have more than 2 square feet per bird, considerably more than industrial layers in so-called free-range housing.

The earth floors of the stalls are covered with a thick litter of wood shavings to absorb and compost the droppings. Like other galliforms, game birds dust-bathe in order to prevent infestation by lice and mites. The birds in Denton's stalls have plentiful access to dust from the floor underneath the litter and have exhibited no problems with external parasites.

As for other health issues, game birds are subject to the same diseases as their domesti-

Started Birds

A couple of young friends in my area started an interesting business: riding the current wave of interest in keeping home flocks, by furnishing started birds in small lots. They brooded chicks—three hundred to four hundred per batch in homemade brooder sheds, mostly traditional dual-purpose breeds, though they raised some of the hybrid layers as well—to sell in small ready-to-go lots. It was a win–win for customers, who didn't have the trouble of brooding them-

cated gallinaceous cousins, but Denton has had no health problems to date. He is careful to buy his stock only from two sources he knows and trusts, never from open auctions. He is equally careful not to bring in used equipment from untrusted sources. Periodically he tests for avian influenza in his flocks.

Though some growers of game birds for market feed different formulations keyed to stage of growth, Denton feeds all three species, from brooder to slaughter, the same mix. He never starts his chicks on the medicated game bird *starter* formula (containing coccidiostats) but uses rather the second stage game bird *grower* feed, boosting protein content to about 25 percent with the addition of soybean meal. This feeding program gives him finished carcasses with plenty of fat on the breasts, which is what his restaurateurs demand.

The one exception to the use of the single mix is the feeding of mature coturnix layers, reserved for an additional two weeks to boost egg production. These hens are fed a commercial chicken layer mash, with the addition of soy meal to boost protein content, plus free-choice oyster shell and granite grit.

Denton makes millet hay for feeding his cattle and horses, cutting just after the seed heads have ripened but before they have begun shattering, and while the stems are still green. All his game birds *love* the millet hay, thrashing about in it, tunneling through it, obviously having the time of their lives. They eat not only the millet seeds, but the green stems as well, leaving only a pulverized residue that becomes part of the litter.

Freestate Farms is based on diversity and relationship marketing. Denton meets with his chefs regularly to plan production cycles for the game birds and other dressed meats he supplies them. Chefs at better restaurants love to vary what they offer, just as they love the added cachet on their menus of high-quality ingredients produced locally—food with a story, food with a face.

Such marketing is an evolving, people- and quality-driven adventure. Denton's chefs have urged him to produce rabbits for them, so he recently bought his first batch of rabbits. They want to offer duck on their menus as well, so he is seeking enough Pekin duck stock for an experimental grow-out to discover a working production model. One of his restaurateurs wants to contract with Denton for a regular supply of his products, something like: a veal calf in week one, fifty rabbits in week two, fifty game birds in week three (any mix of the three species), a pig in week four, fifty ducks in week five, then rotate again through the cycle. Denton figures that if he could make the same arrangement with a second, and at most a third, restaurant, they would among them absorb *all* the production from his farms, and he would not have to scramble about seeking markets elsewhere.[14]

selves and could buy just the few needed for a small flock—rather than twenty-five, the minimum for shipping day-old chicks. Also, customers had their choice of breeds from among those available in the brooder at the time—an exciting option for any children involved. My friends also custom-brooded to order—for example, fifty brown-egg layers for a customer who operates a winery tasting room in which she likes to offer "country eggs" to sippers on a day outing. Started birds remaining unsold by early maturity were no problem—they simply butchered them (exchanging their labor helping at a larger operation for the opportunity to process their own birds with their equipment) and took them to farmer's markets, where they had no problem selling out. As one of them summarized: "There's always something you can do with a bird."[15]

Value-Added Sales

Correspondents who have tried it inform me they make a lot more money on value-added products—chicken potpies made from culled cockerels or old hens; pound cake or brownies or quiches made with lots of eggs; broth made with the feet and other cast-offs of butchering—than they do on their prime-ingredient items themselves (eggs, broilers, and the like). I was especially impressed with the strategy of one of my APPPA correspondents, following publication of an article I wrote on the moral quandary of contributing to the euthanizing of hundreds of thousands of cockerel chicks as a result of our so-convenient all-pullet chick orders.[16] His solution, like mine, was to make straight run orders only in the future—and to butcher the excess cockerels for making potpies to sell in his farmer's market. Now, isn't that an interesting idea—a change of practice taken for reasons of virtue turns out to generate more income as well.

If you do try to turn a profit adding value to your poultry products, be prepared: "Did someone say *regulations*?"

Custom Processing

Efficient butchering in quantity requires equipment that is either expensive or time consuming and challenging to make. Once you've made the investment, you might offer custom processing of others' broilers or turkeys, either as an on-farm service (you come to me) or with a mobile processing unit (I come to you).

Some states place a limit on the number of birds not raised by the farmer that can be custom-processed on-site. You didn't hear it here, but one workaround that has been used is the "sale" to the farmer of birds grown on another farm; their processing as the farmer's own; and their subsequent "sale" to their original owner, all transactions neatly documented. Such is the nature of the bureaucratic rat race.

I don't know many producers who want to do a lot of custom processing—butchering all day is grueling, and most of my correspondents don't want to make a career of it—but I do know producers who have made an MPU for their own use and then rent it out to others for a fee (see figure 30.8).

Custom Milling

As feeding requirements grow, you might start dreaming of Gehl grinder-mixers for milling your own feeds. Again, once the operation is tooled up in this way, custom milling for others in the area could become an option, in cooperation perhaps with partner farmers growing feed grains. At present the structure of the agricultural economy discourages such possibilities, but as our financial, energy, and related crises continue to squeeze, sensible local production options may come to seem—well, sensible.

Fig. 30.8 Mike and Christie Badger use this mobile processing unit to help small-scale producers in north central Pennsylvania butcher their broilers on-farm. Stainless-steel equipment by Eli Reiff is laid out for maximum efficiency in a small space, with the flow of processing clockwise from the killing cones on the right to the scalder and plucker and worktable, ending at the chill tank. Two onboard propane tanks provide heat for the scalder; connections to water and electricity must be furnished on site. PHOTO COURTESY OF MIKE BADGER

Finding Your Niche Market

It can get frustrating trying to compete with the Big Boys who are wooing customers with their cheap chicken and eggs. It might pay off to find niches in the market they have no interest in targeting. Put another way, your ideal market is one that *may not get served* if you do not serve it.

Restaurants

Especially at high-end restaurants, chefs tend to be fanatics about quality of ingredients—at least as much a key to winning awards as their skills in the kitchen or the number of French words on their menus. It's easy for them to get whatever ingredients they need from industrial food distributors. But they know how mediocre most of those are—what they really want is primary ingredients as good as those that show up every day on your table. Those who have tried them know that pastured eggs mean more taste appeal in baked goods, which stay fresh longer, for greater net profit despite the higher cost of pastured eggs; that dressed poultry with gourmet quality doesn't need to be hidden under exotic sauces; and that they cannot make incomparable broth as a foundation for soups and sauces from anything other than chicken feet and stewing hens raised on pasture.

Of course, restaurants operate on thin profit margins, so some chefs are unwilling to pay the producer a worthwhile price. In such cases of price inflexibility, producers may find all the same that, while a restaurant will not pay enough for their broilers, for example, it may prove a good outlet for small-volume, high-impact ingredients like liver and feet.

A high-end restaurant may be a better bet for sales of specialty poultry such as guinea, duck, and quail with gourmet appeal than the general market.

Other Markets

Some of my correspondents have done well selling into cross-cultural markets for chickens and other fowl. Immigrant customers are sometimes willing to buy live and do their own slaughtering, having grown up in countries where small-scale flocks are more the norm than industrial poultry. Many of my correspondents find that customers steeped in other food cultures are eager for "real" (traditional-breed) chickens, however pointy-breasted, and for spent laying hens.

Some producers find that members of local chapters of the Weston A. Price Foundation[17] come looking for chicken feet and necks and spent stewing hens to make the fabulous chicken broth they've come to think of as a foundation of good health. Seek out local foods groups in your area—many of them publish lists of local producers of farm foods for thoughtful and demanding consumers. A number of conservation organizations have published such lists as well, having concluded that a key to land and environmental conservation is good farming, the key to which is the patronage of good farmers.

The Joys of Diversity

A final thought about serving local markets: All the small producers I've talked with have no interest in becoming the next Frank Perdue. Most of them have found their broiler and turkey operations quite profitable, for example—yet I haven't encountered a single one who wanted to specialize with either. If you haven't tried it, butchering poultry all day is *hard* work. Producers I know take in stride the pulses of intense effort required but feel it would be deadening to make that same all-out effort, day in and day out. They prefer instead to *diversify* their operations, to mix it up, keep things interesting. Intuitively they understand that for a farm, just as in nature, diversity is the key to robustness and stability. Most expand to their comfort level of profitable production of broilers or turkeys, then increase no further, instead adding new components to the mix, such as lambs, dairy goats, or honey production. Many are even glad to mentor new producers entering their expanding markets.

Turning Waste into Resource

Vermont Compost Company (VCC) has made compost since 1998, currently 10,000 to 12,000 cubic yards per year on two sites near Montpelier, Vermont. Composting at one site is based on manure from a dairy farm with 900 cows. At the "home" site, which I visited in May, 2011, the compost is based on food residuals from area restaurants and institutions, wood chips from Vermont's extensive forest industry, spoiled hay—and hundreds of hard-working chickens. The composter chickens receive no purchased feeds whatever, foraging their food in the compost heaps entirely.

VCC's owner Karl Hammer blends the components with a front-end loader into starting compost heaps 8 feet high—some of the ridges of aging compost I climbed must have been 12 or 15 feet. If you know the chickens are getting their food from the heaps entirely, you might assume that they're eating mostly food residuals, which must make up the bulk of the heaps. Actually, Karl makes his heaps with only five percent food residuals—ten percent at the most. Even this small percentage contributes a lot of energy for the intense biological activity in the heaps, which is what really feeds the chickens.

Karl stresses that the chickens enhance the quality of the compost. In good measure that enhancement has to do with their own manure, which the chooks turn into the mix as well. VCC's compost is alive with beneficial microbes and thus is vastly superior to dried poultry manure from battery houses, often sold as fertilizer, but containing mineral content only.

VCC has used flocks of a few hundred up to twelve hundred chickens to work the compost heaps—even in winter, when the heat of decomposition keeps the composting process going, and the chooks continue feeding on the various decomposer organisms and their metabolites. Production of eggs was fifteen thousand dozen in the peak year. Which should we say the layer flock offers VCC's operation for free—the egg production or the labor of working the compost?

A neighbor objected to VCC's use of food residuals because crows, opportunists extraordinaire, sometimes land on the compost to grab scraps and fly to a hidden place on the adjacent property. If interrupted, say by a dog, before they finish lunch, they leave these scraps behind. In response to the complaint, the Health Department issued a citation, and for a while it looked as if VCC would have to cease operations, at least as far as the intake of food residuals was concerned.

But VCC has done its best to resolve concerns of neighbors. Karl now assembles the initial heaps containing fresh residuals in two 30-by-48-foot hoophouses, erected last year. Netting over the ends of the hoophouses discourages entry by crows, who in any case are reluctant to enter such an enclosed space. The chickens, on the other hand, are able to move under the netting and range from the newly assembled heaps inside to more developed heaps outside and even on the pasture at the top of the ridge.

Though the struggle to come to terms with the powers that be has been frustrating—and expensive, given the lawyer fees—those same powers that be have begun to appreciate that VCC is performing a valuable service to the community. Since Vermont adopted a goal of "zero waste," officials there have become more keenly aware of the challenges of disposing of the mountains of food wastes that citizens scrape from the communal plate every year. If VCC doesn't take care of these organic wastes, the municipal district will have to do so—at considerable cost to the taxpayer and increased complexity of government agencies and infrastructure. Consonant with resolving citizen concerns and making sure there is no health issue, they see that VCC's successful operation is in the public interest.

Karl made this fascinating observation when I visited him: The poultry industry houses and feeds three hundred million hens in battery cages to produce the nation's egg supply. Meanwhile, we create endless problems for ourselves with the huge quantities of food wastes we send to landfills. As it happens, the two about exactly match each other—if we released those three hundred million chickens on the nation's food scraps, using Vermont Compost Company's model, the chooks would pretty much clean them up with no other feed needed. In terms of the energetics involved, *America could thereby get its entire egg supply for free.*

VCC produces a small portion of its own replacement stock, using hens in the flock inclined to broodiness. For the rest Karl relies on purchased chicks, some of them hatched locally. The growing chicks mostly eat as the adults do—with older hens and cocks in the brooder house teaching them how—and complete the brooder phase with very little purchased feed.

Karl has a deep appreciation of the role of the cocks in his flock, who help organize efficient foraging and provide protection—he is convinced they even keep weasels at bay. Three guardian dogs protect the widely-ranging flock from coyotes and foxes, and even from bears, who sometimes take out entire flocks in the area in one raid.

Fig. 30.9 Chicken-powered composting at Vermont Compost Company. The chickens forage *all* their feed in the highly bioactive compost–a mix of food residuals, wood chips, and hay–while enriching the final product with their droppings.

Fig. 30.10 The amount of compost that can be generated by hardworking chickens is considerable, as we see in this mountain of finished compost ready for screening, blending, and packaging.

Fig. 30.11 This 10-by-40-foot winter hoophouse holds eighty layers, for a total space per bird of 5 square feet. Wooden sides allow for pine shavings and sawdust litter up to 20 inches deep. After cleaning out the litter for composting in the spring, the owner will use the ground for starting early tomatoes for her market garden. PHOTO COURTESY OF COZI FARM, COZIFARM.COM

Market-sized laying flocks often get housed in eggmobiles—recycled recreational vehicles or school buses, or rolling chicken coops mounted on something like an old hay wagon chassis—pulled when needed to greener fields of plenty by tractor, pickup, or ATV. Remember the option for stacking mobile flocks with other livestock species, as in the now classic Polyface Farm model of following beef cattle with layer flocks. In the winter some producers park their eggmobiles and give the birds access from them onto a winter feeding/exercise yard such as the one I use, or even into an attached hoophouse. Deep litter allows for chicken-powered composting through the winter; the vacated hoop house may be used in the spring for an early start for warm-season crops; and the hoophouse can be dismantled and the enriched ground used to grow vegetables (see figure 30.11).

If you are serving a vegetable CSA, adding eggs to your weekly mix would be especially easy. Most producers who make this choice offer the eggs as a separate add-on to the basic subscription. Indeed, some producers have begun offering meat CSAs. Obviously, the greater the diversity of your food products, the more intriguing the mix of CSA foods on offer to tempt potential members.

Note the possibility of developing your own loyal community, whose members revere you as "our farm," and supplying them the full diversity of the foods you produce, changing throughout the year—while keeping in touch via a website, e-mail list, or newsletter to keep them advised about what's coming up and to make it easy for them to place orders.

An interesting thing about diversity is what it does to thinking about the bottom line. As I said before, some producers find a lot of resistance in their markets to paying a fair price for their eggs. Think what a disaster that would be for a producer who had put all his eggs in one basket—that is, whose income depended solely on egg sales. One correspondent in California finds the same resistance but takes it right in stride. With reference to his large layer flock, he says, "Our profit is in the shit." That is, rotating his layers over sections of his farm gives the soil such a fertility boost—for two full years—for his vegetable CSA, it's worth it to his total operation to merely break even on egg sales.

And don't think of diversity in terms of your farm products only. Try to find ways you could put in place some of the integrating practices described in this book, but at the farm scale: stacking your poultry with other livestock species; using market-sized flocks in soil fertility and insect-control projects; cultivation of worms or soldier grubs on a large enough scale to replace a good portion of purchased feeds, or finding local sources of food residuals as a resource for feeding, using the Vermont Compost Company model; experimenting with winter access to the outdoors that benefits the birds, the soil, and the environment. Perhaps these ideas for becoming more independent of purchased feeds and using chickens to do real work can become part of the profit equation in your market flock.

EPILOGUE | THE BIG PICTURE

. . . how we eat determines, to a considerable extent, how the world is used.

—Wendell Berry

The small-scale flock generously furnishes the table, and helps with the work of producing foods other than eggs and dressed poultry. The more uncertain the future, the more welcome the flock will become as partner in working toward food security. But keeping a small-scale flock has wider effects, expanding outward like ripples from the pebble that go all the way out to the edge of the pool.

I am pessimistic that any of us are going to do much about a destructive agriculture on steroids by writing Congress or carrying protest signs. Despite the reality of half-billion-egg recalls and contamination of more than half the dressed chicken in supermarket coolers, I doubt a remedy will be forthcoming that depends on heightened inspection or more complex regulations. I hope for no more profound remedy for the disasters that industrialized agriculture and ersatz food have become than the determination, adopted one family at a time, "There is some shit I will not eat!"[1]—and a consequent growth in farmer's markets and backyard flocks to feed families, rendering the domination and enormous manipulative power of Big Ag and Big Food irrelevant. It is unquestionably true: *One dollar, one vote*—and the votes, though still a trickle, are coming in. With more recalls, more environmental disasters, and less oil, those votes could become a flood. I believe Margaret Mead's observation was correct: "Never doubt that a small group of thoughtful, committed people can change the world. Indeed, it is the only thing that ever has."

But it's not just in the marketplace we are making a difference. Though the differences I make producing some of my own food are small, they are unquestionably real. The omelet or roast chicken on my plate *in fact* reduces by some small fraction the amount of runoff pollution to the nation's waters and the flood of antibiotics into the environment. That I employ chicken power as a key part of my soil fertility program *in fact* sequesters carbon, in however small amounts, into my soil, rather than releasing it to the atmosphere.

The practice of a responsible poultry husbandry leads us into pathways of sustainability as we understand ever more clearly the ecological web in which that husbandry takes place. The truth in the Wendell Berry quotation is profound: All the Big Problems of our time—the energy crisis, climate change, loss of species diversity, soil destruction, pollution—have to do in some way *with what and how we eat*. World-friendly means of producing our own food in our own backyards—and buying it face to face from producers we know and trust—are realistically the only opportunity most of us have to change how the world is used.

For most of us, *sustainability* is a buzzword—we

know it's a good thing but don't really have a clear notion how it works. Having a small-scale flock teaches us what it takes to *be* sustainable, to *create* sustainability. In this great work, this "ministry," to use Joel Salatin's term, we make a beginning every day. A lifetime should suffice.

APPENDICES

A | MAKING TRAP NESTS

Note that the following instructions could guide the assembly of ordinary nestboxes mounted on the wall. Further details on that option follow the description of the trap nest project.

I've seen a number of designs for trap nests, including one in Rolfe Cobleigh's *Handy Farm Devices and How to Make Them*, a useful book for the homesteader first published in 1909.[1] Modern homesteaders could enter "trap nest" into an online search engine and find workable designs.[2] My own design emerged in one of my father's visits. After mulling over a couple of designs I showed him, he concluded that he didn't much like either and proposed, "Why don't we do it the way Granddaddy used to put together his rabbit boxes?" I remembered the live rabbit traps ("rabbit gums," as the old-timers called them) my grandfather made to trap hundreds of rabbits in his fields and woods, and despite some skepticism about turning a rabbit trap into a trap nest for hens, I agreed to give it a try. As it turned out, the design has worked very well for me.

Materials

The materials list is sufficient for two nest units, each containing two separate nestboxes:

- One sheet of plywood, CDX (½-inch or ⅝-inch, depending on which you are confident you can edge-nail effectively)
- Small nails for edge-nailing the plywood and nailing on strips (I used 4d 1⅜-inch and 6d 1⅞-inch coated sinkers)
- Eight #10 2- or 2½-inch self-tapping screws (i.e., not requiring a pilot hole, such as deck screws)
- Two 8-foot 1×4 pine boards
- Two 24×12-inch pieces of ¼-inch mesh hardware cloth
- An additional few small pieces of hardware cloth, any mesh (optional)
- Eight open screw hooks (about 2½ inches or so)
- One ½-inch dowel, 36 inches long

The materials listed assume starting from scratch, but of course many flocksters will have on hand enough scrap from other projects to piece together what is needed—the tracking strips, for example, I cut from "1-by" scrap. No problem altering suggested dimensions to accommodate material you're working with, so long as sufficient interior space is allowed for the laying hens.

Assembly

Cut two 12×24 pieces of plywood for the backs and six 12×18 pieces for the sides and middle partitions (figure A.1).

I cut little windows into the sides of the nests (to

increase ventilation and decrease the trapped hen's sense of isolation), a step that's probably not really necessary. I cut them in the four exterior side pieces only (not in the interior partitions), making the openings about 4×9 (figure A.2).

Cut two 16¼×24 pieces of plywood for the tops. Drill two ⅝-inch holes for the trigger sticks in these top pieces, 6 inches in from the sides and 6 inches in from the back. To ensure a nice sharp edge to engage the notch in the trigger stick, be sure to drill from the side of the plywood top that will be to the *inside* of the nest. For good ventilation, you can add a few other ⅝-inch holes in the top as well (figure A.3).

Edge-nail the top onto the back and exterior sides, using the smaller nails (4d 1⅜-inch), aligning as in the picture (figure A.4). Then nail in the interior partition in the middle of the box thus formed (figure A.5).

Switch to a somewhat larger nail (6d 1⅞-inch) for some of the nailing in the next several steps. Cut one of your 1×4s in half crosswise, then rip one of the resulting 4-foot pieces in half. From the ripped pieces, cut pieces that will stop the doors when they drop. Study figure A.6 carefully, because you can do it more easily than we did. We cut individual pieces that we *inset* into the spaces between the sides and interior partitions. Not only did that mean more cutting and fitting, but we had to toenail one of the inside ends. You're much smarter than that, so you'd probably want to cut a single ripped piece to 24 inches, and nail it *underneath* the front ends of the exterior sides and the interior partition. (When you look at the picture, imagine the two interior crosspieces replaced by a single piece, and the ends of the exterior sides and interior partition sitting on top of it.)

Nail another of the ripped pieces, cut to 24 inches, across the bottom front, aligning as in the picture, to make the perch on which the hen stands as she checks out the interior—the perfect place to lay an egg. Remember to round off the sharp edges with a wood rasp so it's more comfortable for the hen's feet (figure A.7).

Cut nest-front pieces from (unripped) 1×4. The length will be: 24 inches, minus the thickness of your sides and interior partition, divided by 2. I was using ¾-inch stock, so I cut mine to 10⅞ inches. Nail these pieces in place to serve as fronts for the nests. Precise placement is not critical—I set mine 5 inches from the front edge—but we want the nesting material to be held behind this piece, *so the hen will completely enter the nest before triggering the door.* (If it triggers early and whacks her behind, she will become shy of the nest.) Note that in this case there is no alternative to toenailing the end of one of these pieces where it butts to the interior partition. If you have cut windows in the sides, at this point you can staple hardware cloth over the windows to keep smaller hens from escaping (figure A.8).

Rip your remaining 1×4 stock into strips as needed. I used ¾-inch and ½-inch strips for the door tracks and for fastening hardware cloth onto the bottom of the nests, respectively. (Thus if you are using 1×4 nominal stock, you will end up with strips of actual ¾×¾, and ½×¾.) You will need approximately 14 feet of ½-inch strips and approximately 22 feet of ¾-inch strips (figure A.9).

Turn nest unit upside down. Cut ¼-inch mesh hardware cloth to 24 inches long and wide enough (about 12 inches) to span the bottom of the *nesting* areas (*only*). Secure the hardware cloth to the underside edges of the nest areas, using small nails and ½-inch strips cut to needed lengths. (Note that I always use ¼-inch hardware cloth for the bottoms of nestboxes, never solid bottoms. Finer, dustier material sifts out through the wire, I renew the nests with fresh straw from above, and the nests remain largely self-cleaning.) (Figure A.10.)

Cut ¾-inch strips into sixteen 16-inch pieces. Nail into place as in the picture, aligning the front strip with the front edge of the side or partition and using a 1-inch spacer to ensure that the back strip is exactly parallel to the front strip. These pairs of vertical strips will define the track through which the door will fall. (Strips need only be lightly tacked into place using the smaller nails—they will bear no load.) Attach some sort of bumper over the doorstop so the door won't bang down too loudly and panic the hen. I used strips cut from an old bicycle inner tube (figure A.11).

Mounting the Unit and Installing Doors

The unit is now ready to install. If you attach a cleat to the wall (say, a scrap piece of 2×4), it will be easy to install it by yourself. Otherwise, get a buddy to assist (figure A.12).

Attachment to the wall must be rock-solid. I used four #10 2½-inch self-tapping screws for each unit (figure A.13).

Cut doors 12 inches high from plywood or any stock that you have that is ¾ inch thick. The width is determined by the thickness of material you have used for sides and partition, but you should allow for about ¼-inch clearance on each side of the door in its tracking slot. I cut mine to 10½ inches. Drill holes in the exact center of the top edge of each door, and screw in a hook. Position another hook overhead, screwed into any accessible structural member of the building. The position of this hook is the most critical alignment in the whole setup: I used a plumb bob to ensure that the hook was precisely vertical over and in line with the door hook (figure A.14).

Cut a ½-inch dowel into 9-inch pieces for the trigger sticks. Use a saw to cut 5⁄16 inch through the dowel, 3½ inches from what will be the lower end. (This cut is made with a saw because the "shoulder" we are shaping here must be well squared, in order to engage the edge of the trigger hole without slipping.) Drill a 3⁄16-inch hole near the other end of the trigger stick. Using a knife, whittle a notch that starts about 1½ inches toward the upper end and comes down to the inside of the shoulder of the notch (figure A.15).

Time to put it all together. Tie a string between the hole in the top of the trigger stick and the hook in the top of the door, running it through the hook vertically above the door. (Strong braided string that will not stretch, such as mason twine, is best.) The length of the string will depend on the position of the upper hook, of course. Hook the edge of the notch in the trigger stick into the trigger hole, and suspend the door from the upper hook, hanging within its tracking slot. The door should hang slightly above the upper edge of the top of the nestbox, so the hen's back will not bump it as she enters the nest, triggering premature release. Note that the weight of the door keeps tension on the string to keep the trigger securely notched in the trigger hole (figure A.16).

Figure A.17 shows the lower end of the trigger stick in the "set" position from the inside. As the hen begins settling into the nest, she is certain to bump against the trigger stick, knocking it loose and allowing the door to fall into blocking position.

Adding an additional hop-up perch, shown in figure A.18, reduces the chance that a more forceful landing of a hen on the entry perch will jar release of the door prematurely.

It took less than half an hour for my new nest to attract the Cuckoo Marans hen in figure A.19. If I'd had the door and trigger in place, she would have been trapped until I could release her and record her achievement.

Troubleshooting

If the upper hook is not precisely positioned overhead, the door will not fall straight and may jam in the tracking slot. If changing the position of the hook doesn't solve the problem, you can box in the tracking slot with thin strips, ensuring that the door cannot veer to one side as it falls.

If you find it difficult to get the trigger stick to stay locked in the trigger hole, try passing the string from the trigger stick through an additional hook overhead (before it continues on to the hook over the door). In this case, however, you do not want the overhead hook to be precisely in line with the trigger hole—position the hook at enough of an angle to notch the trigger stick securely in the trigger hole.

Doors can be left in the closed position at night if chickens like sleeping in them. A slanted cover can be put in place on top of the nest unit to prevent the birds' roosting there.

If you want to make ordinary nestboxes mounted on a wall, simply alter the construction in these ways: (1) Cut the tops to the same width as the side and middle partitions. (The partitions will not protrude beyond the top as shown in figures A.5 and following.) (2) Omit the additional nest-front pieces shown installed in the interior of the box as in figure A.8—the nesting space will simply be the entire interior, with the nesting material retained by the entry perch strip in front. (3) Thus the hardware cloth shown in figure A.10 will cover the bottom of the entire unit. (4) Omit installation of the tracking strips and doors.

Fig. A.1 Cut two 12×24 pieces for the backs and six 12×18 pieces for the sides and middle partitions. (Only four sides are shown in the picture–we later cut the additional two pieces for the middle partitions.)

Fig. A.2 We cut little 4×9 windows in the exterior sides, a step that's probably not necessary but can be a nice addition.

Fig. A.3 Drill two 5/8-inch holes for the trigger sticks, one over each nest, 6 inches in from the sides and 6 inches in from the back. *Be sure to start drilling from the side that will be toward the inside of the nest.* Drill a few extra 5/8-inch holes for ventilation.

Fig. A.4 Nail on the top, aligning the back and exterior sides as shown.

Fig. A.5 The top nailed on (left) and a better view of the interior partition with the unit upside down (right).

Fig. A.6 Adding the stop (1×4 stock ripped in half) for the falling door. *Note that it would be simpler to nail a single piece underneath the ends of the sides and interior partition,* rather than insetting two separate pieces as in this picture.

Fig. A.7 Nail on another piece of the ripped stock, cut to 24 inches, for the perch from which the hen will enter the nest.

Fig. A.8 Cut and nail into place 1×4-inch pieces that will retain nesting material in the interior of the nest. (Note that by this step we have stapled hardware cloth over the windows.)

Fig. A.9 Cut the remaining 1×4 stock into ½-inch and ¾-inch strips as needed in the following steps.

Fig. A.10 For a self-cleaning nest, use ¼-inch hardware cloth rather than solid bottoms. Secure with small nails and ½-inch strips. (Note that I nailed a strip onto the bottom of the interior partition as well, after this picture was taken.)

Fig. A.11 Tack vertical strips into place to form the tracks through which the doors will fall, using a 1-inch spacer to keep the strips precisely aligned. Fasten rubber bumpers over the doorstops.

Fig. A.12 If the wall of your coop has no structural piece to support the back of the unit, nail a cleat to the wall or call in a buddy.

Fig. A.13 Four well-spaced #10 2½-inch self-tapping screws provided rock-solid attachment to the wall.

Fig. A.14 Screw hooks into the center of the top of each door. Position a similar hook on an overhead rafter or beam, precisely aligned to the door hook. Note how the doors fall between the retaining vertical strips.

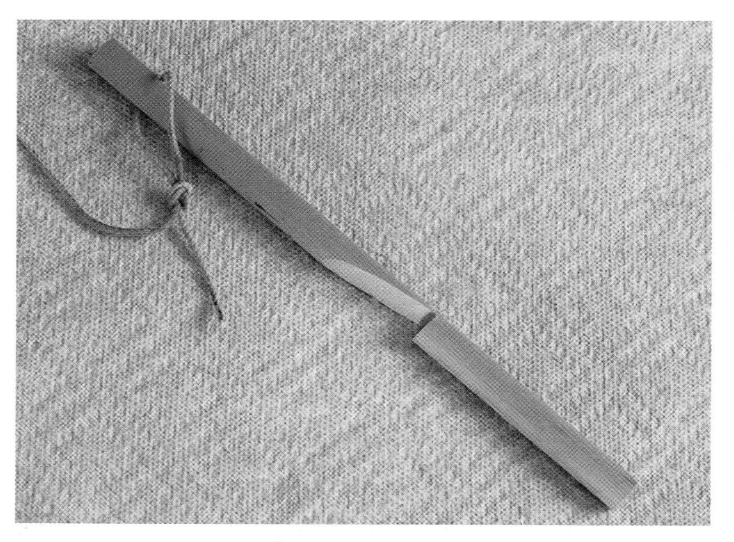

Fig. A.15 Using saw and knife on the 9-inch dowel pieces, make a notch with a perpendicular edge on its lower end. The bottom edge of the notch is 3½ inches above the lower end of the trigger stick. Drill a 3/16-inch hole near the other end of the trigger stick.

Fig. A.16 The weight of the door keeps the trigger stick notched in place in the trigger hole . . .

Fig. A.17 . . . until the hen's settling-in ritual knocks it loose, and the door drops.

Fig. A.18 If the nestbox is set well above the floor, it's better to give the hen a step or two on which she can approach the nest. If she has to fly rather than hop up onto the perch at the entrance to the nest, she may jostle the lower edge of the door, triggering premature release.

Fig. A.19 Success!

B | MAKING A DUSTBOX

If it's too wet outside or the litter inside is not fine enough for dust bathing, a backup dustbox in the henhouse ensures that the chooks can keep themselves free of external parasites. I've made several versions, and I advise you not to make one too small or too shallow—the chickens would kick out too much of the dusting materials as they bathe, requiring more frequent refilling. The dustbox I now use is 24×24×16, with a 2-inch lip around the top edge. Of course, you may have scrap on hand that dictates a different design from the one detailed here.

If you're starting from scratch, note that you can make two dustboxes of this design using a single sheet of plywood. The cutting instructions below assume that you will start by cutting up one sheet of plywood to make all the pieces needed for two dustboxes—step 1 through step 4. Assembly of one of the boxes is illustrated with pictures in step 5 through step 7.

For this project I chose to use ⅝-inch CDX, since I could not find a thinner plywood that was 5-ply, and I preferred to save time and effort by edge-nailing. If ⅝-inch plywood seems like overkill to you, go with whatever thickness you are confident you can edge-nail securely (or attach nailing cleats and use thinner plywood).

In these instructions I am not allowing for the saw kerf. I want to keep the instructions simple, and ignoring the kerf is no great problem—we're not making a kitchen cabinet here. I used a table saw (except for the initial cut), but you could certainly use a handheld power saw for the entire project.

Step 1: Cut the sheet of plywood across the 48-inch dimension into four equal pieces (each 24×48 inches).

Step 2: Cut one of the 24×48-inch pieces in half, to yield 24×24-inch pieces for the two bottoms.

Step 3: Cut the remaining three 24×48-inch pieces 16 inches wide, across the 24-inch dimension, to yield nine pieces, each 16×24 inches. (Eight of these pieces will be sides for the two dustboxes, while the last one will be cut into lip strips for the top edges.)

Step 4: Rip one of the 16×24 pieces cut in the above step into eight strips, 24 inches long and 2 inches wide (to be precise, 2 inches minus the kerf)—these will be used to make the lip around the top edge. Leave four of the resulting strips at 2×24 inches. Cut the remaining four strips to a length of 24 inches minus twice the thickness of your plywood. I used 5/8-inch plywood, so I cut mine to 22¾ inches.

Step 5 (figure B.1): Edge-nail one of the bottom (24×24) pieces onto two of the side (16×24) pieces, aligning them as in the picture. I used 4d 1⅜-inch coated sinkers.

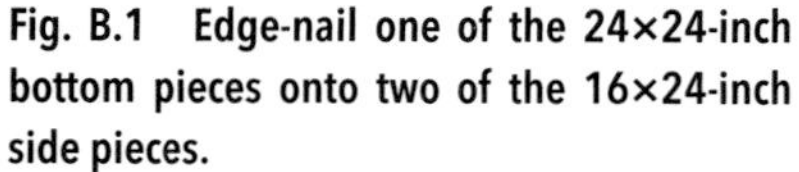

Fig. B.1 Edge-nail one of the 24×24-inch bottom pieces onto two of the 16×24-inch side pieces.

Fig. B.2 The assembled box. (Note the offset top edges.)

Fig. B.3 Nail 2×24-inch strips to make a lip around the inside of the top edges, noting alignment.

Fig. B.4 Close-up of the alignment of the lip strips.

Fig. B.5 Sifting for a fine dusting mix.

Step 6 (figure B.2): Nail the other two side pieces to box in the open ends created by the above step; that is, these side pieces cover the edges of both bottom and first two sides. (The offset of the top edges you see in the picture is intentional.)

Step 7 (figures B.3 and B.4): Nail the 2-inch strips cut in step 4 to form a lip around the inside of the top edges of the box.

Step 8: Assemble the other dustbox from the remaining pieces as above.

Step 9: Put 4 to 6 inches of dusting materials in the bottom of the dustbox. The ideal mix is loose, fine, and easily fluffed by the birds up under their feathers. In figure B.5 I am combining peat moss and wood ash, in a ratio of about 6:1, sifted through a quarter-inch-mesh compost riddle. You could also add dried and sifted clay soil, diatomaceous earth, or elemental sulfur (pure sulfur as a fine yellow powder). *Wear a dust mask!*

C | MAKING A MOBILE A-FRAME SHELTER

Consider an A-frame for your mobile shelter—combining rigidity and reasonably light weight, with more stability in the wind than more boxy designs. Figure C.1 shows my most recent—of several—A-frame, the most successful to date as an all-purpose shelter. At 8 by 9 feet, it would comfortably accommodate up to fifteen layers if confined full-time—I've used it as a "bedroom" only for up to fifty young

Fig. C.1

growing birds who had plenty of space in an electronet enclosure during the day. The roof/sides are solid, providing the flock both shelter from rain and shade if there is no tree cover, but the ends have plenty of wire-covered open framing to permit free flow of air.

This appendix shares my reflections on materials and design choices, materials list, and step-by-step construction of a proven shelter design. Modifications you could make are endless.

My previous most successful pasture shelter was the one shown in figure 11.10. The biggest design change from the old shelter was the choice of baked-enamel steel roofing instead of 24-mil woven poly to cover the shelter. It bothers me to think that the plastic we use with such convenience today could well remain somewhere on the planet in the time of our great-great-great-great-grandchildren, so I now avoid its use as much as I can. Still, woven poly fabric may be the best choice for some. It is much lighter than metal roofing—the metal roofing on my new A-frame added about 80 pounds, while a piece of 24-mil woven poly, custom cut to cover this same shelter, would weigh about 10 pounds. Of course, up to a point, the added weight can be an advantage—there is no question that my new, heavier shelter is more stable in the wind. Cost might also be a factor in deciding between these two coverings: I spent $112 for the metal roofing for my new shelter, including a piece of ridge vent to cap it—a piece of 24-mil white/silver woven poly sized to fit would have cost $78. A final consideration is durability. My previous A-frame with poly cover is still going strong. Based on the current degree of degrading from weathering, I would estimate a total service life of up to fifteen years. The galvanized steel roofing with baked-on enamel paint that I used comes with a twenty-five-year warranty, so I expect total service life to be considerably more than that.

Another change I was glad to make: In the past I have used 8-inch wheels on my pasture shelters, with the 1/2-inch bolts that serve as the wheels' axles set in the exact center of the bottom rails. On my pasture this configuration results in too narrow a gap between the ground and the rear rail, which tends to catch on every bump or tussock when I move the shelter. For the new shelter I bought a set of 10-inch wheels. Instead of drilling through the center of the rail for the axle bolt, I offset the hole toward the bottom of the rail by an inch. The combination of the greater wheel radius and the lower axle hole gives me an additional 2 inches of clearance when moving the shelter. Based on your own particular needs, you may *prefer* a smaller clearance—to prevent escape of birds from inside while moving, for example—but I *love* the way my new Chicken Ferrari rolls across the pasture.

Materials

- Six 6-foot sheets of Fabral's "Grandrib 3" painted steel roofing (see note 1)
- One 10-foot, 6-inch Fabral's ridge vent
- Fabral's painted screws with neoprene washers (optional) or a box of neoprene washers (see note 2)
- Four 12-foot construction grade 2×4s (see note 3)
- Six 10-foot 2×4s
- Four 8-foot 2×4s
- Assorted deck screws (see note 4)
- Four to eight metal corner braces (see note 5)
- Four ½-inch carriage bolts, 5 inches long
- Four each flat washers, lock washers, and hexagonal nuts for the carriage bolts
- Four wing nuts for the carriage bolts
- Four wheels with ½-inch axle bores (see note 6)
- One 10-foot roll 1-inch mesh poultry wire, 48 inches wide
- Small fence staples (for poultry wire)
- Two pairs 1½-inch utility hinges
- One 1½-inch barrel bolt
- Four 4½-inch open screw hooks ("ceiling hooks")

- One sheet ½-inch CDX plywood, optional (see note 7)
- Small roll of ¼-inch hardware cloth, optional (see note 8)
- 1 quart or more wood sealer (see note 9)
- A short length of twisted wire cable and a scrap of old garden hose

Notes on Materials

1: I ordered the metal roofing through my local farm co-op, custom-cut to my order. There are other options for metal roofing than Fabral's Grandrib 3, of course. Check out the possibilities with your local farm supply. The width of coverage of the Grandrib 3 is 36 inches, so the choice of three pieces per side dictated a total length for the shelter of 9 feet. As for the length of the roofing pieces, I used the Pythagorean theorem (the square of the length of the hypotenuse of a right triangle equals the sum of the squares of the lengths of the other two sides) to calculate the 6-foot length, based on a bottom rail of 8 feet 3 inches, and on a preference for cutting angles on the ends of the rafters at 45 degrees. If you are making a smaller or larger shelter, or prefer a different profile with reference to its height, use the good old Pythagorean theorem to recalculate the length of your rafters (the hypotenuse of a right triangle) and the angles at which to cut the ends of your rafters.

2: For fastening Grandrib 3 onto roof framing, Fabral offers a hex-head painted screw complete with neoprene washer. Since I had some left over from a larger roofing project, that is what I used. They are quite expensive, however—up to 16 cents per screw. You could use decking screws (see note 4) for this part of the job as well, adding small neoprene washers to seal the screw holes from the rain.

3: For the framing, I bought all 2×4s and ripped them down as needed. You could of course buy lumber already cut to your needed dimensions (2×2, 1×2, and so on) if you prefer.

4: *Do not use nails* for a pasture shelter—they will work loose as the shelter is jerked about in moving. A coarse-thread deck screw, galvanized or coated against weather, will provide much more durable framing joints and will not require the drilling of pilot holes (except at the ends of pieces to be joined). I keep a supply of a wide variety of lengths and shank sizes on hand at all times. For this project, I used, as appropriate to the join being made, all of the following: #7, 1 and 1⅝ inches; #8, 1¼, 2, 2½, and 3 inches; #10, 3½ inches. Note that, since completing this project, I have begun using star-drive stainless-steel decking screws by preference. They are considerably more expensive than ordinary decking screws, but the star-drive head makes screwing them in much easier, and they last even longer than the galvanized in prolonged exposure to weather.

5: I had a number of 3½×¾-inch steel corner braces ("corner irons") on hand, so I used two on each corner. If you can get larger corner braces, you should need only one per corner.

6: The size wheel for your shelter is up to you. For most pasture shelters you want a solid (non-pneumatic) wheel, available from a garden or tractor supply. See the discussion of wheel size above.

7: If you have some scrap wood on hand, use it to assemble a nestbox. If you do not, you might find it easiest to make one from plywood. If you buy a sheet of plywood, you should have at least half the sheet left over after making your nestbox.

8: I always use ¼-inch hardware cloth (welded wire mesh) for nest bottoms that are more self-cleaning than solid bottoms, so I had on hand the small piece required to floor the nestbox in this shelter. If you can't find a source that will custom-cut to your length, you will probably have to buy a 10-foot roll and keep the remainder on hand for other projects.

9: The sealer I used was Cabot Waterproofing 1000, a clear silicone sealer for wood. I didn't coat the interior parts of the frame that will be completely sheltered from blowing rain. I applied several liberal applications, however, to the bottom rails and end framing and the end grain of stringers and rafters. I used about 2 quarts of sealer.

Construction

Step 1: Set the rip fence of a table saw or handheld power saw to 1¾ inches, and rip all four of the 12-foot 2×4s in half. Cut each of the resulting pieces in half crosswise—that is, into 6-foot lengths. Reserve fourteen of these pieces for the rafters of the shelter, and set aside the other two for other uses.

Step 2: Cut 12 inches off four of the 10-foot 2×4s, to yield four 9-foot 2×4s. Set aside two of these 2×4s to use for the side rails. With the rip fence still set at 1¾ inches, rip the other two 9-foot 2×4s in half. Set aside the resulting four 9-foot pieces, for use as stringers.

Step 3: With the rip fence still set to 1¾, rip one more 10-foot 2×4 in half and set aside for use as collar ties and end framing.

Step 4: Reset the rip fence to 2¼ inches and rip the last 10-foot 2×4. Set aside the 1¼-inch-thick piece for later use. Cut the 2¼-inch piece to 9 feet, for use as the ridge pole.

Step 5: Assemble the bottom rails.

While I did say in chapter 11 that it is possible to cut down on weight by ripping the bottom rails at 2¼ inches, in the present case I was adding a lot of weight because of the choice of the metal roofing as cover. Therefore, I used full 2×4s for the bottom rails. No other parts of the framing needed to be full 2×4—various structural members were 1¾×1¾, 2¼×1½, and 1¼×1½ as specified in the steps below.

Lay out the bottom frame, with the two 9-foot 2×4s previously cut (step 2) to the *outside* and two of the 8-foot 2×4s to the *inside*. Study figure C.2 carefully and make sure that the 9-foot rails set the *length* of the bottom frame at exactly 9 feet; while the 8-foot 2×4s, set on the *inside* of the outer rails, make for a *width* that is 8 feet, 3 inches (8 feet plus the thickness of one nominal 2×4 times two). Stated another way: The 9-foot dimension will become the *sides* of the shelter; the 8 foot, 3 inch dimension, the *front and rear*. You may of course change the size of your A-frame and recalculate lengths and angles, but Pythagorus dictates that you adhere precisely to these dimensions for the bottom frame if you plan to cut and fit your rafters as directed below.

We doubled up on our smaller corner braces and used #10×3½-inch screws for drilling into end grain and #7x1⅝ when drilling into cross-grain. (#8×1⅝ would have been even better.) As noted above, single larger corner braces would be preferable to doubling smaller ones.

Note that as we begin assembling the frame, we want to be as square as we possibly can. If you have a large enough completely flat surface, such as a garage floor, to work on, that is the best choice. That was not an option for us, so I simply used the most level section of lawn I have. For squaring up the corners, I laid each of the corners in turn on a sheet of plywood when joining them with the corner braces. Square your corners as well as you can (measure on the diagonal from opposite corners—the two measurements should be the same if the frame is square) before the next steps, which will lock in the structure.

Step 6: Cut 45-degree angles on both ends of the fourteen rafters cut in step 1. Be careful with this step: You want to end with rafters that still measure a full 6 feet on the *top* edge, with the angles coming *in* from each end.

Step 7: I hope you have a buddy willing to help set the first two pairs of rafters onto the ridge pole—doing that single-handedly would be more challenging than juggling balls walking a tightrope. Screw the ends of all the rafter pieces solidly onto the ridge pole and the side rails, figure C.3, setting the rafters at 18 inches on center. (If you choose a different roofing material, it might require a different spacing of the rafters.) When driving screws into the ends of the rafters—or near the ends of any other pieces in the construction—I first drill a pilot hole to prevent splitting the end. (When driving deck screws into the middle of the work piece, no pilot hole is necessary.)

Step 8: Attach two of the 9-foot stringers cut in step 2 to connect the rear and the front rails (figure C.4). I came in 32 inches from the right and left ends of the front and rear rails to set the stringers. Vary that distance depending on the size door, and access to the nestbox in the rear, that you prefer.

Step 9: Reset rip fence to ¾ inch. Rip the two remaining 8-foot 2×4s to give eight pieces ¾×1½ inches. (No, they're not all precisely ¾ inch thick, but we don't want to go crazy figuring the kerf here.) Set aside four of these pieces for later use. *Check a final time that the bottom frame is square.* Trim the remaining four pieces to make diagonal cross-braces, from each lower corner up to the top in the middle of the structure, attaching to the *undersides* of the rafters wherever they cross, as in Figure C.5. Though fairly lightweight, these cross-braces add tremendously to structural integrity.

Step 10: Using the 1½×1¾-inch stock cut in steps 1 and 3, cut pieces to frame for a door and access to the nestbox on the ends. Study figure C.6 for the general layout. You'll have to do a bit of trial-and-error to establish the correct angles, unless you're a lot better carpenter than I am.

Step 11: Using the same 1½×1¾-inch stock, cut five additional pieces to use as collar ties, which join opposite pairs of rafters (figure C.7). The collar ties add to structural integrity, but they can serve double duty as roosts for your birds. Exactly where you set them is up to you: The lower you set them, the longer they are and the more useful as roosts; the higher, the easier it is to get around inside the structure when you need to do so. A good compromise for us (based also on best use of the precut stock used in this step) was to make the collar ties 29 inches long (measured across the bottom side), which set the top 13½ inches from the ridge pole.

Step 12: To complete the framing, we added the remaining two 9-foot pieces cut in step 2 as an additional pair of stringers, 16 inches above the first pair, connecting the uprights in the framing of the end openings. These two stringers are not essential structurally, but we wanted them to provide a comfortable amount of roosting space for the birds. Remember the 1¼×1½ stock cut in step 4? We cut pieces from it to make two center supports for the pairs of stringers on each side. See figure C.8.

Step 13: I'm going to leave it to you to design your own nestbox (figures C.9, C.10, and C.11) and end door if you want them. I rarely use a door, since I use electronet to protect the flock, but always want the option of shutting in the birds, for a census or selection or whatever—you can see the one I made in figure C.1. For these two projects I used the ¾-inch stock generated in step 9 and the remaining 1¼-inch stock from step 4, plus—for the nest and its hinged access door—some scrap ½-inch CDX plywood. I used some ¼-inch hardware cloth on hand for the bottom of the nestbox. I used more scrap plywood to give additional protection to the ends of the nestbox from blowing rain.

Step 14: While all parts of the frame are still accessible, coat all surfaces that could possibly be reached by blowing rain with a good wood sealer. Don't forget the bottoms of structural pieces: The frame at this point is rock-solid—there is no problem flipping it on its side to get to any surface needing application.

Step 15: Cut the 1-inch mesh poultry wire for the mostly triangular spaces to be sealed off in the front and rear, and staple into place. Staple some over the door frame as well. You could use pieces of hardware cloth instead.

Step 16: We attached the metal roofing using four screws per rafter (figure C.12). As said in note 2 under the materials list, we used some leftover painted screws with neoprene washers—you could use deck screws with neoprene washers instead. The Grandrib 3 we used is designed to be laid down over horizontal roof purlins, but the addition of purlins is not needed in this simple structure. Therefore, we simply attached the screws into the rafters themselves, occasionally doing so through a ridge in the Grandrib 3—not recommended when installing on a house roof—but mostly through a flat section of the profile as recommended. We then attached a ridge vent over the top, which you can see in figure C.1, to protect the interior from rain.

Step 17: Figure C.13 shows the hinged access door we added for collecting eggs. I happened to have some scrap 24-mil woven poly on hand, which I used to cover the hinged top of the door to shed rain.

Step 18: Drill ½-inch holes into the side rails, front and rear. I drilled mine 10 inches in from the end, and 1 inch up from the bottom. Insert the ½-inch carriage bolts through the holes, from the inside of the rails. Place a flat washer, then a lock washer, then a hexagonal nut on the bolt, and tighten until the square shoulders of the carriage bolts bite firmly into the wood (figure C.14). Now simply pop the wheels on the bolts and lock them down with wing nuts (figure C.15). If installing permanently, use more hexagonal nuts instead—the wing nuts make it convenient to put the wheels on and take them off with ease, allowing the use of one set of wheels on multiple shelters.

Step 19: Drill pilot holes, then screw the four open hooks into the front and back rails. Make a pull using a length of wire cable, twisted into a strong loop at each end, and a scrap of old garden hose as padding for your hand. Loop the ends into the hooks, and pull to move the shelter. Now you're rolling!

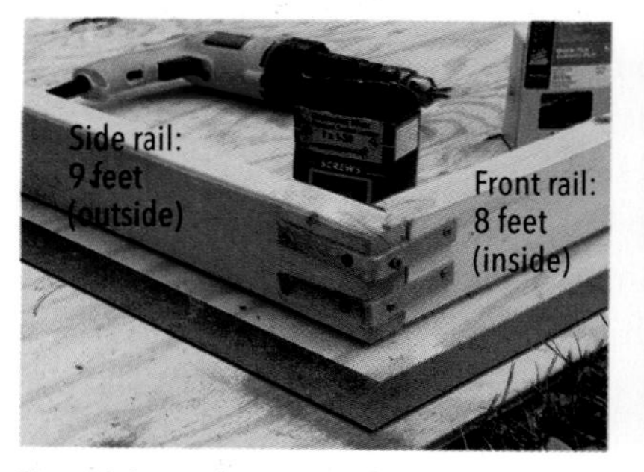

Fig. C.2 Pythagorus insists that the alignment of the pieces for the bottom rails, shown here, is essential if the rafters, as specified, are to fit. The 9-foot side rail is on the left, the 8-foot front rail on the right.

Fig. C.3 Rafters attached to side rails and ridge pole.

Fig. C.4 First pair of stringers attached front to back.

Fig. C.5 Attach four diagonal braces, screwing them onto the underside of the rafters at every crossing point.

Fig. C.6 Frame in openings into the interior, front and back.

Fig. C.7 Collar ties for additional rigidity, positioned to do double duty as roosts.

Fig. C.8 An additional pair of stringers provides more roosting space. Note the two uprights in the middle that add support for the two pairs of stringers.

Fig. C.9 With the end framing and top stringers to attach to, it is easy to make a simple nestbox. The floor of this one is ¼-inch hardware cloth.

Fig. C.10 Here I have added strips of scrap plywood as sides to retain nest materials in the nest. Note the addition of a strip of pine to the top edge in front, to serve as a landing perch for hens approaching the nest.

Fig. C.11 Scrap plywood panels, together with an access door to be added, will keep the nest dry in blowing rains.

Fig. C.12 Attaching the metal roofing.

Fig. C.13 I made the access door to the nest from a piece of scrap ½-inch CDX plywood. A stapled scrap of 24-mil woven poly seals against rain. Note the backing of 1-inch wooden strips to keep the ½-inch plywood access door from warping.

Fig. C.14 Lock ½-inch carriage bolts in the side rails, fore and aft, to serve as axles for the wheels. Note that the lower the placement of the bolt in the rail, the greater the clearance when the wheel is mounted.

Fig. C.15 Note the wing nuts, which make setting and removing the wheels a snap.

D | DUCK CONFIT, CONVENIENCE FOOD EXTRAORDINAIRE

Recipe contributed by Ellen Ussery

Is duck confit fast food or slow food? I'd have to say it's both. It takes careful, patient preparation, but in the end you will have ready at hand the makings of a delicious and deeply satisfying meal that you can put together in no time at all.

Once you have duck confit (pronounced *con-FEE*, it means "preserved duck") in your fridge or cellar, you're ready for anything. You may find yourself tired and hungry at 6 PM and realize that you've given no thought to dinner. You can whip out some confit, heat it gently, toss a salad, add some crusty sourdough bread (or if you happen to have some leftover boiled potatoes, brown them quickly in some of the fat from your confit), and have your dinner. After such a meal your bodily strength will be fully restored.

Any dinner guest presented with your confit will feel honored indeed, certain that such depth of flavor and silky texture could only be achieved by some very hard work on the part of the cook, and that something quite precious is being shared.

Actually, though, making duck (or goose) confit isn't a lot of effort—you can schedule it to fit into other activities and then find that you have spent the last twenty-four hours doing any number of things and, oh, by the way, producing this spectacular food.

So don't be put off by the length of these instructions. Basically there are three steps: Cure the duck pieces in salt and herbs for sixteen to eighteen hours; poach them very gently in rendered fat for three to four hours; and allow them to cool slowly to room temperature in the cooking fat.

The Ingredients

Harvey cuts up the ducks after butchering, and I freeze the wings, thighs, and legs until I have the time and inclination to make confit. When freezing the pieces, I put into each package exactly enough to fit snugly in one layer in the cooking vessel I'm going to use to cook them.

Rendered fat is the other main ingredient. I always make this as soon after butchering as possible. It will keep in the fridge for a long time and in the freezer at least a year. It is crucial to have enough fat to fully cover all your meat when poaching it. Sometimes your ducks will not provide enough fat. If you have geese, their fat is excellent for use in confit as well. Chicken fat is not an acceptable substitute, but good-quality lard is, especially when mixed with some duck fat. I cannot tell you how much you will need. It depends on the size of your birds, the size of your cooking vessel, and what shape container you will use to store the confit. But for a batch cooked in a 12-inch pan, I would have on hand 5 cups of fat at a minimum.

The only other ingredients in my confit are *salt*,

Fig. D.1 **Place the duck pieces in a glass casserole dish with salt, garlic, and thyme.**

Fig. D-2 **Place the cured pieces in a single snug layer in the bottom of the cooking vessel.**

Fig. D-3 **Cover with fat, then with a piece of parchment paper cut to fit.**

Fig. D-4 **Pour in the clear duck fat.**

Fig. D-5 **Cover completely with solidified fat.**

thyme, and *garlic*. There are many recipes calling for more complicated seasonings, but we prefer to accentuate the flavor of the meat itself. I have always used coarse Celtic sea salt. Recently I ran out and used kosher salt, which is not as coarse. It seemed to penetrate the meat more completely than the coarser sea salt I was used to, so if I were to use it again, I would use slightly less. The amount again depends on the size of the container in which you will cure the meat. But for a recipe to be cooked in a 12-inch cooking pan, about a quarter cup should do. I usually have fresh thyme out in the garden until late autumn, and that is what I use, but I have used dried thyme with good results. As for garlic, doesn't everybody have a guest room closet full of garlic, like us? The more the better, I say.

The amounts that follow are what work with my 12-inch copper rondeau, pictured in figure D.2:

8–10 pieces of duck leg, thigh, and wing
¼ cup coarse Celtic sea salt
15 cloves garlic, peeled but left whole
15 sprigs fresh thyme
5 cups (or more) duck and/or goose fat
Parchment paper cut to fit atop the layer of duck pieces

Curing

Using a glass casserole large enough to hold all your duck pieces snugly in one layer, scatter enough of the salt to barely cover the bottom of the pan, then half the garlic and half the thyme. Lay in the duck pieces, skin side up. Sprinkle the remaining salt, garlic, and thyme on top. Cover tightly and refrigerate sixteen to eighteen hours. I usually do this around five or six in the evening so that I am ready to start cooking next morning.

Cooking

Rinse off all the salt from the duck pieces and dry thoroughly. Place them in your cooking pan in one cozy layer, skin side up. A metal pan is better than glass. Add about half of the garlic from the curing process. Pour in the rendered fat, which you have gently melted in a saucepan, being sure to completely cover the duck. Cut the parchment paper for a snug fit in the pan directly on top of the duck—it helps keep the duck covered in fat (see figure D.3). Place in a 200-degree preheated oven. You may have to raise the heat a bit. *What you want is for a tiny bubble or two to rise from the fat every few seconds.* You do *not* want a simmer—that is too hot. So you must monitor it carefully until you get a steady series of bubbles. After three hours, check it by piercing with a fork. In most cases it will be tender, but not falling off the bone. That is the point at which you should stop cooking. But if it still feels quite tough, then continue for up to another hour.

You can also poach your duck in a slow cooker, such as a Crock-Pot. The advantages are that it will cook at the proper low temperature without using as much energy as an oven and, if you are doing this in warm weather, you won't heat up the kitchen. In my case the disadvantage is the size of my slow cookers: Neither of the two I own holds as much in one layer as my rondeau.

Cooling and Storing

Immediately upon removing the duck from the oven, use tongs to place the pieces in the storage container. Then strain the fat into a heat-proof glass jar or measuring pitcher. Do this in two stages. In the first, stop well before any of the cloudy contents of the pan are poured out, to yield a jar of clear fat. Pour this immediately over the duck. It may not cover it completely, but it should be close (see figure D.4). Let cool gently at room temperature—a slow cooling is as important as the slow cooking for developing the flavor and texture of the meat.

Now pour the rest of the fat through the strainer into your pitcher. You will have a bottom layer of meat juices and a top layer of fat. Set the pitcher in the refrigerator for several hours until the fat solidifies. When it does, spoon it out and cover the bare spots of your duck before you store it in the fridge or cellar. *Complete coverage by the fat is essential for preserving the cooked duck* (see figure D.5). The salty meat juices left in the bottom of the pitcher would sour the confit as it ages, so it's important to exclude them in this step. Use them instead for soup or sauce making.

I generally use a flat glass casserole for storage,

since that makes it easier to remove a few pieces at a time. However, it takes more fat to fully cover the duck. By using a widemouthed mason jar, you can squeeze the duck pieces more closely together, and end with a thicker layer of fat on top. It can be difficult to remove the pieces of duck from a jar without mangling them, so I set it in hot tap water till the fat softens enough to remove the pieces I want. Or I might plan to use the whole jar at one time. Another option is storage in a crock.

Most people will store their confit in the refrigerator. But traditionally in France, confit was stored all winter in a cellar with a constant cool temperature. So this would be a safe choice, as long as you are certain that there are no hot-water pipes or the like raising the ambient temperature.

Though you can eat your confit as early as the next day, the flavor improves with aging in the fat. I have kept confit in the refrigerator for as long as three months. But since I don't have a usable cellar, and refrigerator space is at a premium, I find that it works best for me to eat it up in a few weeks and make successive batches. An advantage is that I reuse the fat for the second batch. I won't use it a third time for confit but will cook potatoes or greens with it. Since the fat was never heated to a high temperature, it is perfectly safe to do this.

Once you have learned how to fit confit into the rhythm of your life, you won't want to be without it.

E | A FEED FORMULATION SPREADSHEET

Readers who use electronic spreadsheets will find it a trivial matter to do as I have done—design a calculating spreadsheet to help formulate homemade feeds. Those who are relatively new to the game may benefit from this description of putting together a simple calculator spreadsheet. Note that I'm keeping it simple in this one, which is intended only to illustrate how to set it up. It would be easy to add columns for calculating percentages of fat, carbohydrate, cost of various mixes, or anything else you want to track.

The basic formulation calculator is shown in table E.1. Enter in column A all the feed ingredients on hand to use in various formulations. Make separate listings of the fine, powdery ingredients such as fish meal and mineral supplement, you will make up into a premix for convenience—and of the bulkier components such as corn, peas, and small grains. For any ingredient containing protein, I like to note its percent protein—for example, "Wheat (0.15)." This is just a convenient reminder and has no calculating

Table E.1 Basic Feed Formulation Calculator.

	A	B	C	D
1	INGREDIENT	#LBS/100	% PROTEIN	/25 lb
2	**Premix:**			
3	Aragonite/Feeding Limestone			
4	Nutri-Balancer			
5	Kelp			
6	Fish meal (.60)		0.00	
7	Crab meal (.25) (*Not>2.5/100!*)		0.00	
8	Cultured yeast (.18)		0.00	
9				0.00
10	**Grind/Whole portion:**			
11	Corn (.09)		0.00	0.00
12	Peas (.22)		0.00	0.00
13	Wheat (.15)		0.00	0.00
14	Oats/Barley (.11) (*Not>15%!*)		0.00	0.00
15				
16	Total:	0.00	0.00	0.00

The basic design of the calculator spreadsheet. As values for possible amounts of particular ingredients are entered into the yellow cells, the formulas in the green cells automatically recalculate the results of the changes–for example, the percentage of protein in the mix.

function. In the same way, I might add a note such as "(*Not>2.5/100!*)" to remind myself not to use more than 2½ pounds of crab meal per hundredweight of feed (to avoid excess selenium).

Note that I have highlighted in the other columns two different classes of cells. The highlighting is simply for illustration—once you understand the function of the two types of cells, there is no need to

Table E.2 Entering Formulas in the Calculator.

	A	B	C	D
1	INGREDIENT	#LBS/100	% PROTEIN	/25 lb
2	Premix:			
3	Aragonite/Feeding Limestone			
4	Nutri-Balancer			
5	Kelp			
6	Fish meal (.60)		=B6*0.6	
7	Crab meal (.25) (*Not>2.5/100!*)		=B7*0.25	
8	Cultured yeast (.18)		=B8*0.18	
9				=SUM(B3:B8)/4
10	Grind/Whole portion:			
11	Corn (.09)		=B11*0.09	=B11/4
12	Peas (.22)		=B12*0.22	=B12/4
13	Wheat (.15)		=B13*0.15	=B13/4
14	Oats/Barley (.11) (*Not>15%!*)		=B14*0.11	=B14/4
15				
16	Total:	=SUM(B3:B14)	=SUM(C3:C14)/100	=SUM(D9:D14)

Here is the same calculator spreadsheet, showing the formulas to enter into the green cells that will automatically perform calculations based on values entered into the yellow cells.

Table E.3 A Starter Mix for Chicks on Pasture.

	A	B	C	D
1	INGREDIENT	#LBS/100	% PROTEIN	/25 lb
2	Premix:			
3	Aragonite/Feeding Limestone	1.00		
4	Nutri-Balancer	3.00		
5	Kelp	0.50		
6	Fish meal (.60)	6.50	3.90	
7	Crab meal (.25) (*Not>2.5/100!*)	2.00	0.50	
8	Cultured yeast (.18)		0.00	
9				3.25
10	Grind/Whole portion:			
11	Corn (.09)	23.00	2.07	5.75
12	Peas (.22)	30.00	6.60	7.50
13	Wheat (.15)	24.00	3.60	6.00
14	Oats/Barley (.11) (*Not>15%!*)	10.00	1.10	2.50
15				
16	Total:	100.00	0.18	25.00

The values (weights) of ingredients must be in reasonable proportions, as discussed in chapter 17. As they are entered the spreadsheet calculates the new values for percent protein in the mix, and amounts of ingredients per 25-pound batch.

Table E.4 A Grower Mix for Chicks and Young Waterfowl on Pasture.

	A	B	C	D
1	INGREDIENT	#LBS/100	% PROTEIN	/25 lb
2	Premix:			
3	Aragonite/Feeding Limestone	1.00		
4	Nutri-Balancer	3.00		
5	Kelp	0.50		
6	Fish meal (.60)	5.00	3.00	
7	Crab meal (.25) (*Not>2.5/100!*)	2.25	0.56	
8	Cultured yeast (.18)	2.25	0.41	
9				3.50
10	Grind/Whole portion:			
11	Corn (.09)	24.00	2.16	6.00
12	Peas (.22)	28.00	6.16	7.00
13	Wheat (.15)	24.00	3.60	6.00
14	Oats/Barley (.11) (*Not>15%!*)	10.00	1.10	2.50
15				
16	Total:	100.00	0.17	25.00

As the requirements for the mix vary by age and species of birds we are feeding, it is easy to use the calculating spreadsheet to reformulate.

use color highlighting in your own spreadsheet unless you prefer it.

The yellow cells are those into which I enter possible values (weights) for the different ingredients as I design a feed mix. The green cells contain mathematical formulas—that is, instructions to the spreadsheet program as to the calculations to run on the figures entered in the yellow cells. Note that the cells containing formulas do not show the formulas themselves, which are hidden in the background. Instead, each one shows "0.00," since at this point no values have been entered into the yellow cells.

Table E.2 shows the formulas I added to the spreadsheet. (*Please note* that all the formulas used in this illustrative table are in the syntax required for my spreadsheet program. It is possible that your spreadsheet program requires formula syntax different from these examples. Check the Help section in your program.) I only need one formula in column B: In cell B16 I enter "=SUM(B3:B14)"—that is, this cell will calculate the total of all the cells containing values entered into any or all cells B3 through B14. Note that the title of column B is "#LBS/100." I like to calculate my formulations on a basis of 100 pounds total weight, so that the percent protein is obvious. Thus when I am entering values in cells B3 through B14, I make sure that the resulting total in B16 always comes out to 100.

Column C calculates the weight of protein in the total mix contributed by the weight of ingredient entered in column B. For example, cell A6 lists fish meal, which contains 60 percent protein. Thus the formula to enter in cell C6 is "=B6*0.6"—that is, the weight of fish meal entered in B6 is automatically multiplied by its percent protein to return in C6 the amount of protein it contributes to 100 pounds of feed mix. Similarly, C11 will calculate the protein contributed by corn (A11), which is 9 percent protein, so the correct formula is "=B11*0.09," or the weight of corn entered in B11 multiplied by 9 percent.

The last cell in the column is different. Its formula "=SUM(C3:C14)/100" instructs the program to total all the above cells in the column, C3 through C14, then divide by 100, to give the total percentage of protein in 100 pounds of the proposed mix.

Designing the formulation on a hundredweight

basis is convenient for showing total protein as a percentage, but when we actually mix the feed by hand, making it in batches of 25 pounds is likely to be more practical. Column D will automatically calculate the amount of each ingredient to weigh out per batch. Note that, since we make the premix ahead of time, so it can be handled as a single ingredient, its calculation in D9 must be a bit more complex, "=SUM(B3:B8)/4;" that is, the program will first sum the weights of all the premix ingredients (B3 through B8), then divide by four (there are four 25-pound batches per 100 pounds), to show the amount of premix to weigh out per batch.

Other formulas in column D are more straightforward: A formula such as "=B12/4" will calculate the value (weight) of peas entered in B12, divided by four, to show the amount of peas to use in a batch.

With the spreadsheet ready to calculate, it's time to populate it with some values in order to design a formulation. Table E.3 shows the weights of available ingredients I might use to make a starter feed for chicks on pasture. I entered reasonable proportions of ingredients in column C, and the spreadsheet recalculated with each entry until I hit a target protein content of 18 percent. (That is low by the standards of commercial feeding, but this feed is for chicks on pasture with mother hens, who find high-protein animal foods like worms and insects for the chicks.)

Now suppose I want to reformulate that recipe as the chicks grow older, and that I'm now running them on the same pasture with some growing ducklings and goslings. I can cut back a little on protein, but I want to increase B-vitamin content, since waterfowl need more B complex in the diet. Dried cultured yeast is a good source, so I include some in the new formulation. The revised mix is shown in table E.4.

Tim Koegel of Windy Ridge Natural Farms offers a similar feed calculation spreadsheet, but one tailored to market operations more likely to require feed in ton lots. The sheet has recipes for a chick starter, a layer feed, and a broiler grower ration, each with cost-tracking functionality. Free for download (Microsoft Excel format) at www.windyridgepoultry.com/tools1.htm.

F | SPREADSHEETS FOR TRACKING EGG AND BROILER COSTS AND PROFITS

Tim Koegel, producing member of American Pastured Poultry Producers Association, shares with us a couple of spreadsheets similar to the ones he uses to track production costs and profits in his egg (table F.1) and broiler (table F.2) operations at Windy Ridge Natural Farm in western New York State. To illustrate the usefulness of the spreadsheets for tracking profitability, he has entered figures for four scenarios each for egg and broiler operations, varying with regard to scale of the enterprise, sale prices, labor costs, and so on.

In the presentation of my feed formulation spreadsheet, I discussed some of the details of designing electronic spreadsheets. I will assume that the reader wanting to use these two spreadsheets is well versed in spreadsheet use, especially the entering of calculating formulas. If you study the cells carefully (again, yellow cells are for input of data from the operation, green cells contain the formulas that calculate the input and show the results), it will be clear which formulas are needed in each calculating cell. Once you have entered the needed formulas, you will find that both these spreadsheets are well thought out, and that they are excellent tools for tracking your costs and profits.[1]

Table F.1 Egg Production Profitability Calculator

Windy Ridge Natural Farm
www.windyridgepoultry.com

Egg production profitability calculator

		Scenario #1	Scenario #2	Scenario #3	Scenario #4
total number of hens you want to raise->		250 birds	2000 birds	2000 birds	2000 birds
	Total Pullet cost				
total feed to raise a pullet to laying age->	feed total/lbs	25	25	25	25
cost of grower ration per pound->	feed-$/lb	$0.32	$0.32	$0.32	$0.32
	Total feed $	$8.00	$8.00	$8.00	$8.00
cost of individual chick with shipping->	chick $	$1.25	$1.25	$1.25	$1.25
	Cost ea	$9.25	$9.25	$9.25	$9.25
Chick Mortality->	Mortality %	10%	10%	10%	10%
(True cost of mortality depends on age at loss)	Mortality loss $	$231.25	$1,850.00	$1,850.00	$1,850.00
total cost to raise a pullet to laying age	Flock Cost	$2,543.75	$20,350.00	$20,350.00	$20,350.00
cost of layer ration per pound->	feed-$/lb	$0.32	$0.32	$0.32	$0.32
feed consumption per hen per month->	lb/month/bird	8.7	8.7	8.7	8.7
	lb/month	2175	17400	17400	17400
	feed month/$	$696.00	$5,568.00	$5,568.00	$5,568.00
lay rate (factor in loss here)->	lay rate	89.50%	89.50%	80.00%	89.00%
	eggs / day	224	1790	1600	1780
	eggs / day	18.65 doz	149.17 doz	133.33 doz	148.33 doz
	eggs / wk	131 doz	1044 doz	933 doz	1038 doz
	eggs/month	559 doz	4475 doz	4000 doz	4450 doz
	eggs/year	6806 doz	54446 doz	48667 doz	54142 doz
selling price per dozen of eggs->	retail $/doz	$4.50	$4.50	$2.07	$1.93
wholesale discount (if any) ->	resale disc	0%	0%	0%	0%
	net/doz	$4.50	$4.50	$2.07	$1.93
	ttl $ month	$2,517.19	$20,137.50	$8,280.00	$8,588.50
	net / month	$1,821.19	$14,569.50	$2,712.00	$3,020.50
	net / year	$21,854.25	$174,834.00	$32,544.00	$36,246.00
Number of hours/week->	Labor-hrs/wk	10	30	0	0
Total cost/hr of labor (inc benefits,taxes,etc)->	Labor-$/hr	$13.00	$13.00	$13.00	$13.00
	Ttl labor/yr	$6,760.00	$20,280.00	$-	$-
Cost to package 1 doz eggs	Cartons/each	$0.12	$0.12	$0.12	$0.12
	Total pkg expense	$816.69	$6,533.50	$5,840.00	$6,497.00
Additional costs for your operation->	Capital exp/yr	$1,000.00	$4,000.00	$4,000.00	$4,000.00
Additional costs for your operation->	Electric	$100.00	$800.00	$800.00	$800.00
Additional costs for your operation->	Bedding	$150.00	$1,200.00	$1,200.00	$1,200.00
Additional costs for your operation->	Feed extras				
Additional costs for your operation->	Machines	$200.00	$200.00	$200.00	$200.00
	Overhead	$9,026.81	$33,013.62	$12,040.12	$12,697.12
	Pullet cost	$2,543.75	$20,350.00	$20,350.00	$20,350.00
(includes pullet costs)	Net profit 1st year	$10,283.69	$121,470.38	$153.88	$3,198.88
(does not include any pullet costs)	Net profit 2nd year	$12,827.44	$141,820.38	$20,503.88	$23,548.88
Sales of spent hens at end of 3rd year	Sale price each	$2.00	$2.00	$2.00	$2.00
Revenue of spent Hens (at 0% loss)	Total of spent hens $	$500.00	$4,000.00	$4,000.00	$4,000.00
(includes spent hen revenue)	Net profit 3rd year	$13,327.44	$145,820.38	$24,503.88	$27,548.88

Table F.2 Broiler Production Profitability Calculator

Windy Ridge Natural Farm

www.windyridgepoultry.com

Broiler production profitability calculator

	125 birds	250 birds	500 birds	1000 birds
Feed total/lbs	15	15	15	15
Feed-$/lb	$0.32	$0.32	$0.32	$0.32
Total feed $	$4.80	$4.80	$4.80	$4.80
Chick $	$1.25	$1.25	$1.25	$1.25
Cost of each broiler	$6.05	$6.05	$6.05	$6.05
Mortality %	5%	5%	5%	5%
Mortality loss $	$37.81	$75.63	$151.25	$302.50
processing ea	$1.85	$1.85	$1.85	$1.85
Total processing	$231.25	$462.50	$925.00	$1,850.00
Batch cost	$1,025.31	$2,050.63	$4,101.25	$8,202.50
Avg dressed wt/bird lbs	4.50	4.50	4.50	4.50
cost/lb meat	$1.82	$1.82	$1.82	$1.82
Sale $/lb	$3.99	$3.99	$3.99	$3.99
Batch gross	$2,244.38	$4,488.75	$8,977.50	$17,955.00
Wholesale disc%	0%	0%	0%	0%
Wholesale disc$	$-	$-	$-	$-
Batch net	$1,219.06	$2,438.13	$4,876.25	$9,752.50
Batch Labor-hrs/wk	0	0	0	0
Labor-$/hr	$13.00	$13.00	$13.00	$13.00
Batch Growout/weeks	8.0	8.0	8.0	8.0
Labor cost/batch	$-	$-	$-	$-
Capital exp/alloc	$50.00	$50.00	$50.00	$50.00
other	$-	$-	$-	$-
other	$-	$-	$-	$-
other	$-	$-	$-	$-
Total overhead	$50.00	$50.00	$50.00	$50.00
Total lbs produced / yr	4,275	8,550	17,100	34,200
Total # birds produced/yr	1,000	2,000	4,000	8,000
Total lbs feed used / yr	15,000	30,000	60,000	120,000
Total feed cost / yr	$4,800.00	$9,600.00	$19,200.00	$38,400.00
Batch Net profit	**$1,169.06**	**$2,388.13**	**$4,826.25**	**$9,702.50**
Batches/yr	**8**	**8**	**8**	**8**
Annual Net Profit	**$9,352.50**	**$19,105.00**	**$38,610.00**	**$77,620.00**

G | NATURAL EGGS AND INDUSTRIAL EGGS COMPARED

In 2007 *Mother Earth News* compared nutrient profiles of eggs: conventional averages for supermarket eggs, as reported by the US Department of Agriculture, compared with eggs from fourteen producers of eggs from pastured flocks in all parts of the United States. The results, summarized in table G.1, speak for themselves.[1]

All values are per 100 grams of egg.	Vitamin E (mg)	Vit. A Activity (IU)	Beta Carotene (mcg)	Omega-3s (g)	Cholesterol (mg)	Sat. Fat (g)
Eggs from Confined Birds (per USDA Nutrient Database)	0.97	487	10	0.22	423	3.1
Free-range Egg Averages Mother Earth News, 2007	3.73	791.86	79.03	0.66	277	2.4
Red Stuga; Topeka, KS; Welsummers	3.35	790	73.8	0.69	350	2.07
Polyface Farm; Swoope, VA; Mixed Non-Hybrid Breeds	7.37	763	76.2	0.71	292	2.31
Shady Grove Farm/American Livestock Breeds Conservancy; Hurdle Mills, NC; Buckeyes	2.68	683	42.0	0.59	321	3.16
Norton Creek Products; Blodgett, OR; Mixed Breeds	2.68	781	102.0	0.55	272	1.88
Skagit River Ranch; Sedro Woolley, WA; Mixed Breeds	4.02	1013	99.6	0.74	335	2.68
Spring Mountain Farms; Lehighton, PA; Red Sex-Links	5.36	813	90.0	0.68	231	1.99
Harmony Hill; Troutville, VA; Rhode Island Reds	1.34	700	69.6	0.49	286	3.38
Rocky Run Farm; Dunnville, KY; Brown Leghorns	2.68	Not Tested	Not Tested	0.64	301	2.48
Misty Meadows Farm; Everson, WA; Red Sex-Links	3.35	Not Tested	Not Tested	0.85	283	2.19
Sparkling Earth Farm; Burnsville, NC; Bovans Browns	3.35	Not Tested	Not Tested	0.80	218	2.68
Windy Island Acres; Dayville, CT; Mixed Breeds	6.03	Not Tested	Not Tested	0.52	271	2.60
World's Best Eggs; Elgin, TX; Bovans Browns	4.02	Not Tested	Not Tested	0.46	246	2.01
Longbranch Farm; Fair Play, SC; Araucanas/Ameraucanas	1.34	Not Tested	Not Tested	0.87	271	2.37
Springfield Farm; Sparks, MD; Red Sex-Links	4.69	Not Tested	Not Tested	0.60	201	1.83

Table G.1 Nutrient profiles of pastured and industrial eggs compared.

H | RESOURCES

Below are some resources that I have found useful regarding not only poultry husbandry but also the food independence project as a whole, the coming changes in the national and global economy, and the need for a more sustainable and regenerative agriculture.

Organizations

- **American Pastured Poultry Producers Association:** Consider joining the APPPA for access to a world of information on the whole range of topics involved with practical poultry husbandry. Though it is an organization dedicated to small farmers producing for local markets, APPPA offers much of relevance to the serious poultry husbandman who wants to go beyond the hobbyist level. If you wish to join, go to their home page (http://apppa.org) and click on "Membership" in the left navigation bar. On the linked page you can choose your level of membership. If you join at the "Producer Plus" level, you will have access to the APPPA discussion list—the very best listserve I participate in. Discussion is serious and to the point, and members do not waste my time with idle chitchat.
- **American Livestock Breeds Conservancy:** To learn more about older breeds of poultry and other livestock, join ALBC, dedicated to preserving traditional and historic breeds. ALBC seeks to conserve those breeds, many of which are threatened with extinction, but believes that the key to doing so is using them for the economic qualities for which they were bred—that is, for their fit in the effort to produce food for the family or local markets. www.albc-usa.org; PO Box 477, Pittsboro, NC 27312; 919-542-5704. See the ALBC's list of rare and endangered poultry breeds at www.albc-usa.org/cpl/wtchlist.html.
- **Society for the Preservation of Poultry Antiquities:** The SPPA is another national organization dedicated to perpetuating and improving rare breeds of poultry. (They don't have their own website but share a corner of Barry Koffler's excellent Feathersite.com: www.feathersite.com/Poultry/SPPA/SPPA.html.) Membership in either ALBC or SPPA will put you in touch with other breeders from whom you might get stock. SPPA's *Breeder's Directory* is especially detailed and well organized.
- **Weston A. Price Foundation:** If I could convince you to check out only one source for food and health issues I've touched on briefly in this book, it would be the WAPF, an organization dedicated to reminding the eating public of the profoundly important relationship between traditional foods (as opposed to their industrial look-alikes) and

health, based on the revolutionary work of Weston Price in the earlier half of the last century. The site is a tremendous resource on nutrition and health and could profoundly alter some of your perspectives on these issues. If you like what you see, please become a member—support the foundation's work and receive its excellent quarterly *Wise Traditions*. www.westonaprice.org; 4200 Wisconsin Avenue, NW, Washington, DC 20016; 202-363-4394.

Books and DVDs

Food, Nutrition, and Health Issues

- *Nourishing Traditions*, by Sally Fallon, president of the Weston A. Price Foundation (second edition, New Trends Publishing, 2001). This book goes far beyond the conventional cookbook. It is a comprehensive compendium of information on food and health issues, with a focus on the whole, natural foods emphasized by traditional cultures for thousands of years. It is a book that may revolutionize all your thinking about diet and health.
- *The Omnivore's Dilemma*, by Michael Pollan (Penguin Press, 2006). One of the most important books on food, agriculture, health, and public policy of the past decade. It helps us understand how profoundly food has changed in the age of industrial food, becoming in the process more enemy (to our health, ecology, and future) than sacred gift.
- *Fast Food Nation: The Dark Side of the All-American Meal*, by Eric Schlosser (Houghton Mifflin, 2001). A must-read exposé of the realities behind fast-food franchises and other purveyors of food for our fill-'er-up lifestyle; the manipulation of the consumer (especially children) by advertising; the trickery played on our palates by chemical engineering; and much more. The chapter on conditions in industrial slaughterhouses is an excellent update on Upton Sinclair's *The Jungle*. Schlosser's explanation of why meat in the marketplace has become so hazardous is quite simple: "There is shit in the meat."
- *Food, Inc.* (2009), is a DVD featuring the criticisms of Eric Schlosser, Michael Pollan, Joel Salatin, and others of an industrial food system that is far more about profits and convenience than about nutrition and protecting the integrity of the natural world. It exposes the degree to which production of our food has become invested in a few powerful corporations, to the detriment of our health, the traditional family farm, and the environment in which we all live.
- *The Future of Food* (2005) is a DVD documenting the impact of agribusiness and dangerous technologies such as genetic modification of crop plants.

Sustainability Issues

It is scarcely believable that most of us are largely ignoring the elephant in the room—peak oil and its implications—but the shrinking of hydrocarbon fuel supplies will be the major event of our time.

Some books I can recommend on the subject are James Howard Kunstler's *The Long Emergency: Surviving the Converging Catastrophes of the Twenty-first Century* and Richard Heinberg's *The Party's Over: Oil, War and the Fate of Industrial Societies*; *Powerdown: Options and Actions for a Post-Carbon World*; *The Post Carbon Reader: Managing the 21st Century's Sustainability Crises*; and *Peak Everything: Waking Up to the Century of Declines*.

Particularly valuable analyses of the coming decline in global oil production come from insiders such as oil geologist Kenneth Deffeyes (*Hubbert's Peak: The Impending World Oil Shortage*, Princeton University Press, 2008) and oil investment banker Matthew

Simmons (*Twilight in the Desert: The Coming Saudi Oil Shock and the World Economy*, Wiley, 2005).

If you prefer visual media, see two DVDs: *The End of Suburbia: Oil Depletion and the Collapse of the American Dream* and *A Crude Awakening*. I'm not a fan of the frenetic kaleidoscope of images typical of such presentations, so I found both a bit irritating—but they hit me right between the eyes all the same.

Guns, Germs, and Steel, by Jared Diamond (W. W. Norton Company, 1999). It's hard to characterize Diamond's book. A whirlwind tour of world history it certainly is, but from a perspective you may not have encountered in your history courses. It analyzes, among other things, how changes in the way we produce our food have influenced the rise and fall of empires, the spread of disease, and "the fates of human societies." Even more relevant to the theme of sustainability (or non-) is Diamond's *Collapse: How Societies Choose to Fail or Succeed* (revised edition, Penguin, 2011), a sobering study of numerous previous societies and whole civilizations that collapsed following their overrunning or undercutting of their ecological base (and a few that had the wisdom to avoid such a fate).

Read Wendell Berry! Since he is poet, novelist, and essayist, there will be books of his to suit your particular tastes. They all offer a vision of what has gone so profoundly wrong with agriculture—with culture generally—in our time, but also of the directions we need to take for healing, of both ourselves and the earth. Collections of essays include *The Unsettling of America*; *Sex, Economy, Freedom & Community*; and *What Are People For?*—*Remembering* is a wonderful novel—do find a collection of his poems—it's *all* good.

Suppose you were to adopt the peculiar notion that sustainability in your own corner of the landscape depends on recycling, rather than squandering, your own body "wastes"? The book you'd need for this bit of madness would be *The Humanure Handbook: A Guide to Composting Human Manure* (third edition, 2005) by Joseph Jenkins.

Gene Logsdon, *Holy Shit: Managing Manure to Save Mankind* (Chelsea Green Publishing, 2010) is an informative discussion of the agricultural use of manures. I think Logsdon means both the title and subtitle literally—he forcefully, though wittily, makes the case that the contribution by livestock animals of their manures is a key to sustainable farming.

Frances Moore Lappé's *Diet for a Small Planet* did a great deal to focus our attention on issues of sustainability and waste in agriculture as related to food choice. Unfortunately, she took current industrial agricultural (rather than traditional) practice as the norm and drew the conclusion that direct consumption by humans of grains and legumes is always the more sustainable choice. She overlooked the many possibilities for wise resource use in which animal foods are actually the most sustainable option, not the most wasteful.

That better perspective is provided by Lierre Keith's *The Vegetarian Myth* (Flashpoint Press, 2009). The assumption is almost universally made, even among meat eaters, that basing agriculture and the human diet on plant crops is always and necessarily the more efficient and sustainable way to feed the world. Keith shows just how destructive an agriculture based exclusively on annual crops can be, and argues persuasively that an agriculture that includes livestock animals is not only more sustainable but more compassionate.

Poisoned Waters is a PBS *Frontline* DVD narrated by Hedrick Smith. It is a sobering, indeed frightening, look at the pollution of water systems by human activity, including agriculture and especially the poultry industry.

I highly recommend reading Alan Weisman's *The World Without Us* (St. Martin's Press, 2007), which imagines that *Homo sapiens* as a force in the world disappeared tomorrow. What would be the consequences for the further development of species and natural systems? The reader soon realizes that Weisman's thought experiment is not an idle conceit, but a profound meditation on our impact in the world.

Understanding and Working with Soil

Sir Albert Howard's classics, *An Agricultural Testament* (originally published in 1940, republished in 2010 by Oxford City Press) and *The Soil and Health* (originally published in 1947, reprinted in 2006 by University of Kentucky Press, with an introduction by Wendell Berry) helped inspire the organic movement as a rejection of industrial agriculture. They seem a bit dated today but can still guide our efforts to improve our soil.

Dr. Elaine Ingham has done more than perhaps anyone else to enlarge our understanding of soil life. She wrote most of *Soil Biology Primer* (revised edition, Soil and Water Conservation Society, 2000), an excellent brief introduction for those new to the concept of the soil food web.

Two discussions at greater length of the staggering complexity of soil life and fertility cycles, readily accessible to the layman, are *Life in the Soil* by James Nardi (University of Chicago Press, 2007) and *Teaming with Microbes* by Jeff Lowenfels and Wayne Lewis (Timber Press, 2010).

Building Soils for Better Crops by Magdoff and van Es is a useful overview of soil ecology and soil care.

If vermicomposting for bioconversion of organic "wastes" is new to you, you might check out *Worms Eat My Garbage: How to Set Up & Maintain a Worm Composting System*, by Mary Appelhof (Flower Press, 1982, revised 1997). Something of a classic in the field, with information on worm biology and setup for using earthworms to recycle kitchen "wastes." The basics are quite simple, though—you could easily find all the information you need through an online search engine. Start small, and work up to vermicomposting on any scale you like.

For ideas on vermicomposting at the farm scale, see George Sheffield Oliver, "My Grandfather's Earthworm Farm," available online in its entirety at www.journeytoforever.org/farm_library/oliver/oliver_farm.html—and as chapter 6 of Thomas J. Barrett's *Harnessing the Earthworm* (Bookworm Publishing, 1976).

Poultry

Page Smith and Charles Daniel's delightful *The Chicken Book* (North Point Press, 1982) contains practical information about chickens as a backyard enterprise, to be sure, but is worth reading mostly because of its insights into the role of the chicken, and chicken lore, throughout the ages.

Judy Pangman's *Chicken Coops: 45 Building Plans for Housing Your Flock* (Storey Publishing, 2006) has blueprints and lots of ideas for coops for many flock sizes and management situations.

Fresh-Air Poultry Houses: The Classic Guide to Open-Front Chicken Coops for Healthier Poultry (Norton Creek Press, 2008) is a republication of Prince T. Woods's *Modern Fresh-Air Poultry Houses*, originally published in 1924. It is clumsily edited but is an excellent counter to the common assumption that the chicken coop needs the addition of artificial heat in the winter. Woods tells of one design after another of open-front poultry housing in even northerly climes that actually improved results over tighter housing by increasing airflow through the coop.

A Guide to Raising Chickens by Gail Damerow (Storey Communications, 1995) is a useful guide to many facets of home flock husbandry. And Damerow's *Chicken Health Handbook* (Storey Publishing, 1994) is the most complete treatment of its subject you will find, aside from highly specialized scientific works.

Dave Holderread's *Raising the Home Duck Flock* and *The Book of Geese* (Hen House Publications, 1978 and 1981, respectively) are excellent introductions if you are interested in raising waterfowl.

Chris Ashton, *Domestic Geese* (Crowood Press, 1999), is a useful guide to keeping geese.

Carol Deppe's *The Resilient Gardener: Food Production and Self-Reliance in Uncertain Times* (Chelsea Green Publishing, 2010) is about more than poultry, but I especially recommend her thoughts on using an all-purpose duck flock to fit into the total gardening project in the context of becoming more food-independent as the industrial food system comes under ever greater strain.

Andy Lee's *Chicken Tractor: The Gardener's Guide to Happy Hens and Healthy Soil* (Good Earth Publications, 1994) has excellent ideas for using chickens in the garden—without trashing the garden—with the use of small garden-bed-wide mobile shelters he calls "chicken tractors."

If you're contemplating an egg or poultry marketing venture, by all means read Joel Salatin's now classic *Pastured Poultry Profit$* (Polyface, Inc., 1993, distributed by Chelsea Green Publishing). You should read as well his *Everything I Want to Do Is Illegal* (Polyface Press, 2007, distributed by Chelsea Green Publishing). You will be amused by Joel's trademark humor, but boiling mad as you realize that many of the regulations for the sale of food products have more to do with favoring the profits of gigantic food corporations than the safety of what we put into our mouths.

J. Russell Smith's *Tree Crops: A Permanent Agriculture* (Devin-Adair Company, 1950) is not a poultry book, but I list it here because it suggests a number of foods that ranging flocks can forage from trees—acorns, beechnuts, mulberries, and more.

Other Homesteading Topics

Rolfe Cobleigh, *Handy Farm Devices and How to Make Them* (first published 1909 by Orange Judd Company, republished 1996 by The Lyons Press) has many useful ideas for things you can make yourself for success on the homestead and small farm.

Dave Jacke's *Edible Forest Gardens* in two volumes, *Vision & Theory* and *Design & Practice* (Chelsea Green Publishing, 2005) is an introduction to a fascinating subject, the forest garden, which will suggest many possibilities to the creative flockster. The extensive appendixes are alone worth the price of the books.

Equipment/Accessories

- The **dust mask** I've used that is both the most effective and the most comfortable, excellent for the killing and plucking phase of butchering (when a good deal of poultry dander gets kicked into the air), is the Respro Sportsta, which I buy from Allergy Control Products, www.allergycontrol.com; 800-422-3878.
- If you get serious about making your own feeds, the **feed mill** I use is made by C. S. Bell Co. of Tiffin, Ohio, and can be purchased from Lehman's, their Item #2360, available at www.lehmans.com/store/Kitchen___Grain_and_Grain_Mills___High_Speed_Grain_Mill___highSpeedMill?Args=, priced as I write at about $600. I have been using mine for well over ten years now.
- **Electric net fencing** is a fundamental tool for management of my poultry flocks. My preferred source for electronet and accessories is Premier: www.premier1supplies.com; 800-282-6631. Everyone on the staff at Premier uses electric fencing to manage their personal livestock, so products sent their way get rigorous real-world testing, and technical advice is well grounded in experience. Friends whose judgment I value also recommend Kencove: www.kencove.com/fence; 800-536-2683.
- If you butcher a lot of birds, or if you plan to serve a broiler or turkey market, it might pay to tool up with a **mechanical plucker** and a **scalder** at a minimum. If you want to make your own, Herrick Kimball offers *Anyone Can Build a Whizbang Chicken Scalder* and *Anyone Can Build a Tub-Style Mechanical Chicken Plucker* at http://whizbangbooks.blogspot.com. If you prefer to buy assembled but affordably priced quality equipment, check out David Schafer's Featherman Equipment

Company at www.featherman.net, which offers a scalder and several pluckers. For top-of-the-line stainless-steel equipment designed for pastured poultry operations, see Eli Reiff's Poultryman pluckers and scalders at http://chickenpickers.com/page10.html; or call Eli at 570-966-0769.

- A serviceable **caponizing kit** is available from Nasco (www.enasco.com/farmandranch/Poultry+Equipment/Caponizing/?ref=index).

Online

A great online resource is The Modern Homestead (http://themodernhomestead.us), with information on many facets of home food production, especially the mixed poultry flock. The site will continue to grow as I share more of my articles, thoughts, and experiments in the future.

"Pastured-Raised Poultry Nutrition," by Jeff Mattocks of Fertrell Company, originally written for Heifer International in 2002, is now widely available as a free download in PDF format. One link is http://attra.ncat.org/attra-pub/PDF/chnutritionhpinew.pdf. Key "jeff mattocks poultry nutrition" into a search engine to find numerous other links to this excellent resource.

There are hundreds of breeds of chickens and other domestic fowl. An excellent online introduction to the bewildering diversity of breeds is http://feathersite.com, complete with thousands of photographs. Lots of information and links regarding all things poultry. A site you will come back to again and again.

Journey to Forever has a lot of interesting information for the flockster, homesteader, and small farmer. Its online library at www.journeytoforever.org/farm_library.html contains whole books free for the reading, including some of those listed above—Sir Albert Howard's *An Agricultural Testament* and *The Soil and Health*, J. Russell Smith's *Tree Crops*, and Rolfe Cobleigh's *Handy Farm Devices and How to Make Them.*

Another excellent collection of books available online is the Soil and Health Library at www.soilandhealth.org.

Kate Hunter writes an excellent and highly readable blog, Living the Frugal Life (http://livingthefrugallife.blogspot.com), in which she shares a wealth of information about homesteading on a small scale, including ongoing experiments of her own to boost food independence, for example by integrating a micro-flock of chickens into the process, using principles similar to those in this book.

"Is American Agribusiness Making Food Less Nutritious?" by Cheryl Long and Lynn Keiley, Mother Earth News, June–July 2004, is an overview of the evidence that the nutrition in the national diet has been steadily declining for many decades. See www.motherearthnews.com/Real-Food/2004-06-01/Is-Agribusiness-Making-Food-Less-Nutritious.aspx, or key the title into a search engine for a copy elsewhere.

The peaking of petroleum production will bring in its wake enormous changes in the way we organize our economy, and especially in where and how we produce our food. Post Peak Living at http://postpeakliving.com is an online site that takes seriously the breadth and depth of those changes. It offers persuasive and well-presented analyses of peak oil and related problems for your consideration, plus a series of online courses at reasonable tuition on topics such as "Introduction to Sustainable Gardening," "Sustainable Post-Peak Livelihoods," "Navigating the Coming Chaos of Unprecedented Transitions," and "Chickens 101."

Magazines

Backyard Poultry is a bimonthly magazine that should be useful to both beginning and more experienced poultry people (145 Industrial Drive, Medford WI 54451; 800-551-5691). I have had articles in almost every issue since it began publication in 2006.

Mother Earth News is a long-standing bimonthly magazine that addresses homesteading issues, broadly defined.

Countryside & Small Stock Journal is another long-established bimonthly on homesteading endeavors.

GLOSSARY

bantam: A miniature breed of chickens. A few bantam breeds such as Sebright are naturally small in size. Most bantam breeds have been miniaturized by crossing in the bantam gene and selection.

beak: The bony structure on the head of a bird, consisting of an upper and a lower jaw, which it uses for eating, catching and killing prey, manipulating objects, feeding young, and preening.

bile: An intensely green, bitter-tasting liquid produced in the liver, and essential to digestion of fats in the bird's food and absorption of fat-soluble vitamins.

breed: A group of domestic fowl with the same general conformation, carriage, and size, as a result of consistent selection for the same traits.

broiler: A table chicken slaughtered young enough to be tender enough for dry-heat cooking methods such as frying, broiling, grilling, and baking.

broody, broodiness: The terms refer to the instinctive behaviors associated with incubating and hatching eggs and nurturing chicks. *Broody* can be used as an adjective ("a broody hen") or a noun ("a subflock of good broodies"). A hen is said to "go broody" when entering this special state of mind, "broodiness."

cannibalism: Injurious pecking of one another among chicks or chickens, brought on by stress or inadequate protein in the diet. Well-managed flocks are unlikely to exhibit this behavior.

castings: Excreta of earthworms, an excellent soil fertility amendment.

CDX plywood: Construction grade plywood designed for exterior use (the "X" in its name).

ceca (singular cecum): Long blind pouches attached to where the large and small intestine of a chicken join, serving various functions in the digestive process.

chook: An imported word from Australia and New Zealand, literally meaning "chicken," often used by flocksters playfully or affectionately.

cleat: A strip of wood used to strengthen or provide support to the surface to which it is attached. A cleat can be used to stiffen a span of plywood, or be fastened to the edge of a piece of plywood to receive screws or nails that attach another piece of plywood, when the plywood pieces are too thin to edge-nail.

cloaca: The stretchy membranous pouch at the end of a bird's digestive tract, in which the feces collect before being expelled. The reproductive tract also exits through this area, for expulsion of the egg in the case of the hen, and of semen in the case of the cock.

coccidiosis: A disease in which chickens suffer from overwhelming numbers of coccidia, up to nine species of microscopic parasitic protozoans. Chickens with healthy immune systems develop immunity through moderate exposure to coccidia.

cock: The adult male chicken.

cockerel: A male chicken up to one year of age.

comb: The fleshy appendage on top of a chicken's head, varying in shape and size by breed, larger in the cock. Radiates heat to help cool the body and probably, in the case of the male, serves as a sexual attractant.

confit: Meat, such as duck or goose, that has been cooked and preserved in its own fat.

creep feeder: Any structure that—through use of slats or mesh size—gives access for feeding to younger and therefore smaller livestock or poultry, while physically excluding adults.

crop: A stretchy membranous pouch, an enlargement of the esophagus, just to the right of the base of a chicken's neck. A storage pouch into which to put a lot of food in a hurry, and in which a start is made on digestion by softening the contents with digestive fluids.

CSA: "Community Supported Agriculture," an arrangement in which consumers pay for a season's subscription to a farm's produce up front and receive in return a weekly "basket" of produce, the exact makeup of which changes as different crops are ready for harvest. CSAs in the past have typically furnished vegetable and fruit produce only, but many now offer eggs and even fresh-dressed broilers as well.

debeak, debeaking: To debeak a chick is to clip off part of the upper beak to prevent cannibalism. Such alteration should never be needed in a well-managed small-scale flock.

diatomaceous earth: A powder consisting of siliceous remains of diatoms, ancient one-celled algae that formed shells from silica. Particles, sharp and cutting at the microscopic level, have a lethal effect on insects. For this reason DE is an effective nontoxic insecticide in the feed bin or dustbox, or as an emergency treatment for infestation by lice or mites.

drake: An adult male duck.

duck: Generically, a duck is a waterfowl, *Anas platyrhynchos*, descended from the wild mallard. If the reference is to gender, the *duck* is the adult female (as opposed to the male *drake*).

duckling: Generically, the young of ducks, up to one year of age. If it is necessary to distinguish gender at this stage, a duckling is female, a drakeling male.

duodenal loop: The upper part of the small intestine, where digestive fluids are added to the ground and mixed food passed into it from the gizzard.

dust bath, dust bathing: Like some wild birds, chickens and other gallinaceous fowl make dusty hollows in soil and fluff dust up under their feathers to kill external parasites. As well, dust bathing is an activity they clearly enjoy.

dustbox: A box the flockster builds with dusty materials inside, to ensure adequate access to dust bathing. Alternatively, a bit of ground outside with a cover to keep it dry for dust bathing even in rainy weather.

eggmobile: A movable shelter for a flock of pastured layers, complete with roosts and nestboxes.

electric net fencing: A lightweight mesh fence for confining and protecting pastured flocks within a defined perimeter, made of plastic strands interwoven with fine stainless-steel wires that carry an electric charge.

electronet or electric netting: See *electric net fencing*.

flockster: A word created by the author since no exact one existed: A person who keeps a flock of domesticated fowl, usually based on the conviction that they have a lot to offer as working partners for a more self-sufficient homestead or small farm.

Freedom Rangers: A marketing term for a *type* of pastured broiler bred for pastured production in the French *Label Rouge* certification system. A contrast with the Cornish Cross, the preferred broiler of the poultry industry and many pastured producers as well: colored rather than white, more robust, takes advantage of foraged foods, has a longer grow-out by ten days to two weeks.

gallbladder: A small pouch attached to the liver, the size and shape of a caterpillar, which stores and releases bile.

Galliformes: The avian order to which domesticated

chickens belong (along with turkeys, guineas, pheasants, quail). While Galliformes is the scientific name for the order, we can use the common word *galliforms* to refer to this group of galliform fowl.

game: Refers to breeds of chickens the cocks of which are used in the "sport" of cockfighting, illegal in most jurisdictions. Game breeds have been used in the development of many modern utilitarian breeds.

gander: An adult male goose.

gizzard: Sometimes referred to as the "mechanical stomach." Made up of two sets of powerful muscles around a pouch containing food that has been passed into it by the proventriculus, plus grit, small pieces of stone swallowed by the bird. The working of these muscles, together with the grinding action of the grit, pulverizes and mixes the food and prepares it for further digestion.

goose: Used generically, refers to a domesticated waterfowl descended from wild geese. When the context has to do with gender, *goose* means the adult female (as opposed to *gander*, the male).

gosling: Generically, the young of geese, up to one year of age. If it is necessary to distinguish gender at this stage, I suppose you could use "ganderling" for the male and "gosling" for the female, though such usage is not supported by any dictionary I know of.

grit: Small bits of stone picked up by fowl for furnishing the gizzard with a grinding agent. Crushed granite grit is readily available to purchase as a supplement, in sizes appropriate to the age and species of fowl being fed.

guinea: A galliform fowl, *Numida meleagris*, long domesticated, originally from Africa.

hackle: The rear and side neck plumage of a chicken or other fowl.

hardware cloth: Welded wire screening useful for making protective barriers in the henhouse and self-cleaning nest bottoms. Available in different mesh sizes, half-inch and quarter-inch being the most common.

hen: An adult female chicken.

hock: On a chicken or other galliform fowl, the joint between the lower thigh (what we call the "drumstick" at the table) and the shank.

hoophouse: A shelter, either permanent or temporary, made of plastic sheeting, opaque or transparent, over an arched frame, usually of metal or plastic pipe.

humus: The final carbonaceous residue of organic matter added to soil. Enhances water retention, assists uptake of nutrients by plant roots, and in other ways improves soil quality.

keet: The young of guineas.

Marek's disease: A highly contagious disease in chickens, involving six different herpesviruses that affect especially the nerves but may affect visceral organs, muscle, and skin as well. Though the associated viruses are almost universally present where chickens are raised, well-cared-for flocks with robust immune systems are unlikely to succumb.

Muscovy: A domesticated waterfowl, sometimes called "duck," but with a different wild ancestor, *Cairina moschata*, from that of true ducks.

nestbox: An enclosure provided for laying fowl to retire for egg laying.

nest trap: To use a trap nest to track egg production of individual hens.

ovary: In avian species the female reproductive organ that secretes hormones related to sexual function and produces ova, which start as single female sex cells and develop into yolks.

oviduct: The long convoluted tube in a female bird in which the egg is made and prepared for expulsion.

pancreas: A glandular organ that produces digestive enzymes involved mostly in digestion of proteins.

pasting up, pasty butt: A condition in which viscous feces expelled by a chick stick to the down around the vent, occluding it. It can be caused by chilly conditions in the brooder or poor feed. It's unlikely if brooder is well managed and natural feeds are given from the first day.

pinion: The third and final segment of the wing.

pipping: The breaking open of the eggshell from the inside by the chick, allowing it to hatch.

poult: The young of turkeys.

preening: Care of its feathers by a chicken or other fowl. The chicken expresses oil from its preening or uropygial gland with its beak and applies the oil to its feathers in order to clean, maintain, and partially or fully waterproof them.

primaries: The long, stiff outer feathers of the wing, growing from the pinion, also known as the flight feathers.

proventriculus: In avian species the glandular stomach, between the end of the esophagus and the gizzard, in which hydrochloric acid and digestive enzymes are added to the food the bird has ingested, and digestion begins.

pullet: A female chicken up to one year old.

rendering: Gently heating body fat from slaughtered ducks, geese, or chickens until it liquefies and can be saved as a cooking fat.

rooster: A euphemism for the cock (male chicken), as silly as it is prudish—female chickens *roost* as avidly as males!

saddle: The rear of the back in the cock, extending to the juncture of the back and the tail, covered with the long pointed saddle feathers that are a distinctive part of his plumage.

set, setting: Traditionally, a hen who has "gone broody" is not said to "sit" on her clutch of eggs, but to "set." She is referred to as a "setting hen," not a "sitting hen." That is the usage followed in this book.

sexing: The separation of hatchery chicks by gender, making possible the shipment of all-pullet or all-cockerel orders (as opposed to straight run orders). *Sexing* also refers to the visual inspection of the genitalia of waterfowl (ducks and geese) to determine gender.

shank: The part of the leg between the hock and the foot.

sickles: The long curved tail feathers of the cock, a distinctive part of his plumage.

soil food web: The complex set of living organisms, visible and microscopic, in the top layers of the soil profile, which break down organic residues on or in the soil and make them available as plant nutrients.

soldier grubs: The larval stage of the black soldier fly, which can be cultivated in a bin using organic "wastes" and given as high-protein feed to poultry.

spleen: A small organ beside the liver with important functions in regulating red blood cell supply and the immune system.

spur: A horny projection out of a cock's shank, above the rear toe, short and blunt in most farm breeds, long and sharp in games. Used by cocks as a weapon when fighting to assert dominance. Occasionally found in game hens.

straight run: Hatchery chicks can be sexed (separated by gender) in order to furnish orders of all pullets or all cockerels. Alternatively, the chicks can be sent *straight run*—that is, in the natural gender ratio, which as in humans is approximately half and half.

trap nest: A special nestbox with a door that drops into place when the hen enters, trapping her until released by the flockster, for tracking egg production.

uropygial gland: See *preening*.

vacuum-seal waterer: A waterer that automatically replenishes as the birds drink from it, consisting of a reservoir of water over a base with a narrow lip to prevent wading in the water. A hole in the base allows water to run out of the reservoir. When the hole is covered by the rising water level, a vacuum forms inside the reservoir, preventing additional flow until the hole is once again exposed as the birds drink. Metal and plastic versions are available, from a quart up to 7 gallons or more. You can make your own, so long as the placing of the hole in the base allows for the proper vacuum/release cycle.

vent: The anus of a fowl, through which feces—and in the case of a female, eggs—are expelled.

vermicomposting: The composting of organic "wastes" using earthworms, yielding castings (earthworm poop) as a soil fertility amendment, and potentially a harvest of worms to feed poultry, pigs, or farmed fish. Also called vermiculture.

wattle: The fleshy appendage hanging down from the upper throat of a chicken, varying in size by breed and gender, the male's being significantly larger. Like the comb, the wattle helps radiate heat to cool the body.

ENDNOTES

Chapter 1. Why Bother?

1. The poultry industry feeds its growing broilers a steady dose of antibiotics in order to make them grow faster, and generate more profits. But given the conditions in which the birds are raised, the daily dosage is actually a necessity: None of them would survive to reach slaughter size without them. To my mind feeding antibiotics implies the industry's admission up front that all their birds are sick, and have to be propped up by a diet of medicine long enough to make it to the consumer's plate. If your preference, like mine, is to eat healthy animals rather than sick ones, too bad—the industry is not going to cater to that preference anytime soon. Buy your chicken from someone you know and trust—or better yet, grow your own.
2. This industrial horror story first came to my attention in the form of the news article and video at www.huffingtonpost.com/2009/09/01/chicks-being-ground-up-al_n_273652.html. As you can see in the article, the poultry industry does not deny the "euthanizing" of live chicks in its hatcheries and indeed defends the practice as a necessary part of egg production.
3. The study was done in November 2009 and published in the January 2010 issue of *Consumer Reports*. See a summary of the report at www.consumerreports.org/cro/magazine-archive/2010/january/food/chicken-safety/overview/chicken-safety-ov.htm and the related editorial on the results at www.consumerreports.org/cro/magazine-archive/2010/january/viewpoint/overview/lax-rules-risky-food-ov.htm. A study at the Emerging Pathogens Institute at the University of Florida, published April 28 , 2011, found that salmonella is responsible for more foodborne illness than any other pathogen in the United States. As a food-pathogen combination, campylobacter in dressed poultry tops the list of food-related illnesses, sickening 600,000 citizens each year. Download the full report, "Ranking the Risks: The 10 Pathogen-Food Combinations With The Greatest Burden on Public Health" by Batz, Hoffmann, and Morris at http://www.epi.ufl.edu/?q=RankingTheRisks.
4. This trenchant summation is from Eric Schlosser, *Fast Food Nation* (Houghton Mifflin, 2001), page 197, with reference to equally high-speed—and filthy—processing of supermarket beef. Read *Fast Food Nation* as an excellent update for our day of Upton Sinclair's famous exposé of the meat industry, *The Jungle.*
5. The Union of Concerned Scientists estimates that about 70 percent of microbial drugs used in the United States are used on animals, not people. The Institute of Medicine estimated in 1998 that the resulting antibiotic resistance probably costs the nation as much as five billion dollars annually, a figure that is likely higher now. See "Getting Real About the High Price of Cheap Food" by Brian Walsh, August 21, 2009, at http://www.time.com/time/health/article/0,8599,1917458-1,00.html.
6. Roxarsone is the trade name for the organic arsenic compounds added to broiler feeds. (For reasons not fully understood, arsenic boosts rates of growth.) The industry emphasizes "organic" (it is inorganic forms of arsenic that are toxic) and argues that Roxarsone is biologically inert if present as residues in chicken on the consumer's plate—a point on which I remain skeptical. What is certain is that most of the Roxarsone fed broiler flocks—2 million pounds per year in the United States—is excreted; and broiler house litter is spread far and wide as fertilizer on croplands. Roxarsone is soluble; that is, highly mobile in the environment, leaching to surface and groundwater systems. Many environmental factors degrade Roxarsone to inorganic forms of arsenic—mostly arsenate (toxic), but some arsenite (highly toxic)—sunlight (seen any of that around?), bacteria, presence of nitrates (plenty of that in chicken litter), and especially anaerobic conditions, as found in big wet piles of litter, subsoils, and sediments

under rivers and lakes. The full extent of real-world conversion of Roxarsone to toxic arsenic species is unknown—just as we don't know the full long-term effects of broadscale release of antibiotics into the environment. You and I are the guinea pigs in a colossal scientific experiment to answer these questions—along with countless other species living in our agricultural pollutants downstream. David Kirby's study of CAFOs (confined animal feeding operations), *Animal Factory: The Looming Threat of Industrial Pig, Dairy, and Poultry Farms to Humans and the Environment* (St. Martin's Press, 2010), pages 380ff., includes a discussion of the feeding of Roxarsone to poultry and related issues of contamination of supermarket broilers with arsenic, possible links to cancer in humans, and increase of toxic arsenic species in the environment.

7. Gene Logsdon, *Holy Shit* (Chelsea Green Publishing, 2010) is an informative discussion of an important subject—indeed, so important that Logsdon gave his book the subtitle *Managing Manure to Save Mankind*—how can you pass it up? Written with Logsdon's characteristic wit.
8. The Cornucopia Institute (www.cornucopia.org) states its mission as: "Seeking economic justice for the family-scale farming community. Through research, advocacy, and economic development our goal is to empower farmers—partnered with consumers—in support of ecologically produced local, organic and authentic food." Rely on it to help sort out questions about organic labeling. See, for example, its action alert "Organic Egg Business Being Hijacked by Corporate Agribusinesses," posted to its website September 26, 2010 (www.cornucopia.org/2010/09/organic-egg-business-being-hijacked-by-corporate-agribusinesses-help-reverse-this-scandal). Particularly useful is its "Organic Egg Scorecard," a guide to eggs throughout the United States that truly deserve the organic label—and those whose use of it is nothing but a joke.
9. "Meet Real Free-Range Eggs," by Cheryl Long and Tabitha Alterman, was published in *Mother Earth News*, October–November 2007. It is now available at www.motherearthnews.com/Real-Food/2007-10-01/Tests-Reveal-Healthier-Eggs.aspx and is well worth reading. The article concludes by citing seven studies demonstrating the nutritional superiority of "the real thing" over the supermarket imitations, including *Mother Earth*'s report on samples of eggs from fourteen pastured poultry producers from around the country, prominent among them Joel Salatin's Polyface Farm (Virginia) and David Smith's Springfield Farm (Maryland). One of the studies cited also compared nutritional values of pastured and "conventional" broilers—which indicated as well that your family really does deserve the additional nutrition in the former.
10. Weston A. Price Foundation maintains a list of their chapter leaders by state at http://westonaprice.org/find-a-local-chapter. Most chapter leaders can help you find local food producers. Check out FarmFoody (http://farmfoody.org): A search dialogue on the home page allows you to key in a search for sources of locally produced foods within the radius you specify. Conservation groups, such as Piedmont Environmental Council in my area, are increasingly realizing that land and habitat conservation, environmental protection, and good farming go hand in hand. Many of them publish lists to help consumers find producers in their area. PEC's "Buy Fresh, Buy Local" list for the state of Virginia is a good example (www.buylocalvirginia.org/index.cfm).
11. From a post on American Pastured Poultry Producers Association's online discussion list, based on statistics pulled from a variety of sources by Matthew O'Hayer of Vital Farms in Austin, Texas (http://vitalfarms.com).
12. Michael Pollan demonstrates in *The Omnivore's Dilemma* (Penguin Press, 2006) that, while agricultural subsidies purport to benefit the farmer, in truth farmers are merely the funnel that directs those subsidies into the coffers of agribusiness corporations.
13. See www.trucost.com/news/100/putting-a-price-on-global-environmental-damage for an attempt to price spillover effects of the global economy, to which industrial agriculture contributes a large and ubiquitous share. Since it's hard for the mind to get any purchase on costs measured in trillions, it helps to know that it has been estimated that, when the environmental and health care costs are included, the price of a fast-food burger is actually around $200. I have not been able to find an equivalent for a fast-food chicken meal, but there is no reason to think it would not be the same. The estimation was in Raj Patel's *The Value of Nothing* (Picador, 2009), pages 44ff, citing Nancy Dunne, "Why a Hamburger Should Cost 200 Dollars—The Call for Prices to Reflect Ecological Factors," *Financial Times*, January 12, 1994.
14. As in many of our most heated debates, taking the extreme points of view on either side misses the complete picture—so too in the shouting match between "vegans" and "meat eaters." The best book I know of to sort out the confusion is Lierre Keith's *The Vegetarian Myth* (Flashpoint Press, 2009). Keith is as outraged as I am at the way animals are abused in our heartless, high-confinement meat production systems. But she argues—persuasively, to my mind—that eating animals and their products is necessary (in her

sad experience) to optimum human health, and that including livestock animals in farming is the key to a regenerative, sustainable agriculture.

15. Particularly valuable analyses of the coming decline in global oil production come from insiders such as oil geologist Kenneth Deffeyes (*Hubbert's Peak: The Impending World Oil Shortage*, Princeton University Press, 2008) and oil investment banker Matthew Simmons (*Twilight in the Desert: The Coming Saudi Oil Shock and the World Economy*, Wiley, 2005). Overviews of the economic, political, and social consequences of peak oil include those of Richard Heinberg (*The Post Carbon Reader: Managing the 21st Century's Sustainability Crises*; *The Party's Over: Oil, War and the Fate of Industrial Societies*; *Peak Everything: Waking Up to the Century of Declines*; and *Powerdown: Options and Actions for a Post-Carbon World*) and James Howard Kuntsler's *The Long Emergency: Surviving the End of Oil, Climate Change, and Other Converging Catastrophes of the Twenty-first Century* (Grove Press, 2006).
16. There are several real-time versions of the US debt online, among them www.brillig.com/debt_clock and www.usdebtclock.org. According to the latter, the current national debt per taxpayer is almost $129,000.

Chapter 2. The *Integrated* Small-Scale Flock

1. Sir Albert Howard, *The Soil and Health* (originally published in 1947, reprinted in 2006 by University of Kentucky Press, with an introduction by Wendell Berry), page 11.

Chapter 3. Your Basic Bird

1. See "Farm Chickens' DNA Traced Back to Red Jungle Fowl" by Karen Kaplan (*Los Angeles Times*, March 13, 2010, now at http://articles.latimes.com/2010/mar/13/science/la-sci-chickens13-2010mar13) for a brief overview of this research. The paper on which it was based, "Whole-Genome Resequencing Reveals Loci Under Selection During Chicken Domestication," was published in the online version of *Nature*, March 10, 2010, at http://www.nature.com/nature/journal/v464/n7288/full/nature08832.html.
2. See the discussion by Page Smith and Charles Daniel of the origin of *rooster* and the symbolic role of the cock in various periods and cultures, in their delightful *The Chicken Book* (North Point Press, 1982), pages 51ff. H. L. Mencken discussed the euphemizing compulsion in his *The American Language* (Alfred A. Knopf, second edition 1921), page 149 and elsewhere. Even words with the slightest hint of sexuality were subject to censorship, a bull, for example, being referred to as a *gentleman cow*!
3. Studies of vocalizations of Red Junglefowl have found them to be largely the same as in domestic chickens, with at least two dozen distinct calls, which seem to combine into more complex signals coded to environmental and behavioral context. See, for example, "The Vocal Repertoire of the Red Junglefowl: A Spectrographic Classification and the Code of Communication," by Nicholas E. Collias in *The Condor* (89: 51524), journal of the Cooper Ornithological Society, available at www.jstor.org/pss/1368641.
4. Temple Grandin and Catherine Johnson, *Animals in Translation* (Scribner, 2005), in the section "Rapist Roosters," pages 70ff.

Chapter 4. Planning the Flock

1. You will often see references to this breed as "Araucanas." Do not be deceived. The true Araucana is a small rumpless breed (lacking the protuberance at the end of the spine that supports the tail feathers of most breeds) originating in Chile (bred by the Araucana tribe there). The novelty of its pastel-tinted eggs was its main appeal when it was introduced to the United States in the 1930s. Unfortunately, the gene for rumplessness in this breed carries a lethality factor: When two fowl with the gene mate, one quarter of the resulting embryos die in the shell. Since the gene for the color of the eggshells is dominant, it was easy to retain it in various crosses between the true Araucana and other breeds. These crosses were standardized in the 1970s into the Ameraucana—larger than the Araucana and having a rump, making it easier to breed (because of the absence of the lethality factor). True Araucanas are extremely rare and difficult to find, but the misnomer *Araucana* for *Ameraucana* is extremely common. I do not know of any commercial hatchery that offers Araucanas. I carefully questioned a representative of one hatchery advertising them who insisted that their stock really was the true Araucana—even referring to the distinctive Araucana characteristics: ear tufts and rumplessness. The chicks I gullibly ordered turned out to have muffs and beards (in lieu of ear tufts) and rumps—that is, they were in fact Ameraucanas. If you want to raise true Araucanas, breeders are listed in the Breeders Directory of the Society for the Preservation of Poultry Antiquities (see appendix H). Actually, Ameraucanas are likely to be a better fit in the productive home flock, but let's refer to them with their proper name, shall we?
2. To help in your search, check http://feathersite.com for the most complete set of pictures I know of hundreds of breeds of all sorts of poultry—not only chickens, but ducks, geese, turkeys, pheasants, peafowl, swans, and more. Another good online resource is the Henderson chart at www.ithaca.edu/staff/jhenderson/chooks

/chooks.html, a compendium of characteristics of dozens of chicken breeds.

3. The catalog for Glenn Drowns's Sand Hill Preservation Center has a wealth of information about traditional breeds of chickens, ducks, geese, and turkeys. See www.sandhillpreservation.com/pages/poultry_catalog.html or send for a print catalog from 1878 230th Street, Calamus, IA 52729. Other extensive lists of poultry breeds, their history, characteristics, and productive uses, are found at www.poultrypages.com/poultry-breeds.html and www.ansi.okstate.edu/breeds/poultry. The latter has many excellent color photographs and illustrations.
4. The American Livestock Breeds Conservancy (ALBC) Conservation Priority Breeds List can be found here, in full, including information on other animals besides poultry: www.albc-usa.org/cpl/wtchlist.html#chickens.
5. Do remember if you want to try "Freedom Rangers" that the term refers to a type of colored broiler bred for better production on range, not a traditional breed or even a single strain. The source I've used, with good results, is J. M. Hatchery in Pennsylvania: www.jmhatchery.com.
6. For a more extensive discussion of the shortcomings of the Cornish Cross and alternative meat-bird choices, see "The Cornish Cross: What is wrong with this picture?!" at http://themodernhomestead.us/article/Cornish-Cross.html and "Sunday Dinner Chicken: Alternatives to the Cornish Cross" at http://themodernhomestead.us/article/Cornish+Cross+Alternatives.html.

Chapter 5. Starting the Flock

1. For more on this issue of "euthanizing" excess chicks, see "Moral Puzzles in the Backyard" at http://themodernhomestead.us/article/Moral+Puzzles.html.
2. Ideal posts this advisory at the top of their online order page at https://secuservices.com/ideal/newideal/checkout1.aspx.
3. The National Poultry Improvement Plan (NPIP) was developed in 1935 to eradicate pullorum disease and fowl typhoid from the nation's flocks. Though participation is voluntary, over 95 percent of the US breeding and hatchery industry participates in the program, which is cooperatively administered by the US Department of Agriculture, state agricultural agencies, and the poultry industries. Testing under the program to ensure freedom from pullorum and fowl typhoid in hatchery breeding stock has virtually eliminated those diseases. Other NPIP disease-control programs now include *Salmonella* enteritidis, several mycoplasma diseases, and avian influenza. More information is available from the USDA's Animal and Plant Health Inspection Service at www.aphis.usda.gov/animal_health/animal_dis_spec/poultry.
4. Quoted from "Built-Up Floor Litter Sanitation and Nutrition," Kennard and Chamberlin, 1949. The entire article appears within the body of an excellent discussion of deep litter at www.plamondon.com/faq_deep_litter.html.

Chapter 6. Housing

1. Judy Pangman's *Chicken Coops: 45 Building Plans for Housing Your Flock* (Storey Publishing, 2006) has blueprints and lots of ideas for coops for many flock sizes and management situations.
2. See *Fresh-Air Poultry Houses: The Classic Guide to Open-Front Chicken Coops for Healthier Poultry* (Norton Creek Press, 2008), a republication of Prince T. Woods's *Modern Fresh-Air Poultry Houses*, originally published in 1924.
3. That is, I tied loops around each shank, loose enough to avoid binding the shank but small enough to prevent slipping off. Then I tied together the two loops. The resulting "hobble" allowed the guinea to walk and forage freely but prohibited chasing after the chicken cocks. I was not pleased with this clumsy solution, but there was no pain for the guineas, and it seemed the only alternative to maintaining a separate flock of only a few guineas.
4. I've seen different recommendations about the waiting period to observe between application of raw manure on soil—deposition by chickens of their droppings in a garden space is such an application—and harvest of vegetable crops, ranging from 60 to 120 days. The National Organic Standards specify a waiting time of 90 days for crops that are not in contact with soil like corn, and 120 days for crops that are, such as lettuces and carrots. Frankly, I doubt there is a threat of contamination from a well-managed homestead flock, and I have never observed a measured waiting time. After removing chickens from garden beds, I simply plant and get on with the season.

Chapter 7. Manure Management in the Poultry House

1. The entire article appears within the body of an excellent discussion of deep litter at www.plamondon.com/faq_deep_litter.html.
2. From "Animal Feed Resources Information System: Manure," at www.fao.org/ag/aga/agap/frg/AFRIS/Data/476.htm.
3. Salatin discusses a "loose housing option" for a layer flock in *Pastured Poultry Profit$* (Polyface, Inc., 1993,

now distributed by Chelsea Green Publishing), pages 259–261, in which he recommends a "target minimum space allotment" of 5 square feet per hen. His observations about stocking density in a flock on deep litter in relation to capping are from several of his presentations I have attended.

Chapter 9. Pasturing the Flock

1. See more pictures of this Cadillac of eggmobiles, together with extensive notes on its construction and use, in a free PDF download at www.windyridgepoultry.com/docs1/WRP_photo_book1.pdf.
2. Joel Salatin has inspired countless producers of pastured broilers with his now famous 10-by-12-foot mobile pens, the design and management of which are described in *Pastured Poultry Profit$*, pages 63ff. Some of my fellow members of American Pastured Poultry Producers Association, which Joel helped found, have switched to electric net fencing to manage broiler flocks. Some, however, continue raising them in Polyface-style mobile pens or modifications thereof.

Chapter 10. Managing the Pastured Flock Using Electronet

1. My preferred source for electric fencing and accessories is Premier: www.premier1supplies.com; 800-282-6631. Everyone on the staff at Premier actually uses electric fencing to manage their own personal livestock. Thus any product sent their way gets rigorous real-world testing before they offer it for sale. Service has always been helpful and courteous, and technical assistance outstanding. I have friends who swear by Kencove as well: www.kencove.com/fence; 800-536-2683.
2. For more details about using electric net fencing and lots more pictures, see "Managing Poultry on Pasture with Electronet" at http://themodernhomestead.us/article/Electronet-1.html.

Chapter 11. Mobile Shelters

1. Find much more information on mobile pasture shelters and "chicken tractors" for the garden in the "Pasturing the Flock" section of my website at http://themodernhomestead.us/article/Pasturing+the+Flock.html. At this writing, there are six separate links offered there: "Designing a Pasture Shelter," "Building a Pasture Shelter," "Going Mobile at the Small End of the Scale," "Mobile Shelter: The Classic Polyface Model," "Chicken Tractor: A Tribute to Andy Lee," and "Pasture Shelters for Market Layer Flocks."
2. Indeed, I avoid plastics just as much as possible since reading "Polymers Are Forever" in the May–June 2007 issue of *Orion* magazine, now available at www.orionmagazine.org/index.php/articles/article/270, an abridged excerpt from Alan Weisman's *The World Without Us* (St. Martin's Press, 2007, pages 112ff). Weisman's book is an interesting thought experiment, well worth reading.
3. In this case I was using a material that actually removes plastic from the accumulating waste stream referred to in the previous endnote.
4. Twenty-four-mil poly sheeting should be available from any greenhouse supply—in 100-foot rolls. The only supplier I know of that will custom-cut to your order is Northern Greenhouse Sales (www.northerngreenhouse.com/index.htm).

Chapter 12. Putting the Flock to Work

1. Quoted from "Conservation and Local Economy," one of eight essays in Wendell Berry's *Sex, Economy, Freedom & Community* (Pantheon Books, 1993), pages 13–14.
2. An excellent short overview of the soil food web is *Soil Biology Primer* (Soil and Water Conservation Society, revised 2000), written mostly by Elaine Ingham, one of the premier soil scientists of our time. Two guides through the staggering complexity of soil life and fertility cycles, readily accessible to the layman, are *Life in the Soil* by James Nardi (University of Chicago Press, 2007) and *Teaming with Microbes* by Jeff Lowenfels and Wayne Lewis (Timber Press, 2010).
3. Logsdon, *Holy Shit*.
4. I hope you'll find "The Joys of Cover Cropping" a useful introduction to the role of cover crops and strategies for their use in a soil care program: http://themodernhomestead.us/article/Cover+Cropping+Part+One.html.
5. Sir Albert Howard, *An Agricultural Testament* (originally published in 1940, republished in 2010 by Oxford City Press), pages 47ff. There is a wealth of information from Rodale, a complex of publishing and experimental efforts dedicated to organic gardening and farming, founded by J. I. Rodale, who was inspired by Howard's work. Visit www.rodale.com and key "compost" into the search field to find as much information as you can use.
6. Since that figure may strain credulity, please understand that it is a conservative estimate, and follow along with the calculation if you like. We laid down a 4-inch layer (0.111 yard) of chicken-powered compost on half the beds in that garden (beds 40 inches wide or 1.11 yard). The combined length of all eight beds was 77 ft. Thus the combined volume of compost applied (0.111 x 1.11 x 77 = 9.49) was about 9½ cubic yards. Since a cubic yard of a manure-based compost should weigh about 1,350 to 1,400 pounds, a ton (2,000 pounds) is about

1½ cubic yards. Our 9½ cubic yards therefore weighed more than 6 tons (9.49 cubic yards divided by 1.5 cubic yards = 6.33).

7. Figures like this can be illustrative only, given the enormous numbers involved. I based them on a report from the US Environmental Protection Agency, "Pesticide Industry Sales and Usage: 2000 and 2001 Market Estimates," available at www.epa.gov/pesticides/pestsales/01pestsales/market_estimates2001.pdf.
8. The flock is not the only ally we have in dealing with crop-damaging insects. Other friends include *more insects*. Doing all we can to encourage maximum insect diversity in the backyard ecology is one of the best ways to get off the pesticide treadmill. For an introduction to the basic concepts, see "Dealing with the Competition—Insect Threats and Insect Diversity" at http://themodernhomestead.us/article/(IH)+Insects.html and "Insects in the Garden: Toward a Balanced Insect Ecology"at http://themodernhomestead.us/article/Garden+Insects.html.
9. Carol Deppe, *The Resilient Gardener: Food Production and Self-Reliance in Uncertain Times* (Chelsea Green Publishing, 2010), pages 191ff.
10. Two excellent brief introductions are "Using Weeder Geese" by Metzer Farms (one of the major waterfowl sources in the United States), www.metzerfarms.com/UsingWeederGeese.cfm; and "Weeding with Geese" by University of Missouri Extension, http://extension.missouri.edu/publications/DisplayPub.aspx?P=G8922#crops.
11. References to both Logsdon's chicken-groomed pig pen and the aquaculture ponds are from *Holy Shit*, pages 106 and 112.

Chapter 13. Chickens in the Garden

1. If you do want to try electric fencing of any sort to deter deer, here's a mean little trick to make it more effective: Bait the fence by folding squares of aluminum foil over the charged wires and smearing with peanut butter. If you're dealing with dogs instead, bait by hanging strips of raw bacon over charged wires. If you can tempt contact by nose or tongue, you'll make a strong impression.
2. I planted two vining bean species, both to create shade and to support insect diversity when they flowered: scarlet runner bean (*Phaseolus coccineus*) and hyacinth bean (*Lablab purpureus*). Next year I will grow only the latter, which was much faster growing and made a much denser shade than the scarlet runner.
3. Again, the specific calculations behind this estimate: We applied a minimum of 1½ inches (probably more like 2 inches) of compost over the greenhouse's two growing areas, each measuring 8 feet by 40. Converting to yards (1.5 inch = 0.042 yd; 8 feet = 2.67 yards; 40 feet = 13.33 yards), we applied about a ton (1½ cubic yards) of compost to each growing area (0.042 x 2.67 x 13.33 = 1.49), for a total application in the greenhouse of about 2 tons.

Chapter 14. A Question of Balance

1. An excellent overview of soil fertility management is *Building Soils for Better Crops*, by Fred Magdoff and Harold van Es (Sustainable Agriculture Publications, second edition 2000). Though the book is targeted at farmers, I highly recommend it as well to homesteaders seeking a deeper understanding of the complexities of soil and crop management.
2. Deppe, *The Resilient Gardener*, pages 188ff.
3. The woodlot area referred to is about a third of an acre. It was heavily grown up in *Ailanthus altissima* ("Tree of Heaven," "Heavenwood"), an assertive invasive from China we unfortunates in the mid-Atlantic have to deal with. I cut down the ailanthus and am trying to establish a new canopy—black walnut, hickory, tulip poplar, and sycamore.
4. My friend Don Schrider read somewhere that, traditionally, the Chinese especially prized the flesh of chickens who had foraged a lot of grubs in rotting wood. If that is true, I'm looking forward to eating cockerels who have been ranging on that field of chips next year!
5. To learn more about an integrating homestead idea with endless possibilities, see Dave Jacke's *Edible Forest Gardens* in two volumes, *Vision & Theory* and *Design & Practice* (Chelsea Green Publishing, 2005). The extensive appendixes are alone worth the (considerable) price of the books.

Chapter 15. Thoughts on Feeding

1. An excellent compendium about feeding chickens naturally is "Pastured-Raised Poultry Nutrition," by Jeff Mattocks of Fertrell Company, originally written for Heifer International in 2002, now widely available as a free download in PDF format. One link is http://attra.ncat.org/attra-pub/PDF/chnutritionhpinew.pdf. (Key the article title and author name into a search engine and you will find many more.)
2. I ordered from "Uncle Blaine" at www.sparrowtraps.net. Extensive testing in its first season convinced me this is an excellent design.
3. The Wikipedia article on atrazine summarizes questions about its dangers and gives links to dozens of scientific papers on the subject: http://en.wikipedia.org/wiki/Atrazine. See a summary of the USGS's findings on water contamination in "Atrazine: Poisoning the

Well—Atrazine Continues to Contaminate Surface Water and Drinking Water in the United States" at www.nrdc.org/health/atrazine.

Chapter 16. Purchased Feeds

1. Mattocks, "Pastured-Raised Poultry Nutrition," pages 9–10.

Chapter 17. Making Our Own Feeds

1. Actually the situation is even worse than that: If you make your own prepared feeds you will still, as a taxpayer, be subsidizing the "cheap" chicken and "cheap" eggs supplied in the supermarket, in effect making a down payment with your tax dollars on industrial foods you're not willing to feed your family. That fact about your ultimate feed costs should get you really upset.
2. I bought my grain mill from Lehman's, their Item #2360, available at www.lehmans.com/store/Kitchen___Grain_and_Grain_Mills___High_Speed_Grain_Mill___High_Speed_Grain_Mill#2360, priced as I write at about $600. It's made by C. S. Bell Co. of Tiffin, Ohio. I have been using it for well over ten years now and recommend it highly. Note that I bought the hand crank on offer for the mill as well, for use in a prolonged loss of electric power.
3. You'll waste a lot of money if you buy diatomaceous earth in small packets. The 50-pound bag I buy from my grain supplier for $25 lasts for years. If your feed supplier doesn't stock it, they can probably special-order it.
4. Countryside Organics, www.countrysideorganics.com; 888-699-7088.
5. Fertrell (www.fertrell.com) offers organic fertilizers and livestock supplements. Jeff Mattocks—vice president and nutritionist and fellow member of American Pastured Poultry Producers Association—is a gold mine of information on poultry feeding and health. I have cited his excellent "Pastured-Raised Poultry Nutrition" above.
6. If you're interested in exploring this subject from the standpoint of human diet and health, see Weston A. Price Foundation's "Soy Alert!" ("Everything you need to know about why you should avoid modern unfermented soy foods") to get you started—www.westonaprice.org/soy-alert.html. As for agricultural and environmental effects of the combination of GMO soybeans and the herbicide glyphosate, see the interview with Don Huber, a plant pathologist for fifty years and professor emeritus at Purdue University, in the May 2011 issue of *Acres USA*, pages 50–58. A letter from Dr. Huber to Secretary of Agriculture Vilsack dated January 16, 2011, advised Secretary Vilsack that a hitherto unknown pathogen seems to be encouraged by the glyphosate used to grow all GMO soy in the United States—which is to say, almost all the soy grown here. The pathogen affects both plants and livestock animals including poultry, through the feeds made with glyphosate-treated soybeans and corn. Dr. Huber, a sober scientist not given to extreme statements, warned the secretary that demonstrated and potential problems associated with the new pathogen are "widespread, very serious," and that they deserve "immediate attention with significant resources to avoid a general collapse of our critical agricultural infrastructure."
7. "Hexane is a byproduct of gasoline refining. It is a neurotoxin and a hazardous air pollutant. Soybean processors use it as a solvent—a cheap and efficient way of extracting oil from soybeans, a necessary step to making most conventional soy oil and protein ingredients. Whole soybeans are literally bathed in hexane to separate the soybeans' oil from protein." Quoted from the Cornucopia Institute (www.cornucopia.org/2010/11/hexane-soy), which notes as well—not surprisingly—that hexane is prohibited in the processing of organic foods.

Chapter 18. Feeding the Flock from Home Resources

1. There is a good deal of confusion about comfrey types and terminology. Since there are indeed some comfrey (*Symphytum*) species that can become problematic, it is important to find one developed for garden use, which will not spread by seeds or runners. Probably the best bet is one of the Bocking clones. Bocking #4 and #14 are available from Richters Herbs in Canada: www.richters.com; 357 Highway 47, Goodwood, ON L0C 1A0 Canada; 800-668-4372 or 905-640-6677. The best (possibly the only) source for garden comfrey clones in the United States is Nantahala Farm & Garden in western North Carolina, which offers Bocking #4 at excellent prices: www.nantahala-farm.com; 828-321-4913; coescomfrey@yahoo.com. Note that many official sources advise against feeding comfrey to livestock. Despite centuries of use as a fodder plant and even as a human food and medicinal, comfrey contains pyrrolizidine alkaloids. Lab rats fed these alkaloids with massive chemically pure doses have developed liver disease. You will have to do your own research and draw your own conclusions. My conclusion is that there is no problem using whole comfrey leaf as fodder for livestock. See Lawrence D. Hills's *Comfrey Report*, and key "pyrrolizidine alkaloids" into an online search engine. See also the discussion of the

subject in an appendix to Lawrence Hills's *Comfrey: Fodder, Food & Remedy* (Universe Books, 1976), pages 229–237.

2. Examples of toxic plants are castor bean (*Ricinus communis*), milkweed (*Asclepias* spp.), immature berries of nightshade (*Solanum nigrum*), oleander (*Nerium oleander*), jimsonweed (*Datura stramonium*), poke-berries (*Phytolacca americana*), and yew (*Taxus* spp.). Such plants are rarely a real threat to chickens, however—they avoid them instinctively. Note that in some cases it is not so much the plant that's the problem as its mature seeds. For example, I no longer use hairy vetch (*Vicia villosa*) as a cover crop here, since their seeds are poisonous. Chickens are unlikely to eat jimsonweed but may eat the seeds. Seeds of both plants have been implicated in actual poisonings of poultry.
3. Deppe, *The Resilient Gardener*, pages 188ff.
4. The more commonly planted is Japanese buckwheat (*Fagopyrum esculentum*), the instant cover crop—thirty days from seed to bloom in my garden. Though sensitive to frost, it's a great weed suppressor, even in hot and dry parts of the growing season. In addition, it produces nutritious seeds the chickens can self-harvest. Tartary buckwheat (*F. tartaricum*) is somewhat more tolerant of cold weather and produces seeds higher in protein.
5. J. Russell Smith, *Tree Crops* (Devin-Adair Company, 1950), page 186.
6. See Kate Hunter's blog entry describing her use of acorns to feed her small suburban flock at http://livingthefrugallife.blogspot.com/2010/10/acorns-as-chicken-feed-revisited.html.
7. A good overview is R. A. Leng, J. H. Stambolie, and R. Bell, "Duckweed—A Potential High-Protein Feed Resource for Domestic Animals and Fish," available at www.lrrd.org/lrrd7/1/3.htm, from *Livestock Research for Rural Development* 7, no. 1 (October 1995); www.lrrd.org.
8. A couple of good sources to get you started are The Free Library's "Small-Scale Silage: Your Yard Can Be Your Field" at www.thefreelibrary.com/Small-scale+silage+your+yard+can+be+your+field.+%28Feeds+%26+feeding%29.-a083553803 and Smallstock in Development's "Little Bag Silage" at www.smallstock.info/tools/feed/silage/lbs02.htm. The problem I have with the methods used in both is the generation of large amounts of plastic waste. I have begun experimenting with silage making in one of my emptied vermicomposting bins, filling it with fresh cut grass and stomping it down to exclude oxygen.
9. For more information on goji, see www.phoenixtearsnursery.com/information.html.

Chapter 19. Cultivating Recomposers for Poultry Feed

1. Quoted from *New Scientist* in "The Myth of the Rabid Locavore," by Kerry Trueman; posted August 20, 2010, to www.huffingtonpost.com/kerry-trueman/the-myth-of-the-rabid-loc_b_689591.html.
2. Mary Appelhof, *Worms Eat My Garbage* (Flower Press, 1982, revised 1997).
3. The description here, of filling the bins with horse manure, is from past practice. Since getting the results of soil tests, related in chapter 14, I am no longer importing pickup loads of horse manure. I am experimenting with other bedding and feeding options, of which there are many. I am as well experimenting with "cooking" feeds for the worms along the lines of "A Solar Cooker for More Efficient Feeds" (in the following section "An Alliance with the Soldier") to increase feeding efficiency and speed up growth of the population in the bins.
4. For more information and pictures, see "The Boxwood Vermicomposting System" at http://themodernhomestead.us/article/Boxwood+Vermicomposting.html.
5. George Sheffield Oliver, "My Grandfather's Earthworm Farm." Available online in its entirety at http://www.journeytoforever.org/farm_library/oliver/oliver_farm.html—and as chapter 6 of Thomas J. Barrett's *Harnessing the Earthworm* (Bookworm Publishing, 1976).
6. See www.journeytoforever.org/farm_poultry.html#flies. Journey to Forever is an interesting site with a lot of creative homesteading ideas—I recommend spending some time exploring it.
7. See the section "Putting the Bluebottle Fly to Work" at www.journeytoforever.org/farm_library/oliver/oliver2b.html.
8. Ibid.
9. Aside from articles in technical and academic journals, most useful material about this species is online. The following two links have lots of solid information, and an abundance of links to information elsewhere: www.thebiopod.com (the website of ESR International, LLC, the company that designed and sells the BioPod for soldier grub composting) and http://blacksoldierflyblog.com (a blog maintained by "Jerry aka GW," who has been cultivating soldier grubs to feed pond fish for several years).
10. BioPod is a trademark of ESR International, LLC. This high-density molded plastic unit is manufactured in Vietnam, using a design by Paul Olivier—who has written a good deal about bioconversion of "wastes" using recomposers like soldier grubs—along with his

son Robert, who helps run ESR International and sells the BioPod (www.thebiopod.com).

11. Solar cookers can be simple or complex. Here are a few sources of information about design if you want to give them a try: www.cookwiththesun.com/wood-box.html; www.cookwiththesun.com/solar.htm; http://solarcooking.wikia.com/wiki/Principles_of_Solar_Box_Cooker_Design; and especially http://solarcooking.org/bkerr/DoItYouself.htm.
12. A well-documented description of this do-it-yourself project is at http://blacksoldierflyblog.com/bsf-bucket-composter-version-2-1, which includes fifty detailed photographs and four instructional videos.
13. Though it probably needs updating to reflect climate change, the USDA's zone hardiness map is available at www.usna.usda.gov/Hardzone/hzm-ne1.html.
14. For a more extensive introduction to the cultivation of soldier grubs with additional pictures, see "Black Soldier Fly, White Magic" at http://themodernhomestead.us/article/Black+Soldier+Fly.html.

Chapter 20. One Big Happy Family

1. My friend Don Schrider notes that in his experience it is more likely to be a toenail than the spur that injures the hen. The problem may emerge more in some breeds than in others: When he was working with Buckeyes, Don trimmed the toenails as well as the spurs on the cocks to prevent injury. In his current work with Light and Dark Brown Leghorns, he trims spurs and toenails if he notices the backs of hens getting bald.
2. If I'm having problems with hawks, I will keep chicks in a movable shelter for a while, until they've grown enough to be less vulnerable. Remember to monitor as well to make sure these still lightly feathered young birds do not get wet in either a light rain or a heavy dew.

Chapter 21. Protecting the Flock from Predators

1. Be aware of laws that protect some predators. Raptors in particular are protected, and killing them may bring severe penalties. I endorse laws supporting raptor populations—I love the hunting birds for their magnificence, and for their indispensable ecological services. But Joel Salatin pointed out in *Everything I Want to Do Is Illegal* that, while the draconian laws protecting raptors were necessary, it's time to recognize that they succeeded—most populations have rebounded robustly. Consequently, today it really wouldn't mean the potential extinction of a species if a farmer takes out the occasional "rogue predator" who keeps coming back. Responses involving light-gauge shotguns come to mind or #1 traps, either concealed in a shallow covered pit or fastened to a perch at the edge of the pasture. Raptors frequently return to a half-eaten carcass—an easy meal in contrast with initiating a new hunt—which can be used as bait in such strategies. Be aware as well that devotees of falconry or hawking (the sport of using live raptors to hunt game) may have permits to live-trap individual raptors for training. If you have a problem with a persistent raptor, try to find a hawking enthusiast in your area willing to set a live trap for your rogue.

Chapter 22. Helping the Flock Stay Healthy

1. I should add a major caveat: I have had dreadful luck trying to brood turkey poults (see "Turkeys" in chapter 24), though started turkeys who had "graduated" from my buddy Mike Focazio's brooder have been easy for me to grow to superstar status on the Thanksgiving table. It's a common observation: Turkey poults are pretty fragile, and losing distressing numbers of them can happen with little clear idea why they're dropping like flies—then, past the brooder phase, they're tough as nails. In any case, I'm not the guy to advise you about health issues encountered when brooding turkey poults.
2. It is common for beginning flocksters to want to take a sick chicken to the veterinarian. If you want to do so, by all means give the vet a try. But in my experience local veterinarians have no interest and little expertise in dealing with chickens. The experts are likely to be poultry scientists in your state's extension office or ag schools, but they are more interested in commercial flocks. If it seems that your flock has a general infection, as opposed to the rare unexplainable death, by all means seek their help, through your local extension agent.
3. Gail Damerow, *The Chicken Health Handbook* (Storey Publishing, 1994), page 207.
4. Damerow's *Chicken Health Handbook*, ibid., is an excellent overview of chicken health and health problems and contains a lot of useful information about treating birds who have become ill. See as well "Remedies for Health Problems of the Organic Laying Flock: A Compendium and Workbook of Management, Nutritional, Herbal, and Homeopathic Remedies," compiled and edited by Karma E. Glos and available for free download at www.kingbirdfarm.com/Layerhealthcompendium.pdf.
5. You don't have to worry about the precise recipe. It's actually sufficient just to coat the shanks liberally with Vaseline or mineral oil, thereby smothering the mites—the strongly antimicrobial qualities of the tea tree oil and/or oil of oregano just make for a stronger assault on the entrenched critters. Make up a thick oily goo of

the mineral oil and Vaseline and add enough of one or both oils—a dropperful or two—to strongly scent it, and your mix will be completely effective. These are not difficult mites to kill.

Chapter 24. Other Domestic Fowl

1. For more on raising waterfowl, see "The Homestead Waterfowl Flock" at http://themodernhomestead.us/article/Waterfowl-1.html.
2. Both Dave Holderread, *The Book of Geese* (Hen House Publications, 1981), and Chris Ashton, *Domestic Geese* (Crowood Press, 1999), describe vent sexing waterfowl—both of ducklings and goslings, and of adult ducks and geese.
3. See "My Long Goose Breeding Saga," published in the December 2010–January 2011 issue of *Backyard Poultry*, now available at http://themodernhomestead.us/article/Goose+Breeding.html.
4. See "The Silver Appleyard: A Great All-Round Duck," published in the August–September 2010 issue of *Backyard Poultry*, now available at http://themodernhomestead.us/article/Silver+Appleyard.html.

Chapter 25. Breeding for Conservation and Breed Improvement

1. I kept Newcomer-strain New Hampshires for years and can recommend them as a sturdy and reliable dual-purpose breed, hens of which are good winter layers and sometimes good mothers. I have just started seventy-five Newcomer-strain New Hampshires in the brooder and plan to begin breeding them again next year. I ordered from Estes Hatchery in Springfield, Missouri (www.esteshatchery.com). Cackle Hatchery of Lebanon, Missouri, also offers Newcomer-strain New Hampshires (www.cacklehatchery.com).
2. The American Livestock Breeds Conservancy (www.albc-usa.org; PO Box 477, Pittsboro, NC 27312; 919-542-5704) bases its work on the proposition that breed conservation depends on the restoration of the economic qualities (productivity) of historic and traditional breeds in danger of extinction. Descriptions of fifty chicken breeds, both utilitarian and exotic, along with their conservation status, are at www.albc-usa.org/cpl/wtchlist.html#chickens. The site has similar information for ducks, geese, and turkeys.
3. As a matter of fact, historic breeds such as Old English Games were used to develop modern breeds. It would be foolish to assume that we only need to keep breeds of immediate use in today's circumstances. Who can say when we will need the widest range of poultry genetics to deal with new challenges and changed circumstances? An example from plant crops illustrates that this is not merely a theoretical concern: By 1970 the hybrid corns that made up the bulk of the American corn crop came from an increasingly narrow genetic base. That year a devastating epidemic of southern corn leaf blight wiped out 15 percent of the US corn crop—50 to 100 percent in some areas—resulting in at least a billion-dollar loss to the agricultural economy. The solution was simple: the breeding in of resistant genes from older varieties of corn. But imagine the dimensions of the catastrophe if no one had taken the trouble to conserve those earlier varieties.
4. If you're interested in working with Will Morrow's Delawares, you can get some of his stock (chicks or hatching eggs) from his Whitmore Farm: www.whitmorefarm.com/store/212; 10720 Dern Road, Emmitsburg, MD 21727. If you are interested in working with the restored Buckeye, I advise contacting the ALBC at www.albc-usa.org (the contact page is www.albc-usa.org/tellus.php) to get in touch with breeders working with the Buckeyes who came out of the ALBC's improvement breeding project.
5. Is toe punching painful for the chick? I can only assume that it is—which might argue for another option such as plastic spiralettes. Not so fast. It would be silly to focus solely on this question of causing pain, and ignore the undeniable fact that the handling of the chick is itself stressful. If you opt for the spiralettes, you will have to change them several times as the bird grows—meaning that, over the course of its growth to full size, multiple handlings will result in greater total stress for it.
6. For more on hand selection for egg production traits, see Don Schrider's "Selecting for Egg Production" at www.albc-usa.org/documents/ALBCchicken_assessment-2.pdf.
7. Jeannette Beranger and Don Schrider detail "Selecting for Meat Qualities and Rate of Growth" at www.albc-usa.org/documents/ALBCchicken_assessment-1.pdf.
8. I have experimented with crossing Old English Game cocks onto proven broodies of larger-bodied breeds. The daughter offspring almost always express the broody instinct, and since they're larger, they can incubate more eggs per clutch, increasing hatching efficiency. It would be possible to continue selecting from the initial crosses to generate not only new strains but even a new breed of hardworking broodies. In such a breeding project, a high level of broody skills should always trump body size.
9. There is a wealth of information in the "Educational Resources" section of the American Livestock Breeds Conservancy (ALBC) site (www.albc-usa.org

/EducationalResources/downloads.html) that will be helpful in a breeding project, including record-keeping forms, lists of hatcheries and private breeders offering heritage breed stock for sale, and other resources. The excellent compendium on breeding—"Chicken Assessment for Improving Productivity"—was written by Don Schrider and is available in three parts in PDF format: "Selecting for Meat Qualities and Rate of Growth," coauthored with Jeannette Beranger (www.albc-usa.org/documents/ALBCchicken_assessment-1.pdf); "Selecting for Egg Production" and "Ongoing Selection of Breeding Stock" (www.albc-usa.org/documents/ALBCchicken_assessment-2.pdf and www.albc-usa.org/documents/ALBCchicken_assessment-3.pdf). See as well Don Schrider's "Productive Purebred Poultry," *Backyard Poultry*, June–July 2007.

10. Joel Salatin, *The Sheer Ecstasy of Being a Lunatic Farmer* (Polyface, Inc., 2010, distributed by Chelsea Green Publishing), pages 195–197.

Chapter 26. Managing the Breeding Season

1. As I proceed with my new breeding plans for Spiral Matings of two breeds, I should in the future enter a normal breeding season with twelve cocks: one older and one younger for each of three "families" of New Hampshire—six breeding cocks—plus the same for Old English Games.

Chapter 27. Working with Broody Hens

1. A note about terminology: Traditionally, a broody hen has been called a "setting hen"—as opposed to a "sitting hen." I continue that usage here.
2. Thanks to Ed Hart, fellow member of the Society for the Preservation of Poultry Antiquities, for pointing me to Cackle Hatchery (www.cacklehatchery.com; PO Box 529, Lebanon, MO 65536; 417-532-4581) as a source of standard Old English Games. They offer eighteen variants of this unusually beautiful breed, from more traditional black-breasted red to Golden Duckwing, Wheaten, Red Pyle, and Crele. Another source of Old English Games is Sand Hill Preservation Center (www.sandhillpreservation.com; 1878 230th Street, Calamus, IA 52729; 563-246-2299).
3. Actually on occasion I have made a daytime graft onto a broody hen of chicks just received in the mail—after all, the chicks have had a long and stressful trip, and I want them in the foster mother's tender loving care as soon as possible. But a daytime graft is chancy, something I attempt only with a seasoned broody I am confident is rock-solid as a willing mother.

Chapter 28. Butchering Poultry

1. I recommend reading Stephen Budiansky's *The Covenant of the Wild* (published by William Morrow, 1992, republished 1995 by Terrapin Press), the subtitle of which, *Why Animals Chose Domestication*, encapsulates its interesting take on the origins of domestication: Certain animal species *initiated* the process even more than did the humans. Budiansky's book is not perfect—for example, he uses a single anecdote near the end of the book, about one of his ewes with a prolapsed vagina who welcomed tight confinement in a stall, as an implied apology for the entire CAFO enterprise (confined animal feeding operation). But it imparts a new and more scientifically based correction to our usual human-centric concept of the origin and nature of domestication.
2. Again I highly recommend reading Lierre Keith's *The Vegetarian Myth*, with its excellent and persuasive argument that an agriculture based on animals and plants is ultimately more regenerative, sustainable, and compassionate than one based solely on plant crops.
3. The one I use is from Allergy Control Products, www.allergycontrol.com; 800-255-3749.
4. If I honor the bird by utilizing it as thoroughly as I can, then the very best practice for disposal of the entrails is to cycle them through my maggot buckets to generate live protein feed for the flock. As noted in chapter 19, I have had problems with that option in the past. But it seems that the generation of botulism had to do with soured feed in the crops of birds not properly starved prior to slaughter. I continue to utilize entrails to feed the maggot buckets, but only when the birds have been starved in preparation for slaughter long enough to clear the gastrointestinal tract entirely. Note that I have also been feeding entrails in small lots to my soldier grub bin, with no problems to date.

Chapter 29. Poultry in the Kitchen

1. An excellent source of information on the important role of cholesterol in human metabolism is Chris Masterjohn's http://www.cholesterol-and-health.com.
2. Published in January 2006 as "Dietary Cholesterol Provided by Eggs and Plasma Lipoproteins in Healthy Populations," now available at http://www.ncbi.nlm.nih.gov/pubmed/16340654.
3. http://blog.cholesterol-and-health.com/2010/11/does-choline-deficiency-contribute-to.html.
4. www.motherearthnews.com/Relish/Free-Range-Versus-Pastured-Chicken-And-Eggs.aspx and www.cholesterol-and-health.com/Egg_Yolk.html.
5. Water glass is available in a 1-gallon size, enough for fifty dozen eggs, from Lehman's (www.lehmans.com).

6. I cannot make the case here, but the Weston A. Price Foundation has done much to dispel the misinformation on this critical subject. If you have not already begun to question the received wisdom about the role of fats (both good and bad) for health, see www.westonaprice.org/know-your-fats.html to get you started—it has numerous links to well-written articles on all aspects of this subject. I also recommend Mary G. Enig's *Know Your Fats: The Complete Primer for Understanding the Nutrition of Fats, Oils and Cholesterol* (Bethesda Press, 2000).

Chapter 30. Serving Small Local Markets

1. The American Pastured Producers Association (http://apppa.org) was founded in 1997 by Joel Salatin and other small farmers basing reasonably scaled commercial production on pastured flocks. Membership is at three levels: Basic, Producer Plus, and Business. I maintain membership at the "Plus" level ($60 annual dues, only $20 more than Basic), since it gives me access to an active Web-based forum that is hands down the best in which I participate. Members do not waste my time with idle chitchat, sharing instead from their abundant experience, experimentation, and curiosity about all things poultry.
2. The work of the Weston A. Price Foundation (www.westonaprice.org) is based on the belief that health can only come from whole, unprocessed, traditional—as opposed to ersatz—foods. WAPF provides invaluable antidotes to much erroneous received wisdom about food and health—the supposed benefits of soy, myths surrounding cholesterol, the "necessity" to pasteurize milk. Check them out, and if you are convinced, support them with your membership.
3. The mission of the Farm-to-Consumer Legal Defense Fund (www.farmtoconsumer.org) is to defend the right of family farmers to sell their foods to consumers, as well as the right of consumers to obtain their foods directly from farmers; and to protect America's small farmers from harassment at all levels of government. Joining and making a donation to FTCLDF could well mean more for sane, sustainable farming, and by extension ecological stewardship, than membership in other worthy organizations such as Sierra Club, Audubon Society, Nature Conservancy, and Slow Food.
4. Read this war story at http://polyfaceyum.blogspot.com/2010/10/war-story.html.
5. A quick overview of management for egg quality and egg washing is at www.plamondon.com/faq_eggwashing.html. That page also has links to a free download of the chapter on the subject from Milo Hastings's classic *The Dollar Hen*, first published in 1909, reprinted in 2003 by Norton Creek Press; and to the USDA's Egg-Grading Manual. (The latter is of special interest because of its many pictures from inside industrial egg factories.)
6. There is a good deal of buzz about these "Freedom Rangers" in broiler production circles these days, so be aware when you hear the term that it isn't the name of a specific strain or "genetic package." It's a marketing name for a type of market broiler: bred to be raised on pasture; colored rather than white (red or bronze, sometimes with white or black flecking); slower growing than the Cornish Cross; with a longer body and a bit narrower breast but larger legs and thighs. I've grown these meat birds a number of times. They are so much more enjoyable to work with than Cornish Cross, and their flavor is markedly superior. (The source I ordered chicks from is www.jmhatchery.com. There are pictures of some of my Freedom Rangers on pasture at the link to my website in the next endnote.)
7. See my article "The Cornish Cross: What Is Wrong with This Picture?!"—first published in *Grit!*, the newsletter of American Pastured Poultry Producers Association in January 2001, and subsequently included in the APPPA's compilation *Raising Poultry on Pasture: Ten Years of Success*—now at http://themodernhomestead.us/article/Cornish-Cross.html. See also my "Sunday Dinner Chicken: Alternatives to the Cornish Cross," first published in the April–May 2009 issue of *Backyard Poultry*, now available at http://themodernhomestead.us/article/Cornish+Cross+Alternatives.html, for a fuller discussion of market broiler options.
8. If you want to build your own high-grade scalder and plucker, check out Herrick Kimball's manuals *Anyone Can Build a Tub-Style Mechanical Chicken Plucker* and *Anyone Can Build a Whizbang Chicken Scalder* (http://whizbangbooks.blogspot.com). A friend of mine made a Whizbang plucker (see my articles "The Whizbang: An Affordable Homemade Poultry Plucker" at http://themodernhomestead.us/article/Whizbang+Plucker.html and "The Whizbang: Materials List and Hints for Building a Plucker from Scratch" by Michael Rininger at http://themodernhomestead.us/article/Building+Plucker.html), and I have corresponded extensively with APPPA producers for broiler and turkey markets who use their Whizbangs for serious production work. If you prefer to buy assembled but affordably priced quality equipment, check out David Schafer's Featherman Equipment Company at www.featherman.net, which offers a scalder, several pluckers, and much more. For top-of-the-line stainless-steel equipment designed for pastured poultry operations, see Eli Reiff's Poultryman pluckers and scalders at

http://chickenpickers.com/page10.html, or call Eli at 570-966-0769.

9. For a fuller discussion of market broiler production, see my "Stepping Up to Production for a Small Broiler Market: Thinking It Through," which relates actual experiences of producing members of APPPA. See as well "Serving a Small Broiler Market: Three Examples," a look at the broiler operations of friends in my area: Chrystal Mehl, Deanna Child, and Matthew and Ruth Szechenyi. Both were first published in the February–March 2008 issue of *Backyard Poultry* and can now be read at http://themodernhomestead.us/article/Broiler+Market.html and http://themodernhomestead.us/article/Small+Broiler+Market.html.
10. Project details are at http://livingthefrugallife.blogspot.com/2009/07/diy-chicken-plucker.html.
11. I bought my caponizing kit from Nasco (www.enasco.com/product/C10606N). A simple instruction manual, originally written for Sears, Roebuck & Co. in 1922, is available online at www.afn.org/~poultry/capon.htm.
12. To expand on these three points: (1) Sewing or binding the incision through which the testes are extracted has never been a part of caponizing. Pull the skin to one side before making the incision, and afterward it will move back into a covering position over the incision site. (2) I scrub my hands and clean the stainless-steel work surface before beginning, and bathe my hands and caponizing tools frequently in alcohol—this level of heightened cleanliness is sufficient, given the robust resistance to infection characteristic of a backyard chicken. (3) An anesthetic? Are you going to duplicate the long training of the operating-room anesthetist? Where will you obtain anesthetic drugs, which after all can kill at levels a little beyond sedation? Administration of an anesthetic is simply not a practical option—the major reason the Animal Welfare Approved standards prohibit caponizing. Just how much pain does a cockerel feel when being caponized? I can't say, though in my experience the cockerel doesn't even blink when I make the incision. On the other hand, a violent spasm as if he'd been hit with a jolt of electricity taught me to avoid hitting a particular nerve inside. In any productive flock excess males must be culled in any case. Which cockerel gets the better deal: the one who is culled at twelve weeks or the one who endures the undeniable stress of caponization but continues a nice life a year beyond that? Has anybody asked the bird?
13. For more information, including reports from my own experience working with capons, see "Caponizing: Reviving a Lost Art" at http://themodernhomestead.us/article/Caponizing.html.
14. A more detailed account of Denton Baldwin's game bird niche market operation was published as "Game Birds: An Income Opportunity" in the June–July issue of *Backyard Poultry*. Now available at http://themodernhomestead.us/article/Game+Birds.html.
15. Read the full story of Leila and Allison's enterprise, "Serving a Niche Market for Started Birds," first published in the June–July 2009 issue of *Backyard Poultry*, now at http://themodernhomestead.us/article/Started+Birds.html.
16. "Moral Puzzles in the Backyard" was first published in the February–March 2010 issue of *Backyard Poultry*—read it now at http://themodernhomestead.us/article/Moral+Puzzles.html.
17. The WAPF can help you find local chapter leaders in your area, who can in turn help you make contact with foodie fanatics like me. See the listing of local chapter leaders by state at http://westonaprice.org/find-a-local-chapter.

Epilogue: The Big Picture

1. The line from "i sing of Olaf glad and big" by E.E. Cummings from *Complete Poems: 1904–1962* by E.E. Cummings, edited by George J. Firmage. Copyright © 1931, © 1959, 1991 by the Trustees for the E. E. Cummings Trust. Copyright © 1979 by George James Firmage.

Appendix A. Making Trap Nests

1. Rolfe Cobleigh, *Handy Farm Devices and How to Make Them* (first published 1909 by Orange Judd Company, republished 1996 by The Lyons Press), page 119.
2. Gail Damerow suggests a couple of simple designs in her excellent *A Guide to Raising Chickens* (Storey Publishing, 1995), pages 172–173.

Appendix F. Spreadsheets for Tracking Egg and Broiler Costs and Profits

1. You can download live-action copies (with formulas entered and ready to calculate) of these two spreadsheets (Microsoft Excel format) at Tim Koegel's website: www.windyridgepoultry.com/tools1.htm.

Appendix G: Natural Eggs and Industrial Eggs Compared

1. The *Mother Earth News* article is "Meet Real Free-Range Eggs," now available at www.motherearthnews.com/Real-Food/2007-10-01/Tests-Reveal-Healthier-Eggs.aspx. The chart presented here is available at the end of the article as a PDF download.

INDEX

Note: Page numbers in **bold** refer to definitions. Page numbers followed by *t* refer to tables.

ABOUT THE AUTHOR

Harvey Ussery has been developing his whole-systems poultry husbandry for decades and has been writing about chickens and other fowl for *Backyard Poultry* since the inception of the magazine in early 2006. He has also written numerous articles for *Mother Earth News* and *Countryside & Small Stock Journal*, and has published in the American Pastured Poultry Producers Association's newsletter, *Grit!*, over the years. Ussery has presented at national and local events on poultry, homesteading, and energy and sustainability issues, and maintains a highly informative website, TheModernHomestead.US. He lives with his wife, Ellen, in Virginia.

About the Foreword Author

Joel Salatin and his family own and operate Polyface Farm, arguably the nation's most famous farm since it was profiled in Michael Pollan's *New York Times* bestseller *The Omnivore's Dilemma* and two subsequent documentaries, *Food Inc.* and *Fresh*. An accomplished author and public speaker, Salatin has authored seven books. Recognition for his ecological and local-based farming advocacy has included an honorary doctorate, the Heinz award, and many leadership awards.

the politics and practice of sustainable living

CHELSEA GREEN PUBLISHING

Chelsea Green Publishing sees books as tools for effecting cultural change and seeks to empower citizens to participate in reclaiming our global commons and become its impassioned stewards. If you enjoyed *The Small-Scale Poultry Flock*, please consider these other great books related to Gardening and Agriculture.

GAIA'S GARDEN, Second Edition
A Guide to Home-Scale Permaculture
TOBY HEMENWAY
ISBN 9781603580298
Paperback • $29.95

THE WINTER HARVEST HANDBOOK
Year-Round Vegetable Production Using Deep-Organic Techniques and Unheated Greenhouses
ELIOT COLEMAN
ISBN 9781603580816
Paperback • $29.95

THE FARMSTEAD CREAMERY ADVISOR
The Complete Guide to Building and Running a Small, Farm-Based Cheese Business
GIANACLIS CALDWELL
ISBN 9781603582216
Paperback • $29.95

THE HOLISTIC ORCHARD
Tree Fruits and Berries the Biological Way
ISBN 9781933392134
Paperback • $39.95

For more information or to request a catalog, visit **www.chelseagreen.com** or call toll-free **(800) 639-4099**.